# The New Social Economy

# The New Social Economy:

## *Reworking the Division of Labor*

Andrew Sayer and Richard Walker

Cambridge MA & Oxford UK

First published 1992

Blackwell Publishers
3 Cambridge Center
Cambridge, Massachusetts 02142, USA

108 Cowley Road, Oxford, OX4 1JF, UK

*Library of Congress Cataloging in Publication Data*
Sayer, R. Andrew.
    The new social economy : reworking the division of labor / Andrew
Sayer and Richard Walker.
        p.      cm.
    Includes bibliographical references and index.
    ISBN 1-55786-278-8 (alk. paper).—ISBN 1-55786-280-X (pbk. :
alk. paper)
    1. Division of labor.    I. Walker, Richard, 1947–    .    II. Title.
HD51.S29   1992
306.3'6—dc20                                                    91-39336
                                                                    CIP

*British Library Cataloguing in Publication Data*

A CIP catalogue record for this book is available from the British Library.

Typeset in Sabon on 10/11.5 pt. by Best-set Typesetter Ltd.
Printed in the USA

This book is printed on acid-free paper.

# Contents

# Preface

This book grew out of a decade's work on the changing nature of production and industrial geography in the advanced capitalist nations. In our efforts to make sense of a range of topics from class and gender, through service industry, the labor process, and industrial organization, to the failure of state socialism, the concept of division of labor repeatedly forced its way onto the agenda. It became apparent that divisions of labor and ways of organizing them were being continually reworked in practice under capitalism and socialism, unleashing profound changes in the workings of these systems that were often attributed to the wrong cause. As these once-tangential efforts began to add up, we began to realize that something was seriously amiss in the prevailing conceptions of industrialism, capitalism, and socialism. What had been backdrop moved into the foreground, as the theme of the division of labor, and its general neglect in social theory, became more and more insistent.

On the strength of our new conviction we decided to pull together a set of essays around this central theme. The result, after substantial rethinking, is a much clearer vision of the role of division of labor in the social economies of the late 20th century. The first three chapters originated with Richard Walker, the last three with Andrew Sayer. They still bear the mark of their drafters, in spite of considerable discussion, critique, and cross-editing; we have not attempted to suppress altogether our differences of style and emphasis. While co-authorship has its stresses and strains, especially when it involves an international division of labor spanning over 5,000 miles (with only three face-to-face meetings), we have greatly enjoyed working on this project and believe it to be more than the sum of our individual efforts.

Andrew Sayer would like to thank Fiorenza Belussi, Steen Folke, Andy Graves, Costis Hadjimichalis, Frank Hansen, Bryn Jones, Dan Jones, Peter Maskell, Doreen Massey, Kevin Morgan, Robin Murray, Michael Storper, and Stephen Wood for their comments on various parts of the book; the Universities of Copenhagen, Roskilde, and Lund, and the Copenhagen Business School for their hospitality; and the University of Sussex for sabbatical leave.

Richard Walker thanks Michael Burawoy, David Harvey, Michael Johns,

Nigel Thrift, Annalee Saxenian, Michael Storper, Michael Webber, and Eric Wright for their input, inspiration, and comradeship along the way; and the Department of Geography at the University of California for sabbatical leave. We together offer our great thanks to Peter Solomon for his efforts to impose clarity and consistency across the gulf of British and American writing styles, Charlie Hadenfeldt and Natalia Vonnegut of the Geography Department in Berkeley for their help with the manuscript preparation, and Chic Dabby for eagle-eyed proofreading.

Finally, we would like to acknowledge our much-loved partners. We hope the periods of distraction have not been too long and that disturbances to our domestic divisions of labor arising from the book have been minor rather than major. We are grateful, too, that boring old "labor" is not all that is involved. Our love and thanks to Hazel Ellerby and Chic Dabby, as ever, and to Lizzie and Nadya for allowing us to rediscover the delights of youth.

*Berkeley and Brighton*

# Introduction

In this era of triumphant individualism and heralding of the free market, the rediscovery of the inherently social nature of modern economies goes against the grain. One expects the reformer, revolutionary or moralist to make the claim for human life and work as a social enterprise, in which cooperation, democracy, and meaningful labor are the hallmarks of a just order. But quite unexpected questions have been raised within the bosom of capitalism, coming from business schools, policy studies, and other unlikely quarters, about the need for greater collaboration, less fragmented work, and more human involvement in the labor process. The new social economy, it appears, is already well upon us. It is manifest in closely orchestrated delivery systems, international networking among business firms, and the shift toward greater "service" labor, among other things. Capitalism is once again offering new possibilities for social labor and human achievement, even if many of those suggestions continue to go unfulfilled. This book explores several of these key developments, and tries to make sense of them in terms of a common axis, the division of labor.

The division of labor is one of the most neglected categories in contemporary political economy and social theory. Compared to the likes of class, gender, or markets, it seems a rather indifferent concept, part of the backdrop rather than one of the key forces of economy and society. The central conviction behind this book is that this is a dangerous assumption, one which produces serious misconceptions about many of the fundamental issues of contemporary economy and society, from the place of the new middle class to the challenges of international competition. While we do not hold that the division of labor gives one a privileged viewpoint, we do believe it provides a fresh angle from which to cut through certain theoretical knots and shine new light on old dilemmas.

The division of labor refers to the system of work specialization that runs through all human societies, from the most primitive to the most advanced. It is not an inert set of specialized slots into which people fit like pegs, but an active force in social ordering, economic development, and the lived experience of the participants. Far from being a simple consequence of more profound social forces, the division of labor has far-reaching effects of its own which are often falsely attributed to other structures. The differences it erects between people are deeply implicated in the emergence and sustenance of gender, race, and class inequalities. The multitude of specialized products it creates are the basis for exchange in markets, and the hierarchies of administration and knowledge it generates within social labor processes are deeply etched in the frames of modern corporations. The productivity it has sustained is a fundamental part of the continuing industrial revolution of the last two hundred years, playing a major role in the global successes of capitalism.

Because the division of labor is so much taken for granted, there is a need to defamiliarize ourselves with it, to treat the seemingly mundane as problematic. This means taking a fresh look at the extraordinary range of specialized tasks and collective activities which make up modern production, as well as the vast geographic scale of a global division of labor that embraces everything from a household in Mexico to a factory in France, from a London bank to a West African cattle range. More than this, we need to grasp the dynamism in the way the fabric of the advanced economies has been enlarged to encompass new sectors, new jobs, and new ways of working, while the surfaces and densities of social labor have been stretched, folded, and refolded in upon themselves like dough, taking shapes previously unimagined.

Rethinking the division of labor also means attempting to grasp the sheer magnitude of millions of commodities weaving their way from one place of work to another, and from those to a vast number of sites of consumption. Along with these substantial things goes the equally vast flux of money, information, and ideas, and the swirl of people flooding in and out of workplaces, cities, and countries. This immense, translucent ballet of people, things, and representations challenges every given conception of production, industrial linkage, and economic coordination. It demands a richer understanding of the social integument of human labor than has normally prevailed in economics, and a fresh look at the problems of industrialization, innovation, and economic growth.

Refamiliarization with the division of labor also means confronting the immense puzzle of how the far-flung pieces of social labor are knit together in space and time to produce intended and useful results. The hidden hand of the market is celebrated as the only mechanism capable of such a task, and the magic of the price system has been unleashed with a vengeance in our time by Thatcherism, the international debt crisis, and the dissolution of the Communist bloc. Yet how can markets cope with all the unruly acts of labor without boundaries and all the untameable flows beyond the space of commodification? At the same time, there are few today who will defend the performance of the Soviet planning system, whose targets and allocations have broken upon the shoals of economies too complex for the most supercharged computers.

The intellectual reduction of economics to markets or state planning is wholly unsatisfactory, for it leaves out almost everything social about the division of labor and modern production: the great varieties of people at work, the social integument that binds together their activities, and the social context in which production and circulation proceed. In effect, this shorthand denies the exigencies of power and domination, coordination and conflict, economic development and social change. On the right this has meant steadfast evasion of the problems of capitalist exploitation, patriarchy, and imperialism, and of the pervasive malfunctioning of capitalist economies due to the division of labor itself, such as the coexistence of unmet needs and unused labor. On the left, it has meant avoiding the unpleasant record of Soviet despotism and mismanagement, and a dogged disinterest in tackling the practical problems of socialism as a form of large-scale economic organization and production, in favor of romantic notions of the associated producers coordinating and guiding themselves. Such ideals skirt all the difficulties of taming the division of labor, such as the clash of interests of different producers, timing and quality of materials flow along supply chains, and meshing rates of investment and innovation with the needs of users. Democracy in the abstract means very little when there's no soap, as Soviet miners have recently discovered.

While we side with the left in its moral determination to find a better way than capitalism, we insist on the need for more imagination about how to organize a socialist economy that works as well or better than a capitalist one. Now, more clearly than ever, the old homilies are not enough. The collapse of existing socialism renders hard testimony to the inadequacies of central planning as an alternative to generalized markets. Yet, despite the noise and the shouting from the bourgeois back-bench as to the triumph of the free market in Eastern Europe, nowhere in the capitalist world does the market in fact function free of regulators, corporations, urban systems, national boundaries, police, and lawyers, or free of the social tissue of trust, coercion, command, and dependence surrounding all production and exchange. And the governments of Britain and the United States, at the high tide of market liberation, are increasingly disconcerted as savings and loans collapse, exports sink, and bankruptcies mount, while better planned and produced Japanese and German commodities flood their home markets with blatant disregard for the fine points of ideology.

The general crisis of communism, then, is matched by a selective unraveling of industry in the capitalist world which lacks only a financial disaster to become quite general itself. In part, the economic woes of capitalism derive from the unruliness of the division of labor, as exemplified by competitive challenges across international divides, radical improvements in complex manufacturing systems, and the rise and fall of industrial sectors based on shifting technologies. As New England sinks into depression along with the fading fortunes of its mini-computer industry, or Liverpool tries to make something of its decrepit docklands, it is cold comfort to be assured that the Polish shipyard and steel towns have it worse.

This book will demonstrate the considerable theoretical and practical significance of the division of labor and offer this as a corrective to certain standard accounts in political economy and social theory. Among economists the division of labor has figured very imperfectly. On the right, neoclassical theorists are obsessed with the formal requirements of timeless and spaceless equilibration through market exchange in a known constellation of firms. For them, the division of labor presents no obstacles to market optimization thanks to convenient assumptions about substitution in production functions, flows of resources across units, and anticipation of future conditions. Austrian economists take the division of labor more seriously – Hayek even considers coordination of the division of labor the gist of the economic problem – but remain equally tied to market exchange as the only solution.

On the left, institutionalists and evolutionary economists weave the threads of the market and economic calculation into a broader tapestry of laws, institutions, and constrained behaviors, allowing the division of labor room to grow and act in unpredictable and disharmonious ways, but do not subject divisions of production themselves to further inquiry. Marxist economics takes very seriously the divisions created within the labor process and their implications for class formation, and by its attention to the "hidden abode of production" ferrets out the unseemly side of class power so fastidiously glossed over by conservatives. Yet by demoting exchange, Marxism is led to neglect the social division of labor and to consign the larger tissue of circulation and economic organization to the nether world of "capitalist anarchy", a step back from the institutionalists if not, indeed, from the Austrians. Feminist economists have pushed the discourse on the division of labor in fruitful directions by looking outside the sphere of capitalist production, into the contortions introduced into supposedly gender-blind labor markets and capitalist forms of exploitation, and the general devaluation and underpayment of women's labor. But they have not been able to supplant altogether the insights of Marxism and mainstream economics into the operation of the capitalist system, and they have chiefly portrayed the division of labor as another dimension of patriarchy (or capitalist patriarchy) rather than a force in its own right.

None of these approaches will suffice on its own. Marxists and feminists are correct to insist on the way power relations affect all areas of the social economy, Austrians and neoclassicals to point to the importance of commodity exchange across divisions of labor, and institutionalists to show how the infrastructure is requisite to production and exchange. But none of them seriously explores the realm of the division of labor itself, either as the full expression of social labor and production as a whole, which Marx's work suggests to be the subject of political economy, or as a source of social difference and domination in its own domain, which is where the logic of the feminist critique of class theory leads. Nor, as we shall see, can some more recent and fashionable academic notions such as flexible specialization, post-industrialism, and class stratification escape criticism: all are found seriously wanting on these grounds.

In tackling the effects of the immense division of human labor in the modern world, we cannot be blind to the illusions it creates. John Berger (1973) has written "it is not time but space which hides consequences from us", and the differences built into human geography are in large measure those of the division of labor. Even within small, seemingly knowable, communities the division of labor works its magic on the consciousness of participants in differentiated societies: the mysteries of things unknown because done by others or misunderstood because known only by others. That powers of human labor can be turned to powers over other people, with the help of mystifications, denials, and exclusions, is evident enough in single households and villages, let alone among nations. At larger scales, the origins of what we consume and the destinations of what we produce are still more obscure. This extraordinary combination of separation and interdependence is so much a part of contemporary life that we are thoroughly inured to it. Yet the observation that household objects may contain parts made on several continents, often under unspeakable conditions, by workers speaking many languages, demands reflection, not dismissal. Nor are those characteristics of the division of labor closer to home any less worthy of a fresh look for the insight they offer into human difference and conflict, human achievement and possibility: the division of waged and unwaged work, the separation of industries, the assignment of men and women to traditionally gendered tasks.

Each of the three major themes in this book raises fundamental questions about the nature of advanced industrial societies, capitalist and socialist, and about social and economic theory as well. The first is the significance of social labor, or the orchestration of complex labor systems; the second is economic change, or the role of the division (and integration) of labor in the repeated overhaul of modern economies; the third is social power, or the place of the division of labor in social inequality and the struggle for democracy.

## Social labor

The growing division of labor implies something more than separation and difference; it speaks to the development of social labor of an ever-increasing scale and richness. At the most abstract level, this connotes the complexity of roles and intricate web of social interdependency in industrial society addressed by Durkheim and the sociologists of modernism, or the transcendent operations of the vast globalized economy in all its myriad parts, as tackled by contemporary theorists of the supermodern, such as Castells. We shall not push things to that level, but seek instead to grasp the more tangible, if still vast, questions of the multiplication of industrial activities and of industrial organization in its fullest sense.

In our view, the economic problem, so called, is not so much the allocation of scarce resources as it is the orchestration of social labor to achieve and sustain the kind of productivity that overcomes classic scarcities, and to put that enormous productive power to effective and good use by meeting the needs and desires of the largest number of people. This is not the formula of the hidden hand, so beloved by conservative economists, but a restatement of a general condition of industrial societies. The expanding division of labor brings with it a correlate economic problem, the integration of social labor. The activity of social labor on an unprecedented scale puts industrial organization at the center of the economic puzzle and makes it key to strategies of development and social harmonization. The unruly division of labor has to be harnessed, made to work for those who control industry and those who depend on it for their livelihood. Yet we find here a field of research long consigned to side branches of economics and sociology, as well as an arch silence on the left about fundamental issues of economic organization under socialism.

The puzzle is this: how are the various specialized activities to be connected so that they function in a reasonably coherent and effective manner? The classic answer has been through firms and markets. The capitalist controls the internal organization of the firm by virtue of ownership and the power it confers, while the price mechanism coordinates relations among many firms having a rough equality before their fellow buyers and sellers. But the matter of economic integration is not exhausted by the simple duality of firm and market. Many other modes of organization exist, from workplaces to territorial complexes, and many hybrid forms, from subcontracting to joint ventures, can be explored under the heading of relational contracting and networks. And there are several means of integration at work in these various institutional frames, besides exchange and hierarchy, including persuasion, reciprocity, extortion, and indicative planning. The result is an organizational problem much richer and more open to fresh analysis than has been previously realized.

An explosion of interest in the new industrial organization theory is now before us, moved by the press of events, particularly the discommodious fate of so many formerly successful firms and local industries in the face of global recession, fierce international competition, and heated financial maneuvering. The most successful capitalists in recent years have been those applying innovative organizational formulae, from the small firm complexes of the Third Italy to the just-in-time system and subcontracting networks of Japan. It has become clear that industrial organization is a key social technology that can confer competitive advantage on certain firms, industries, and countries. More broadly, it is clear that capitalism must not only reproduce its conditions of existence in terms of workers at the factory gate or new surplus value; it must reproduce, extend, and reintegrate its division of labor in the face of a continually expanding and shifting industrial foundation.

The same applies to socialism. The significant problems that arise in connection with the integration of social labor are not reducible to the class character of society, nor does the conventional Marxist antithesis of socialist planning and capitalist anarchy provide sufficient guidance for coping with the complex economic and political needs of post-capitalist societies. Economies are not homogeneous wholes needing a few well secured plans, like wrapping tape; they are highly differentiated as to production, consumption, and inequality, thanks to the division of labor. Moreover, that differentiation – which socialism must deal with (but cannot abolish) – is coupled to the very means of integration created to cope with it. Hence class and uneven development persist even when private property is formally eliminated.

*Economic change*

As Adam Smith so forcefully argued, the continual elaboration of the division of labor has contributed fundamentally to economic development. It is one of the basic forces of production, along with mechanization, technical knowledge and the capacities of workers. Because the division of labor is so deeply implicated in economic change and industrial restructuring, it raises an old dilemma: how do we distinguish structural continuity from social transformations, or, more simply, figure out what changes and what does not in complex and evolving systems? If a growing division of labor has been an essential element of industrialization, it is hardly possible that a shift in that division can, by itself, herald the end of the industrial era – much less the end of capitalism, as the post-industrial theorists hold. Yet it can be central to long-term transformations and upheavals in capitalism and its socialist rivals in a way that has been insufficiently apprehended on either the left or the right of social theory.

The capitalist world, despite eager proclamations of the final triumph over communism and history, has been wracked by discord and debilitation, particularly in the formerly most advanced industrial regions of the United States and Britain: witness a general decline of manufacturing, formerly regarded as the essence of economic growth, and a massive shift towards a "service economy" of low-productivity work and low-paying jobs; the fierce challenge in the international domain from Japanese and German firms employing unfamiliar forms of production, industrial organization, and labor relations; and the failure of many formerly leading companies, sectors, and industrial centers of the so-called "Fordist" epoch to make a good showing in the new environment. This has led to intense speculation about transitions to a new era, called everything from "post-industrialism" to "New Times", implying a remarkable break with the past and its legacies of industrialism, capitalism, and laborism. We shall argue that these developments, while important and disorienting, do not signal radical departures from the known workings of industrial capitalism. Rather, they pivot on the shifting division of labor and new methods of industrial organization – in short, new forms of social economy thrown up by economic development and the initiatives of new contenders for capitalist supremacy.

If the problem of change is a thorny one within capitalist societies, it becomes more complicated still in relation to socialism, or constructing an alternative mode of production. This is a matter over which few today are sanguine, given the legacy of bureaucratic administration and Stalinist politics in the Soviet Union and Eastern Europe. The institutions of centrally planned economies ossified in forms that no longer met the demands of a wealthier and more sophisticated citizenry, measured up to the standards of global competition, or, in extreme cases, held together disintegrating national economies. Yet the socialist ideal still offers the tantalizing promise of a more humane, manageable, and democratic economy. Any effort to reach that ideal puts the Gordian knot of a recalcitrant division of labor at the center of social theory. For if the problem of handling social labor cannot be solved except in a free market, capitalist fashion without generating technical stagnation, deadening labor capacities, and denying the fruits of industrialization to the consuming public, then economic history will indeed be at an end.

### Social power

As economics has been struck by the importance of industrial organization, social theory has become increasingly sensitized to the varieties and pervasiveness of power relations in human affairs. Marxism long played the central role in the critique of social power, but its focus on class has had to be supplemented by a fuller acknowledgement of other forms of domination, such as patriarchy, racism, homophobia, and imperialism. These axes of power have to be recognized for their depradations on immense numbers of people; and the struggles of the oppressed have to be acknowledged as social movements that do not fit classic revolutionary models. Yet an essential figure in the landscape of power has left only a shadowy trace in contemporary social thought.

The division of labor is largely neglected in the literatures dedicated to explaining difference and domination in human affairs. Yet it is so closely caught up in the workings of social domination that it has in the past been recognized as synonymous with some basic forms of inequality, as in the separation of mental and manual labor, the division between town and country, or international disparities of rich and poor. Wherever we look in world history, the division of labor provides a lever by which people can be pried apart, elevated and subordinated, even as they are caught up in relations of mutual dependence and reliance on the collective achievements of social labor. Capitalists, slaveowners, feudal lords, and male masters of the household have all used this tool in rising to power.

While division of labor is frequently taken to imply equal exchange of products in the market, this is by no means guaranteed, whether equality is measured in terms of the nature of the goods and services, the amount of labor embodied in their making, or the social conditions of their production. And while social labor implies mutuality, it also binds people to collectivities, from the family to the factory, without which their labor cannot be gainfully employed and so forces them into compliance with a social order that may be anything but just. Inequality, dominance, and exploitation are thus not exceptions but likely concomitants to the division of labor in society, contrary to the views of social conservatives such as Durkheim, Hayek, and the neoclassical economists.

The division of labor needs to be accorded its fair share of credit for the prevailing social relations of capitalist societies. This does not mean it abolishes or transcends class, race, or gender, but rather that it contributes actively to their making and remaking over time and space. Unfortunately, the conservative political hegemony of the 1980s, and the rout of the communists before the practical criticism of their own citizens, have led to widespread cynicism about traditional theoretical critiques of capitalism. This has led the left to retreat from class analysis just when the empowerment and enrichment of those at the top has been proceeding unabated. The question before us is how to blend the theories of class, gender, and other major forms of social domination into an even more powerful alternative vision.

This cannot be done unless the division of labor is included in the equation of social power. Reformers and revolutionaries have too often relied on ideals of the new economic order that abolish the division of labor with a wave of the hand. On one side stand the utopias of the local and immediately accessible, such as workers' self-management of factories, ecotopias, and communitarian villages, which are supposed to escape from the insidious power relations of larger collectivities; but barring a general return to primitive economies, this is impossible – even nations the size of the Soviet Union and China have not been able to absent themselves wholly from the global division of labor. On the other side stand the utopias of internationalism and world harmony, which are utterly inaccessible to human beings rooted in the places and conflicts of the present. In between stands the bourgeois utopia of the market, which is supposed to reconcile all human difference and abolish all exploitation through the magic web of exchange and the law of value. But social differentiation and the division of labor resist leveling, and neither workers' councils nor the United Nations nor the free market can fully command these forces.

Humankind has developed economic systems so complex and so far-flung that no one solution at the local, national, or global level can work. The capitalist world depends on a vast array of institutions and power-centers to orchestrate and contain the industrial economies. These are in most cases horribly undemocratic, whether one is speaking of General Motors, the US Federal Reserve Bank, or the councils of the Group of Seven. But the road to democracy is not blocked only by the privileges of a capitalist class holding on to the means of production; it travels over the rough terrain of the division of labor – not to say divisions of culture, nationalism, and religion. Democracy is not the first-born of capitalism, despite the ideological claims now being trumpeted across the fallen Berlin Wall. To be sure, it was stillborn under communism; the failures of the 20th century, on both sides of the World Wars and Cold Wars, teaches us that political democracy cannot be achieved where economic democracy is lacking. It seems ever more hopelessly utopian to propose a harmonious resolution of differences on a world scale. Yet any idea of separate and autonomous development has become equally utopian, precisely because of the global interdependence developed through the division of labor; hence the contradiction of speaking simultaneously of the Arab "World" – as if it were separate – and the centrality of its oil to non-Arab economies. Difference and interdependence go hand in hand, and division of labor is a prime structuring force in this turbulent process.

Power, change, organization are three themes addressed in this book, as we seek to restore the division of labor to its status as a major topic in political economy and social theory. To this end, we take on several of the most pressing practical and theoretical topics in the new social economies of capitalism and socialism.

We begin, in Chapter 1, with the intersection of class, gender, and the division of labor, three fundamental axes of social power and widespread differentiation, exploitation, and domination. Each has a distinctive character and plays a particular role in social life, and the failure to distill the discrete effects of the division of labor from those of class and gender has led to no end of confusion in studies of everything from social stratification to women's work. At the same time, these three dimensions of social structuring have been deeply entwined over the course of modern history, resonating in ways that typically reinforce the inequities and insults of everyday life.

In Chapter 2, we examine the shifting social division of labor at a very broad scale in advanced industrial economies. Our target is the thesis that industry has been transcended by "services" which involve qualitatively different outputs, labor inputs, ways of working, and relations between producers and consumers. Capitalist industry is, indeed, significantly altered in all these dimensions, but not in quite the way the service thesis contends, and at the center of these changes lies a core process of deepening and broadening the division of labor. This has taken place through the fine parceling of tasks in the workplace, the vast proliferation of commodities and industrial sectors, the great expansion of every social labor process into realms such as design and administration, and the huge growth in the work associated with moving commodities, money, and information.

Chapter 3 considers the problem of an expanding division of labor from the opposite direction, that of organizing the swarming pieces of complex and multiple labor processes. In it we tackle the transaction cost and corporate management theories of industrial organization, showing the narrowness of their approach to the problem of economic integration and the limits to their solutions. The palate of organization is laid out in all its varied tones so we can paint a picture that fills in the missing middle ground between markets and firms, highlights the multifarious connections among capitalists, and adds the infrastructure of territories and states beneath all strictly micro-economic forms of interaction. Organizational technology is put at the center of industrialization, uneven development, and the shape of capitalist power.

Chapter 4 focuses closely on a special form of industrial organization at the micro-level, the Japanese just-in-time system which has revolutionized mass production and posed such a challenge to Anglo-American methods. We find that the popular conception of this as a way to cut delivery schedules is blind to the profound advances it represents in orchestrating social labor systems within and among factories and firms. Just-in-time systems not only provide a novel way of synchronizing labor to generate cost savings, but also depend on and promote learning processes and continual piecemeal improvements; they are a dramatic illustration of the significance of organization of division of labor. We explain the just-in-time system's preconditions and its likely effects on workforces and on the spatial organization of production at all geographical scales.

Chapter 5 pulls back a step to examine the currently popular thesis of a transition from "Fordist" to "post-Fordist" production and industrial organization. We take to task the validity of the concept of Fordism itself as a way of portraying a distinct age of industrialization, and find that post-Fordism is an equally infirm characterization resting on greatly exaggerating the import of a limited number of cases, particularly the flexible production districts of Italy, at the expense of the much more significant instance of Japanese industrial organization. The global division of labor has, indeed, been radically made over from the post-war era of Anglo-American dominance, but not quite as post-Fordist theory acknowledges. An examination of some aspects of industrial organization in Japan, which have contributed greatly to its success, shows how they both contradict many features of the flexible specialization/post-Fordist literature and, in a more constructive light, suggest other novel forms of industrial organization which demand attention.

In Chapter 6, we return to the questions of power and problems raised by the division of labor for overcoming economic domination and coercion within capitalism and socialism. Here, the focus is on modes of micro-economic integration – the ways in which the vast array of different products and consumers are coordinated in an advanced economy. Neither neoclassical, Austrian, or Marxist theories deal with this satisfactorily. In particular, Marxist accounts typically attribute to class and the social relations of production outcomes which have more to do with the intractability of a complex, advanced social division of labor. Our work leads us to revise the Marxist concept of the social division of labor and this, in turn, prompts a major revision of familiar explanations of domination and economic problems under both capitalism and socialism.

We have been deeply influenced in general approach by historical materialism and critical realism (Sayer, 1984; Bhaskar, 1987). We believe that while all knowledge is in principle revisable, reliable and coherent explanations must recognize that events are invariably outcomes of the interaction of several social structures and mechanisms.[1] To say this is not to fall prey to reductionism and essentialism. On the contrary, it enables us to confront one of the main problems of inquiry in social science – sorting out the range and force of the multiple structures simultaneously present in any situation. It is no easy task to tease out the significance of the division of labor from other social forces and one faces hard judgements about causality. This is a recurrent problem in all kinds of political economy and social theory. How much is the division of labor between men and women a product of patriarchy and how much a product of capitalist mechanisms? To what extent does the division of labor within the workplace reflect class power and to what extent is it a consequence of large-scale production? To what degree are the problems of socialist economies an irreducible function of coordination problems associated with complex divisions of labor rather than consequences of their social relations of production? The chance for misattribution of causality is daunting. Whatever the merits of our analyses we hope that they will provoke others to re-examine the long-neglected powers of division of labor, and their crucial role in the new social economy of today's industrial world.

---

[1] Such talk of structures and causality may be taken amiss by those influenced by post-scientific and post-structuralist thought. As critical realists we are not "foundationalists." But to acknowledge fallibility is to say very little; to say where our substantive analyses are wrong (i.e., where they misrepresent their objects) is to say something useful. Relativists who back off from this challenge are not liberal but dogmatic. The idealisms and relativisms of "anything goes" or "no theory can be shown to be better than any other" or "there is nothing outside discourse to which it can refer" only serve to protect the status quo from criticism.

# 1

# Class, Gender and the Division of Labor

The division of labor is among the most basic social relations governing social life, but its full influence is rarely credited. Class and gender, in contrast, are widely acknowledged as two of the most important areas of social difference, oppression, and exploitation in the modern world. While there are disputes over the priority of class and gender with respect to race, nation, or state, almost no one enters the lists on behalf of the division of labor. On inspection, however, class and gender theory prove to be fraught with disputes over definition, range, and effect. We shall endeavor to show that a considerable burden of unnecessary confusion can be lifted by recognizing the division of labor as a distinct pivot of social development. While our approach is in line with recent attempts to demote class or gender from the absolute primacy ascribed to them by some Marxists and radical feminists, and to open social theory to a wider gamut of forces and voices, it has the paradoxical effect of strengthening those concepts by unburdening them of explanatory weight and political expectations they cannot bear, and which have long threatened to sink them. Cutting loose the division of labor also allows categories like class and gender to become more supple, revealing the coevolution and intertwined workings of structural relations in social affairs.

A few comments on method are in order. Class, gender, and division of labor are social structures. They exist as the result of the social relations in which people are enmeshed, and are not willed directly by any one individual or group. Social structuration is the outcome of repeated human thought and action in the form of widespread, systematic, and sustained practices, such as systems of property ownership or marriage. Structures consist of internally related positions, and associated resources and capacities, at the command of the parties involved. These endure beyond the immediate participants, and may persist despite transformations in specific institutional arrangements.

Why emphasize certain social relations? Because some, such as class and gender, are particularly widespread, durable, and robust in their effects on thought and behavior. Social structures influence but do not determine actions and events, which depend on situation, circumstance, and human agency. Moreover, social situations are invariably codetermined by interacting and intertwined structures. It is impossible to close off one structure and observe the unambiguous effects of its operation. As a result, what is contingent in one context may be essential in another. Much of social analysis involves teasing out the interwoven threads of a stratified reality, determining the coherence of different subsystems, and weighing the strength of their structures, always mindful of the risks of attributing to one structure what is due to another.

One must be particularly alert to the level of abstraction.[1] Most analyses fail to specify whether a concept is meant to be abstract or concrete, whether, that is, it is intended to capture certain restricted aspects of people or to summarize a range of important properties. Talking at cross-purposes is a staple of debates on social structuring. Some use the term class, for example, to describe the way the political economy is structured by the ownership and control of the means of production, an abstract or minimalist category. Others are more interested in social inequality and stratification as shown by income, consumption patterns, and life chances; this is a more concrete and comprehensive, but taxonomic, conception. Unfortunately, many people, usually practicing sociologists, want class to perform both functions: to cut through social complexity *and* to provide a map to concrete differences among individuals. The social taxonomists wish to assign each and every person to a class or class trajectory, a task comparable in futility to describing the fall of a particular tree on the basis of Einstein's theory of universal gravitation. Worse, in overstretching the concept of class, without recognizing the myriad intervening forces at work, they end up casting doubt on the analytic and political usefulness of the concept itself.

The construction and reproduction of structured social systems is a creative process involving human agency. This has two sides, the reproductive and the disruptive. In ordinary circumstances, the vast majority of people behave in ways that are consistent with and supportive of existing social structures. This is not a matter of passive acceptance or of learning set roles, but of coming to understand in a practical way the rules of the game of life, and employing them in a structured but indeterminate way.[2] Structure is, in other words, both limiting and enabling. While structural reproduction depends heavily on strong patterns of thought and action, what Bourdieu (1977) calls the *habitus*, it also allows room for creativity and change in a way that puts to shame the finest formal games. Power, for example, must be exercised in the pursuit of practical interests in order to maintain the conditions from which it flows, e.g., a capitalist must

[1] It is common to refer to structures as lying deep in the social fabric hidden from everyday view, but this does not mean that structural relations cannot be observed in appearance or effect. It is too easy to beg hard questions of explanation by such a device (Parkin, 1979).
[2] Structural relations are too extensive and often act too subtly to be readily comprehensible to the individual or limited social group. Ordinary people may be very much aware of structural influences on their lives, but this is not the same thing as achieving the kind of systematic knowledge in the abstract and well-defended form that constitutes social theory (Giddens, 1979).

control the workers sufficiently to realize profit from capital investment or lose power over them.

Social structuration is a dynamic process of conservation in the midst of change (Piaget, 1970). As human creativity introduces new ways of doing things, old forms of life and thought are transformed. This will very likely not be intended as a break with the past, but rather seen as the consistent application of social rules, as with the transformative effect of technical innovation by capitalists pursuing competitive advantage. Equally important are the contradictions embedded within the logic of structured systems (or between different elements of society), which raise unsolvable puzzles and conflicts and put creative action at odds with seamless reproduction; for example, notions of justice raised by capitalism but not realized in practice for workers, racial minorities, or women (Bowles and Gintis, 1986). The future is forever being born, bearing both the stamp of the past and the imprint of present pressures and limits. Structural relations must be repeatedly formed and reformed under changing circumstances.

For all these reasons we cannot expect societies to break neatly along the fault lines of class, gender, and division of labor. While the structural cleavages, like continental plates, run deep and grind mercilessly, they nevertheless turn and twist against one another, against circumstance, against the willful intransigence of human beings, and against their own internal contradictions. As we proceed to analyze division of labor, class, and gender as distinct structural relations in modern societies, we are well aware of the need for caution. We are particularly aware that the evolution and workings of structural relations depend very greatly on their continuous interplay, so our analysis must consider their dynamic interactions. Even with this move away from a wooden structuralism, any attempt to distill the essentials from more circumstantial elements must recognize that social reality is easily denatured; social inquiry cannot remain at this level of abstraction for long without losing its grip.

## The Division of Labor and Places in Production

At first glance the division of labor seems a plain enough thing. The simple meaning of the term is work specialization, either of the narrow kind between individual workers, such as that within factories and families, or the broad kind involving large groups of people, as between branches of industry or between the household and the formal economy. At this level of abstraction, division of labor exists whatever the definitions of specialization or whoever occupies various positions. But this common definition buries at least three different analytic problems. The first is to determine the technical base of specific tasks to be done, products to be made, services to be rendered. The second is to describe the social fabric inherent in dividing, unifying, and directing the labor of the collective worker. The third is to discern the way work is distributed in society, or the allocation of social labor. These raise difficult questions about the technological necessity of given divisions of labor, the social organization of divided labor, and the social relations of the division of labor and its organizational forms.

The following are some common divisions of labor in modern societies:

- Between family labor attending to children and household, on the one hand, and tending fields or otherwise producing subsistence on the other.
- Between domestic labor in the home, and labor outside the home; or, more broadly, between non-commodity production involving barter of goods and labor-services, and commodity production organized by capitalists.
- Among branches of commodity production ("industries").
- Among stages in the extended production cycle of a commodity, including not only the immediate production process but prior steps (research and development), and subsequent steps (delivery, installation, and repair).
- Between production and exchange (circulation), i.e., between industry and trade.
- Between the circulation of commodities and the circulation of money (between commerce and finance).
- Between organizational command posts and the direct work of production.

The division of labor is both social and technical. In normal usage, technical division of labor is that found within a firm, while the social division of labor is mediated by markets and exchange between firms. This is an important distinction in capitalist societies, but the terminology is somewhat misleading as it implies that the division of labor within the factory or firm is entirely determined by technical considerations, while it is market mediation that gives the inter-firm division of labor its social content. We need to carry the dialectic of technical and social in the division of labor further than this.

The division of labor has an irreducible technical foundation in the nature of the work to be done to produce a desired result from materials provided by nature and history. Every labor process contains a series of ordered tasks as well as particular tools and knowledge. Where these exceed the capacity of an individual worker, who cannot master more than a finite number of tasks, a division of labor will eventuate. As jobs diverge in significant ways, owing to the nature of the specific materials, tools, or products involved, so will the skills and the ways of working of those holding different jobs. This is as true between sectors as it is between tasks within a single factory.

Technologies are socialized to their roots, however, because the practical problems of transforming nature never strictly determine the way people do things (Noble, 1986). Workers who embody the knowledge and competence necessary to handle tasks, tools, and materials are social beings, and their actions can never be reduced to strictly mechanical dimensions. Machines, product designs, and process configurations always result from human inspiration and choice, which follow cumulative and often unpredictable paths despite certain "natural trajectories" in the way technology unfolds (Rosenberg, 1976). The particular configuration of tasks in jobs, jobs into work units, work units into factories, and so on, is a social decision and a social process, in which the abstraction of social interdependency is converted into concrete dependencies among groups of workers.

The division of labor can be considered as part of the forces of production, owing to its close configuration with technologies of work, and especially to its role in raising the productivity of labor (Cohen, 1979; Storper and Walker, 1989). Social labor develops through the division of labor because specialization breeds expertise and fluency in specific tasks and branches of labor, in the manner first elucidated by Adam Smith (1776). Even more, the multiplication of tasks, tools, products, and expertise coalesces into a system of economies and advances in technology that is more than the sum of the parts (Young, 1929). On the other hand, the division of labor contains within it the seeds of definite social relations of production; it is not just a mute force to be harnessed by capital or patriarchy.

One aspect of the social relations of the division of labor is the way specialization divides people experientially, organizationally, and ideologically. There has been a great deal of wishful thinking that such divisions are not socially fragmenting because they are readily overcome by the interdependency of different roles in society (Durkheim) or by the leveling effects of proletarianization and factory production (Marx). Unfortunately, conflict and rupture are endemic to divisions of labor, and to the working class. Oppositions appearing horizontally between people of roughly equal standing vis à vis the means of production, i.e., between those of equal standing in the division of labor, are frequently more immediate and strident than those occurring vertically between classes or genders. It is thus meaningful to speak of division of labor struggles, though the concept is not isomorphic with class struggle.

At the same time, the division of labor necessitates its converse, the integration of labor. One has to speak of social labor, the collective laborer. This raises the question of methods of work organization, and what is also generally called industrial organization. Marx (1863) introduces simple cooperation as a method of developing the forces of social labor antecedent to the division of labor, but fails to take the matter farther. The organization of work is a kind of technology itself, involving mastery of human interaction, labor allocation, and organizations as means of production. We shall have much to say in subsequent chapters about the dialectic of division and integration of labor in the new social economy.

The division of labor also implies a social hierarchy.[3] All large labor processes require direction, independent of the class character of society, as Marx's oft-cited example of the orchestra and conductor indicates, and various kinds of skills, with more or less command over the techniques of production. Command over large organizations, research and marketing activities, or the circulation of money all imply very different levels of control over social production. All this is essentially independent of the class relations in modern industrial economies, and it will not do to treat these hierarchical elements of the division of labor as derivative of separate power relations, whether those of charismatic authority, class, or gender. Without question, the hierarchical relations of the division of labor become heavily infused with class and patriarchal relations, but to discuss those in a clear-headed manner we must first recognize the real and often unpleasant elements of difference, power, and exploitation inherent in the division of labor.

The relations that emerge from the division of labor and the practical tasks carried out in social production are fraught with social differentials and unities. That is, the division of labor provides a material axis around which people not only develop technical capabilities and knowledge, but also positions of power over others, associations that enhance their collective leverage, and even means of exploitation of the labor of others (Rueschmeyer, 1986). Critics of Marx have asked repeatedly how ownership and exploitation can be the only sources and objects of power. The answer is they cannot. Divisions of labor congeal into definite social relations of production, such as the old inside contracting system in the steel industry, where skilled workers hired, commanded, and paid less-skilled workers. In attributing relations of power and exploitation to the division of labor, however, we do not mean to equate or confuse them with consequences of class and gender, which differ both in degree and in kind. The social dimension of the division of labor is distinguished by power relations that are relatively muted and technical relations that are more apparent.

We must here engage Wright's attempt to introduce two new kinds of class exploitation, complementing exploitation through private property, one based on organizational assets, the other on skill assets.[4] Organiza-

---

[3] The distinction between horizontal and vertical dimensions of the division of labor is subject to some confusion. Vertical can either mean work hierarchies or sequential steps in processing; horizontal usually refers to divides between commodity sectors or parallel processes on the shop floor, but might as easily apply to different stages in the production cycle or to work in circulation. While it is hard to avoid use of these terms, they cannot be more than rough and ready pointers.

[4] Wright gets here by way of Roemer's (1982) neoclassical Marxism, on which something must be said. (cf. E. Wood, 1989). His "discovery" that non-human inputs to production are exploited is not news to anyone familiar with Marx's distinction between use-value and labor-value. His

tional assets are positions of command in managerial hierarchies of large organizations (corporate, state, non-profit); skill assets are those scarce capacities that accompany technical labor, usually based on higher education. Both sorts of assets allow managers and technicians to extract a measure of surplus labor from others, tilting the distribution of income their way through inflated salaries, benefits, fees generated by educational credentials, professional monopolies, expense accounts, company cars, and the like (Larson, 1977; Collins, 1979). But none of this constitutes class exploitation, unless we throw the definitional door wide open. The kinds of inequalities and exploitation Wright is after arise principally from the division of labor, not class.[5]

We have been talking up to now in the rather dry terms of social position in the division of labor, rather than of the people who animate those positions. That is, allocating different people to different tasks is very much a part of the division of labor as a structuring process; the creation and filling of positions are deeply entwined. If white men once lay hold of key positions of managerial power and technical skill, for example, while blacks and women are consigned to menial jobs or domestic duties, this will have an enormous impact on the flow of resources and the way new jobs open up for (or are closed to) each group and, conversely, on how the division of labor itself is extended and elaborated.

When social theorists venture far back in time, the division of labor is always close at hand in child-rearing, hunting, gardening, warfare, and the like. As a force of production present throughout human history, the division of labor has been steadily elaborated in number of specializations, enlarged into new areas of human activity, and ordered and reshaped into vastly diverse forms. As a set of social relations it has always played a critical role, as in the social divides between military and civilian life, seafarers and nomads, cotton plantations and wheat farms. These divisions have deep resonance with relations of class and gender, as in the way Germanic warriors were transformed into feudal rulers in northern Europe, agriculture became the province of independent farm families in the northern United States, or irrigated rice growing is being transferred from

definition of exploitation as transfers of (surplus) labor from one person to another is sound, but his rejection of the labor theory of value is based on a failure to apply the concept of socially necessary labor (that is, he rejects Adam Smith's theory, not Marx's). He is correct to argue that exploitation can occur "at a distance", via markets, without direct control of the labor process, owing to unequal distribution of assets. However, his rejection of the labor transfer definition of exploitation on the basis of the silly example of the poor lazy man exploiting the rich industrious one is, well, silly (recall Mandeville's fable of the bees); lazy poor men do not stay in power long. Finally, the game-theoretic approach, defining exploitation as situations where individuals would choose to opt out of the system to improve their welfare, is based on an untenable methodological individualism which annuls the foundations of Marxist social theory.

[5] Indeed, Marxist value theory necessarily implies secondary forms of exploitation by virtue of the mechanism of value-transfers between sectors, regions, or between production and circulation (Devine, 1989). Wright's treatment of value theory and skill exploitation is suspect because he does not factor in skills as part of the socially necessary element of labor time and quality (Itoh, 1987).

women to men in the Gambia. We consider first the intersection of class and division of labor.

## Class and the Control of Production

Class is sometimes defined simply as any relation of power between groups of people (e.g., Dahrendorf, 1959). So universal a definition, however, is insufficient for analytic work. A net cast so wide comes up with any and every form of social domination, such as that exercised by the police over a city or the teacher over a classroom. It is more helpful to give these relations names other than class. Similarly, feminists have rejected the term sex-class as ultimately unhelpful in understanding the specific condition of women (Ramazanoglu, 1989). Class power can best be distinguished from mere domination on the basis of the objective, material benefits it confers on the ruling group, benefits which go beyond a Nietzschean pleasure in controlling other human beings.

The Marxist definition of class, which we adopt, rests on the relations of production, that is, on the social conditions under which the necessary labor of transforming nature to support the populace is undertaken.[6] At the heart of class relations of production is exploitation, a process by which quantities of surplus labor/output are extracted from the direct producers and used to support the dominant class(es). Looking out from capitalist Europe, Marx and Engels pressed home the central role of private property (the form of possession of the means of production), and the degree of domination over production (control of the labor process) in defining production relations. Modes of production, such as capitalism, feudalism, and slavery, were, in their view, particular combinations of relations and forces of production. It is the relation to production which class exploitation holds, and the centrality of production in human life and historical development, that gives the Marxist concept of class its force. This much is common knowledge and common ground among Marxists. At such a level of abstraction the concept of class cannot bear the weight of too much historical specificity, however. The Marxist concept, derived from the study of capitalism, has been applied fruitfully to European feudalism and ancient Greco-Roman slavery (Brenner, 1985; Ste. Croix, 1984), but not without serious dispute (e.g., Bois, 1984; Wood, 1988). Although less germane in non-European settings, it can probably be stretched to cover class systems such as ancient China, where the state extracted most of the surplus and the ruling class included a large, propertyless mandarinate (Moore, 1966). But it is not infinitely elastic or universally applicable.

---

[6] This is not to say that class is the only kind of production relation possible. As we have just argued, the division of labor also implies certain relations of production; so does gender.

Class structures have a substantial degree of historical permanency, even as they are adapted and extended; that is, there can be changes in particulars without necessarily transforming the essential class relations. Nonetheless, social structures cannot be frozen without killing the living tissue of societies developing through time, as Thompson (1964) has observed.[7] While Marx dissected class using his classic triad – extraction of surplus, ownership of means of production, and control of the labor process – he did not stop there. His method was to spiral up and out from a bare bones definition, not just adding elements to the analysis, but recasting the categories as seen from new angles (Ollman, 1971). This is not just a strategy of thickening the categories, but indicative of Marx's awareness of capitalism as a dynamic and self-transforming system of production.

To illustrate, consider the period immediately preceding the rise of industrial capitalism in the United States. There was a class of merchant capitalists defined, in part, by their control over the means of commerce and commercial money, though they did not control production or the means thereof. At the same time a class of small masters existed which did own its tools and directed the work of apprentices. As farmers began to move from the land and artisans to leave their tools with the creation of an army of wage labor, a new element was added to the pre-capitalist system, but this did not immediately usher in a whole new class system. Wage labor, as day labor on docks or building roads, coexisted with petty commodity production well into the 19th century.[8] This was not yet the era of industrial capitalism, in the sense of modern factories and generalized wage-labor, but it cannot be cast into the limbo of a "transition to capitalism", which in the American context has little meaning.

Perhaps the crucial moment was the real subsumption (control) of labor by the rising class of industrial capitalists, which proceeded rapidly in New England in the period before 1850. This stage was still dominated by manufacture (small workshops with a developed detail division of labor but little machinery) in which small capitalists mastered production and hired wage labor, but did not displace the class of merchants on whom they were dependent. Or the critical step may have been when merchant capitalists were displaced, with widespread development of the factory system and credit money, and industrialists achieved financial and commercial independence – a point reached roughly by the Civil War (Porter and Livesay, 1971).

[7] Of course, it is necessary to stop the clock from time to time, and take a cross-section. This is essential to the taxonomic studies of sociologists, which reveal the composition of social forces in their own way. The hazard lies in failing to start the clock again. Perhaps a more graphic metaphor is the tale of the geographer who was too lazy to take a proper living core of a large bristlecone pine and who, on cutting it down to count the rings, discovered to his horror that it was the oldest tree known!

[8] As Przeworski (1985) argues, the formation of classes under capitalism depends as much on the way in which wage workers are absorbed into the new economy as on the fact of their divorce from the means of production.

Wright is the leading Marxist exponent of class theory in sociology today, and his exemplary efforts to define capitalist classes clearly and objectively illustrate the problem of pinning down the precise nature of the concept. Wright is determined to reduce class to a fundamental triad, yet the fundamentals of this triad have changed over the years. This is not a sign of inconsistency or infirmity of judgement, but reflects the mobile nature of class itself. In his first effort, Wright (1976) begins with the labor process and chooses three elements: control of one's own labor, control of tools, and control over the labor of others. Later, Wright (1980) realizes this leaves out control over investment, so the triad becomes: control over money capital, physical capital, and variable capital (labor). In his major work to date, Wright (1985), as noted above, shifts his foundation to private property, organizational assets, and labor skills. But the list could easily be extended to other aspects of the production, circulation, and organization of capital, for instance, to the vital control points of merchant or commercial capital, new scientific and technical knowledge, or today's elaborate securities trading and credit systems. As capitalism has developed as a system of production, so has the meaning of capital been enlarged, deepened, and altered (see chapter 3).

It is easy to see the problem. At which point did the *real* capitalist class stand up? Class is a relation of power which must continually be maintained, extended, and recreated in the face of changing conditions. It is therefore inextricably entwined with the development of capitalism as a system of production, circulation, and exploitation. Carchedi (1977) is right to insist on lodging the definition of class within a framework of the evolution of capitalism. One cannot settle on a tidy definition of class that stops history in its tracks.[9]

Of course, this kind of argument can take us so far away from the core relation of ownership of private property and employment of wage-labor that we lose our grip. Long-term structural continuities within US capitalism can be discerned in the fact that a small group of people persistently control the key elements of capital. We need a concept of structural reproduction, even though that reproduction is neither guaranteed nor fully intentional. Some writers have argued that the information revolution has annulled the old class system, with its antiquated power relations in rusting factories (e.g., Bell, 1973). The answer to this is that capital of the "old" type has been quite successful in capturing the transformative potential of advancing knowledge (though by no means containing it completely) through the tried and true methods of owning, controlling, and bribing the broadcast media, newspapers and publishers, the advertising industry,

---

[9] This is not only a matter of class relations forming through a number of consolidations of power over more concrete elements of production, as Marx's theory of subsumption of labor implies, but of the most abstract foundations of class also being altered – as in the redefinition of private property toward corporate forms, for example (Lustig, 1982).

universities and experts, and so forth (Bagdikian, 1990). Such are the stakes of the class struggle.

Marxists traditionally depict class formation and class struggle as taking place on the foundation of class structure, as in Lukacs' (1922) formulation of class-in-itself/class-for-itself. But the dualization of class structure and class struggle leads friend and foe alike to displace the explanatory failings of Marxian class theory – particularly the failure of the working class to make the socialist revolution – onto the subjective dimensions of consciousness and politics. Przeworski (1985, pp. 66–67) represents this tendency when he slips from the reasonable formulation that "Classes as historical actors are not given uniquely by any objective positions" to the fallacious proposition, "if struggles do have an autonomous effect upon class formation, then the places in the relations of production, whatever they are, can no longer be viewed as objective....", a position that negates the whole idea of class structure. He overlooks the objective process by which class structuration itself is (or is not) altered by class formation and class struggle.[10] One can, for example, think of many modifications of the wage-labor relation that affect not only the experience of the working class but its very class base. Replacing hourly wages with monthly salaries is not merely a change in the wage form; it modifies the relation of wage-labor itself because it firms the worker's tenuous ties to the means of production and gives a greater solidity to the distributional claim over a portion of the final product. Unionization, so often treated as only a mode of working class organization, modified the basic wage relation of 20th century capitalism by altering the legal rights of employer and employee (Clark, 1989).

A persistent source of confusion in class analysis comes from expecting the abstract concept of class to generate broad taxonomic class maps.[11] One cannot infer from the pervasiveness of class effects in society that their cause, a class relation, is equally pervasive. If class is defined in relation to employment, then children, retirees, or the indigent cannot be assigned to the working class except by indirect association (e.g., with a working husband, past or future work life, industrial reserve army, etc.). To say that wage-labor is a restricted experience does not mean that it is not widespread or that its force cannot be felt at a distance, however. Indeed, the power of class lies precisely in the fact that large effects issue from a limited cause, not that everything or everyone is in a class relation.

[10] Worse, he goes on to claim, "The very theory of classes must be viewed as internal to particular political projects. Positions within the relations of production ... are thus no longer viewed as objective in the sense of being prior to class struggles. They are objective only to the extent to which they validate or invalidate the practices of class formation...." This dissolves structuration into class formation as a strictly political process, and annihilates social science in the name of political projects (cf. Burawoy, 1989; Walker, 1989a).

[11] This confuses abstractions with generalizations, in classic positivist fashion (Walker, 1989a). Wright (1990) has finally acknowledged the futility of a Marxist class theorist submitting to the impulse to find a class niche for everyone in society.

The "discovery" by critics (e.g., Marshall *et al.*, 1988) that not everyone has a position in the Marxian "class structure" does not obviate the usefulness of class theory in uncovering important characteristics of capitalist societies (cf. Burawoy, 1989).

The confusion of static versus dynamic or abstract versus concrete class analysis is amplified when one moves to international comparisons of capitalist class systems, as Wright and his colleagues attempt to do. A single class scheme cannot be applied unproblematically to different countries where class relations have evolved in different ways. This point has been sharply illustrated in the transition to agrarian capitalism from pre-capitalist modes of production: witness the contrast between the incorporation of the Prussian Junkers into German capitalism and the revolutionary elimination of the French nobility of the *ancien regime* (Moore, 1966). Similarly, the British ruling class is deeply marked by the failure of the industrial bourgeoisie to displace the mercantile and financial capitalists of the City of London (Anderson, 1987), while the US ruling class cannot be understood without considering the defeat of slavery followed by the curious reinsertion of southern landholders after the Civil War (Mandle, 1978). The diverse forms of capitalist class formation, incorporating elements of state and military, manufacturing and banking, working class struggle, and the like, are not unseemly contingencies of politics but lively differences over time and place that the theory of class is supposed to help one grasp. Class is to be sought in the way production relations are orchestrated and dominated by particular configurations of power, institutions, leadership, and ideology. To be called capitalist, those relations must pivot on private property in a commercial setting, but one looks in vain for a *pure* form of capitalist class structure behind the refractory circumstances in which classes actually form – for an essence that is *at once* abstraction and concrete circumstance.

## Class and the Division of Labor

The division of labor is at the heart of several dilemmas faced by class theorists. As soon as they try to bring in more subtlety, multidimensionality, and concreteness, they begin mixing in elements of the division of labor, either explicitly or implicitly, without any theoretical preparation. Wright's work is symptomatic: the division of labor is merely an "empirical" concept in contrast to the structural force of class (Wright, 1980). Indeed, the term does not even rate an entry in the index of his magnum opus, *Classes*. He fails to see that the quandaries of class analysis cannot be overcome without a judicious grasp of the division of labor as a force in its own right, and of the way class and division of labor coordinate and overlap as systems of difference and power. Without a clear view of these dynamic interactions, class analysis rests entirely on an overburdened and inadequate category of class.

The confusion begins with Marx himself, as Rattansi (1982) has shown. In his early writings, such as the *1844 Manuscripts* and *The German Ideology*, Marx overloads the division of labor as an explanation for the origin of class, gender relations, social development, exchange, and alienation, ending up with the ultimately insupportable claim that socialism must mean the abolition of the division of labor itself, embodied in the infamous statement about being a hunter, fisher, shepherd, and critic all in the course of a day. In his later works, *Grundrisse* and *Capital*, Marx separates class clearly from the division of labor as he penetrates more deeply into the capitalist relations of production lurking behind the market. The clearest example of this distinction is found in his view that the hierarchical relations of management and supervision of collective labor are necessities independent of the capitalist form of modern industry. Marx continues to criticize the way capital distorts and coopts such differences, including the deadening separation of mental and manual labor, but he no longer imagines that socialism means abolishing the division of labor, *tout court*; only its previous capitalist forms. Furthermore, he goes on to analyze the way the labor process, including the division of labor, helps break down distinctions among workers and furthers their alienation from control of their own labor: labor becomes more abstract, more homogeneous, more social, and hence class relations are firmed up. This emerging dialectic of class and division of labor has been lost in subsequent thinking.

### Stratification and the shifting class map

Marxian class theory has foundered for over a century as new occupational groups thrown up by capitalist development failed to conform to expectations as to shared organization, interest, and politics with the bulk of the blue-collar working class (Przeworski, 1985). For example, Kautsky, the leading theorist of the Second International, took note of a new category of proletarians emerging in the late 19th century – educated workers in offices of banks, railroads, government, and the sciences – but dismissed the growing crew of "social parasites" such as middlemen and saloonkeepers, as there is no real class position for them in his framework. These difficulties grew in the 20th century with the enormous expansion of white-collar work and of employment in non-profit institutions and government, as well as the vastly increased numbers of students, retirees, welfare recipients and others outside the realm of paid labor. These groupings were the source of heated debates in the revival of Marxian theory in the 1970s (e.g., Poulantzas, 1975; Carchedi, 1977; Walker, 1979) and the delineation of the true proletariat and the theoretical standing of the so-called new middle class occupy center stage in contemporary debates.

The working class can be defined simply as all those employed as wage-labor, but this includes salaried workers in the highest ranks of management, the professions, or the state who conform in no reasonable way to the vision of an oppositional proletariat. The theorist wishing to maintain political rectitude could fall back on a minimalist conception of the proletariat as manual and factory workers (Gorz, 1982). This is the solution of Poulantzas, who excludes all those who do not create surplus value (and consigns them to the purgatory of the petit bourgeoisie). Przeworski, despite his critical distance from orthodox Marxist interpretations of proletarian politics,[12] also ends up pushing everyone else off into a capacious grab-bag called "surplus labor." Braverman (1974) tried to reconcile these two positions by arguing that capitalist production has required the de-skilling and homogenization of all wage-labor, hence extending proletarianization into the realm of white-collar workers, destroying their middle class pretenses.

Juxtaposed to this long debate is another stream of thought on class, derived from Max Weber (Weber, 1978; Lockwood, 1958; Goldthorpe et al., 1969; Parkin, 1979). Weberian stratification analysis sorts people according to rough equivalence of income, rewards and chance for advancement (market situation), authority and autonomy on the job (work situation), and prestige (status situation). Its theoretical foundation lies in Weber's view that class rests on skills and authority as well as on property; that classes are mediated by market exchange and its distribution of rewards; and that class closure cannot be determined independently of personal mobility, life chances, and life styles.

For the Weberian social taxonomists, Marxian class analysis fails because it cannot reveal classes in a self-evident way through inspection of the everyday social world. They treat class as a strictly empirical category, boxes that hold individuals, with a simple mapping of positions. This misunderstands the purpose of Marx's class analysis, which was to see beyond everyday appearances and discover a more profound dynamic, set in train by relations in the hidden abode of production (Burawoy, 1985). These are opposing visions of the purpose of class analysis: to grasp the structuring of large-scale dynamic social systems or to locate individuals in a stratification system. The division of labor is an important bridge between the two.

In Weberian analysis, the middle classes bulk so large that the bourgeoisie and proletariat are reduced to mere shadows of their Marxian selves. The *working* class includes only manual workers, while an *intermediate* class absorbs all routine non-manual office workers, personal service workers, low-grade technicians, and supervisors (along with small proprietors). The former make up 36.5% of the employed in Britain, at last count, the latter, another 36.3%. Next up the hierarchy is the *service* class of managers, high-grade technicians, and professionals, who constitute fully 27.3% of the employed population (Marshall et al., 1988; cf. Mills, 1951). The *capitalist* class appears as a separate category in some

---

[12] Indeed, he provides a devastating materialist critique of those leftists who have always been uncomfortable with the social-democratic tendencies of the working class.

versions (e.g., Abercrombie and Urry, 1983), but disappears entirely into the service class in others (e.g., Goldthorpe, 1982).

This schema rests heavily on bringing occupational categories, and their work and reward attributes – in other words positions and advantages within the division of labor – into class analysis. Weber's key class elements of skill and authority are similarly constituted, as is his obsession with power as monopolization of opportunities and rewards. Weberian stratification studies, then, can be seen as an effort to combine class with division of labor so as to achieve greater specificity in representing everyday situations. There is much to be gained by such a project, so long as it is not confused with the aims and categories of Marxian theory. Yet that is exactly what has happened, thanks to the conceptual muddle that Weber made of class theory;[13] it can be argued that he returned to the thinking of the young Marx, in which class and division of labor were not yet sorted out, and the market, rather than production, was the principal moment of economic determination.

Neither Marxian nor Weberian approaches think through the division of labor before juxtaposing it to class. The people called white-collar workers or unproductive labor or intermediate or service classes have definite positions in the industrial capitalist economies. They fall largely into three broad categories of labor – circulation, consumption, and indirect – which are examined in the next chapter. For the moment, an outline will suffice. *Circulation workers* are employed in the sphere of distribution, finance, and correlate activities relating to the movement of goods and money. They are unambiguously wage-laborers, entirely homologous to factory workers and manual trades in relation to their bosses, although they work in stores, warehouses, and offices, and frequently work with paper, electronic signals, and other relatively insubstantial things. They are also overwhelmingly women, bringing the gender divide into the thick of class relations (see below). *Consumption workers* are those in sales, consumer services such as entertainment, and social services such as health and education. They, too, are wage-workers in the main, but their workplaces are again distinctive (hospitals, stages, schools, etc.) as is their work, which can involve a good deal of personal contact with consumers and is far from the physical hewing of wood and grinding of metals.

All complex labor processes, whether in production, circulation, or consumption, require large numbers of *indirect workers*, including engineers, repair(wo)men, and janitors, who support, extend, and amplify the labor of those working directly on manufactured goods, at sales desks,

---

[13] The end-products of this are well summarized by Przeworski (1985, p. 64): "The system of stratification distributes people along continuous strata, bulging in the middle to generate the 'middle class.' The resulting consequences are well known: empirical descriptions of 'socioeconomic standings' became independent of any historical understanding; the vision of classes as historical actors became replaced by statistical analyses of distribution of income, education and prestige; the analysis of social differentiation became separated from the analysis of conflict."

or in hospital wards. Many are ordinary wage-workers, with few claims on authority or skill, but it is here that we find the special group at the heart of the new middle class: managers, technicians, and mental workers. These people are elevated by the hierarchical dimension of the division of labor, they become deeply implicated in the design, conception, and control of the collective labor and the work of others, and are rarely free of the taint of ruling class prerogatives, incomes, and ideology. Their position in the class structure is therefore profoundly ambiguous.

Both Marxian and Weberian class analysis also struggle with the problem of where to place those people not directly employed by capital (or the state). Some, such as children, are outside the world of work entirely, as noted before. But others are working outside the sphere of capitalist production or the control of the state, in households and communities, informal and underground economies. In short, they occupy divisions of labor unrecognized by traditional class theory (Pahl, 1984). Przeworski rightly chides Wright for putting aside virtually half the adult population of the United States as economically inactive before getting down to the brass tacks of class analysis. One cannot deny that class relations are attenuated when there is no direct involvement in the *capitalist* division of labor. We shall return to this issue; for the moment, let us remain inside the capitalist sphere.

In Wright's principled defense of Marxian analysis, he tackles the puzzle of ambiguous class positions, especially that of the middle class. His original contribution was the idea of contradictory class locations, straddling class boundaries (Wright, 1976). For example, the intermediate position of managers and professionals is based on their control over the work of others (or of themselves). Wright fails to recognize, however, that these characteristics rest in large part on managers' and technicians' positions in the division of labor, not class relations *per se*. In other words, he lacks a notion of multiple structured systems which must be meshed in order to comprehend the complexity of the social order.[14]

In his later work, Wright (1985) tries to deal with multiple class positions by introducing skill assets and organizational assets, but this does not ultimately prove any more satisfactory in handling the empirical problems.[15] Moreover, he beats a tactical retreat by arguing that skill and organizational exploitation represent elements of post-capitalist modes of production (statist and socialist) in the bosom of capitalism. This only ends up shifting the muddle onto the concept of socialism, which appears

---

[14] In failing to distinguish between structures, Wright, in the end, merely inserts boxes between boxes (MacKenzie, 1982). Indeed, Wright (1980) relegates division of labor to the position of "function" while preserving for class the notion of structure – as if class did not have a function (exploitation) or division of labor a structure.

[15] Moreover, it veers remarkably close to Weber's formulation, as Wright acknowledges (1985, pp. 106–08). For a comparison of Goldthorpe and Wright's class maps in Britain, see Marshall *et al.*, 1988, ch 3. For a discussion of convergence between some Marxist and Weberian class theorists, see Burris (1987).

to be defined as the rule of managers and experts: the much-maligned "New Class" of Eastern European critics of communism (Szelenyi, 1986).

After a very long march round the subject, Wright (1989a), in his latest reconsideration, returns to Marx's two basic classes, workers and capitalists. He now realizes that overloading the little boat of class theory merely swamps it. He sees the complications of stratification analysis as lying outside the core structure of class, and notions of contradictory, multiple, and mediated class locations, of temporal trajectories and secondary exploitations, now fill his vocabulary, which is, as always, admirably clear and sensible. Only one term is still lacking: division of labor.

If we consider class and division of labor together, it is possible to understand why the new middle class holds a contradictory position. While managerial and technical workers do not own the means of production, they do hold a favored situation in terms of degree of control and ability to demand a higher reward than other workers. This need not be attributed to class standing, as it flows from the division of labor itself. Relative advantage may lead such people to keep their distance from manual workers, create further advantage through professional organizations, and cozy up to their capitalist betters, which the latter are happy to encourage through such perquisites as lofty titles, stock offerings, and salary increases. Indeed, many managers and skilled workers advance to the top of the corporate heap or set up their own companies, where they unambiguously join the capitalist class.[16]

In short, there are many reasons for the ambiguous standing – and the ambivalent politics, organization, and consciousness – of the new middle class, but none of them require creating new categories, abandoning the structural analysis of economically based classes, or other intellectual somersaults. We simply need a more amplified sense of social structuring than class analysis alone can provide.

### The ongoing interplay of class and division of labor

It is not sufficient to recognize that class and division of labor are simultaneous features of capitalist production. They are also features of capitalist *development*, interactive and mutually modifying; the two evolve in tandem. As Przeworski (1985, p. 60) says, "the proletariat could not have been formed as a class once and for all by the end of the 19th century because capitalist development continually transforms the structure of places in the system of production and realization of capital. . . ." The reciprocal relationship between capital and the division of labor is by no means simple. Division of labor is not merely a modifier in the grammar of class.

---

[16] On the other hand, the perfidiousness of the so-called labor aristocracy is usually exaggerated, and skilled workers have often been in the leadership of working class movements (Barbalet, 1987).

On the one hand, capitalism clearly drives the development of the division of labor. From the first industrial revolution onward, capital and the capitalists have seized upon existing divisions of labor and profoundly transformed them by destroying household units, gathering workers into workshops and factories, multiplying the detail division of labor, creating new branches of industry, and so forth. In Marx's view, certain aspects of this development color the division of labor as indelibly *capitalist*, especially the infusion of power into the work of supervision and direction, the manic rationalization of the detail division of labor, and the separation of mental and manual labor (Braverman, 1974; Rattansi, 1982).

Causality runs in the other direction as well. As the division of labor has been expanded and elaborated, it has given rise to new dimensions of human activity – new jobs, new processes, new knowledge, new workplaces, etc. – and, consequently, new sources of differentiation, power, and exploitation. These become sites and tools of class struggle and class struggle, in turn, affects the class structure. But all this must be placed on the shifting ground of an unfolding economy. Capital must try to prevail vis-à-vis the most important components of the economy if it is to continue to operate effectively; capitalist class power, like capital, must accumulate in order to be reproduced (Carchedi, 1977).

As Marx showed, once capital became the guiding force of social production, it began to revolutionize economy and society in ways that built a whole new edifice of class onto the elemental base of proletarianization and private property. This process must be tied to the shifting division of labor, which has played a decisive role in reframing the house of class over time. As whole new spheres of work such as engineering and medical technology opened up, they became contested zones of class structuring (Kocka, 1980a). In the United States, for example, engineering became the right arm of capital and engineers frequently moved on to become capitalists themselves (Noble, 1977), whereas in Britain engineers are mostly seen as a kind of skilled worker. This difference is a consequence of contrasting histories, including sharper class lines in Britain, weaker capitalist control of production methods, poorer development of special schools, and more extensive unionization. Doctors have carved out a substantial wedge of power and income in British and American society, by means of a huge research apparatus, professional monopoly, and control of most medical institutions (Brown, 1979; Starr, 1982), and many individuals have been able to parlay these gains into considerable personal wealth and huge staffs, swelling the ranks of the capitalist class. Very few doctors, even those on salaries in the largest hospital or the National Health Service, can be considered working class, or consider themselves so.

With engineers and doctors, new divisions of labor have served to strengthen the capitalist pole of the class structure, while creating a distinctive and potent "new middle class" which has played a decisive role in economics and politics in most 20th century capitalist countries. For example, the new middle class were leading actors in the transformation of capitalist production and government achieved by scientific management in the early 20th century (Veblen, 1921; Ehrenreich and Ehrenreich, 1977) – even if neo-Weberians like Urry (1986) greatly overstate the autonomous role of the "service class" vis-à-vis capitalists in these changes (Walker and Greenberg, 1982). Similarly, the middle class has been pivotal in the contemporary triumph of the New Right (Davis, 1986; Ehrenreich, 1989). Marxism is only weakened by a failure to recognize the place of the middle class in the economy and politics and to situate the leverage of this group in objective, structural characteristics of the capitalist economy. While these rest more on the division of labor than on property, the efficacy of the middle class as a social agent is closely tied to its cozy relations with capitalists within the large corporation and other organizational networks of modern production. We explore the latter shifts in chapter 3.

Division of labor also insinuates itself into the creation of new working class positions. The simple condition of working for wages has lost some of its force. Classic proletarianization continues to affect millions around the world, but wage work only establishes capitalist class relations in a weak sense in advanced economic regions. Hence, Marxists have had to append a second stage of proletarianization, arguing that the capitalist labor process grinds exceedingly fine as it deskills and homogenizes common workers (Braverman, 1974; Gleicher, 1985–86). Yet the deskilling mechanism works most imperfectly, and is balanced by several opposing forces, not least the way the division of labor throws up new kinds of work involving new skills (Walker, 1989b). The working class standing of people employed in these jobs is by no means assured at the level of class structuring, let alone class formation and consciousness – think of the ambiguous histories of class alignment in office work, teaching, or health care. Nurses, for example, are strongly elevated by education, skills and responsibilities, as well as by the professional model set by doctors, but are also demoted by their gender, service mission, and subordination to physicians. The working class consciousness of nurses is much stronger where they have become militant unionists, but many stand aloof from traditional working class organizations and politics. It is best, in such cases, to recognize their ambiguities and contradictions, while stressing the ongoing fact of class power (cf. Callinicos and Harman, 1987).

Transformations of the division of labor (and of attendant spatial divisions of labor) are profoundly disorienting, for the working class especially (Storper and Walker, 1989). Fundamental class transformation has accompanied the transition from every major capitalist epoch to the next, providing an essential foundation for quieting class struggle in Britain after 1850 (Foster, 1974), the disastrous collapse of US unions in the 1920s (Davis, 1986), as well as the massive defeats of the working class in both the United States and Britain in our time (Bluestone and Harrison, 1982). The recent debate over class in Britain has been triggered in no small part by the massive restructuring of industry that has undermined so many bastions of traditional manual working class and Labour Party power, and highlighted the long-term shift into new kinds of work, especially in London and the southeast (Massey, 1984). As Marshall *et al.* (1988, pp. 3-4) observe:

> Critics of class analysis usually begin from observations about the economy and the impact of economic or technological change on the occupational structure [i.e. on the division of labor].... The net effect of these several processes is to fragment traditional "bourgeois" and "proletarian groupings" and redraw the boundaries of common interest in new ways not specifically reducible to established differences of class. Myriad "new middle" and "new working" classes have been created. Their conceptual locations and political proclivities are objects of intense speculation.

This speculation has, of course, been fueled by the enormously successful mobilization of the right under Thatcher and Reagan.

A central element of such disorientations, according to some, is that increasing fragmentation is inherent in an expanding division of labor, and this makes it harder for individuals to perceive class effects as opposed to particularistic interests deriving from occupational groupings (e.g., Hobsbawm, 1981). We find this dubious. The division of tasks and scale of social labor was already enormous in the "Golden Age" of labor organizing and is, over time, balanced by processes of integration and organization such as the growth of large firms, improvements in communications, concentration in larger and larger cities, or the emergence of national unions. One can argue with equal force that international competition and the growth of more flexible forms of production have had a splintering effect on capitalists. In the short term, affairs may well disadvantage the working class but this need not signal the demise of class or class politics.

### Class formation and open class structures

The overdetermination of class outcomes by the mediating impact of division of labor compels us to consider the process of class formation. This term is often laden with voluntarist overtones, as in Przeworski's strategic shift from class structure to class formation in order to evade the political difficulties of incorporating new groups of workers into traditional working-class organizations, parties, and voting patterns. He ends up, as noted above, making political formation of the working class *the* structuring process, with economics and production fading into a nebulous background. The same criticism can be made of Thompson's (1964) over-shift to culture and experience in the making of the working class. Without some kind of structural ordering, class formation becomes so vague a phenomenon as to be meaningless.[17] We have insisted on lodging many of

[17] Or we move to the realm of the absurd, as in Brint's (1984) definition of the "New Class" as sociocultural specialists.

the fundamental problems of class analysis in a more supple understanding of class structuring itself, including the interaction of class and division of labor. With that in mind, it is reasonable to move into the realm of class formation, and subsequent problems of class consciousness, politics, and struggle. This does not mean, however, that class formation can be relegated to a derivative realm of ideological or political superstructure.

Class formation is, in a sense, what people do with the material at hand in particular social situations (Katznelson, 1981). It is a middle-level concept needed to close the model of class society; class formation operates on the vast middle ground between abstract, elemental structures and concrete specifics of everyday life. That is, societies are open systems with indeterminate consequences, thanks both to conscious breaks with the past (the Marxists' class struggle) and to the unpredictable conjunctures of history generated by the complexity of real societies. The elemental production relations of class under capitalism are necessary but not sufficient to determine the formation of capitalist classes (Wright, 1985). People act on the basis of the rules of a class system, and in so doing make more of it than can ever be determined by knowledge of the rules alone – such is the dialectic of possibility and realization in human life. As a result, the elements of class institutions, cultural practices, and ideas, and the way the class struggle is fought out in all its dimensions, affect the very "classness" of society – the strength of the class field of force.

On the basis of abstract class relations we can only say that the working class and capitalist class will never cease to be recognizable so long as capitalist firms employ wage-labor; we cannot ever assure that they will form the homogeneous and self-conscious entities that inhabit the revolutionary dreams of devout Leninists. In his defense of the middle-range tactics of social democracy, Przeworski is quite correct to insist that because class is never a settled matter, political organization and struggle can be as much about the definition of class as about the pursuit of structurally defined class interests. The same is true of the Weberian concern with the closure of class boundaries through family, inheritance, schooling, and the like, since these undoubtedly affect the degree of class solidarity and organization, and hence the class divide in society (Murphy, 1988). Since change is persistent under capitalism, class formation is necessarily partial, and the imprinting of class on everyday life is incomplete. We can acknowledge this sort of feedback on class structure without cutting class formation loose from relations of production.

Class formation traditionally referred to the kinds of unions and parties classes erect on the structural foundation, but since Thompson's influential work, it has been understood to encompass the way people put flesh on that structure, particularly in making recognizable class cultures, class institutions, and class consciousness out of the complex materials of social situations which include language, artistic expression, traditions of popular liberty, and so forth (Bourdieu, 1977). The possible outcomes are many and unpredictable. The openness of history must be faced straight on, and with it the variant pathways to class formation in different countries (Kocka, 1980a). That class organization or class culture are weaker in the United States than in Sweden is more than an incidental difficulty to be overcome by militant, far-sighted leaders; long histories of republicanism, ethnic quarrels, and residential segregation, for example, have deeply colored the class structuring of American society (Katznelson, 1981; Davis, 1986).

## Gender and Patriarchy: Beyond Production

Feminism has shaken the house of social theory to its foundations. It has not only shone a light on the facts of women's oppression, it has forced examination of areas of social life not previously considered central to history and social science, and has destroyed the barrier between the public and private spheres, politicizing things once consigned to the realm of the personal, such as family, sexuality, and emotional life. The impact of feminism on left thinking has been especially profound; awareness of the stark toll of gender oppression on women has altered forever the meaning of historical materialism and human liberation (Hartsock, 1987). How, then, do gender relations compare with class relations and the division of labor as a structuring force on capitalist societies? To answer this, we must first establish the distinctive nature of gender and patriarchal domination, then work out the choreography of these deeply interwoven structures.

Gender relations are, decisively, relations of power. The name given this power of men over women is usually patriarchy, and its attendant violence, sexism, marital relations, childrearing practices, control of women's labor, and so forth can be seen across the widest range of past and present societies.[18] The constellation of patriarchy needs to be sorted out, however. As with class, gender power must have an object and an objective: the capacities of women and the gaining of material advantage by men. The power of men over women involves three basic elements: control over women's labor, control over women's child-bearing powers, and control over women's desires and affections (cf. Connell, 1987). As with division of labor, each of these realms involves separation and specialization. We shall consider each in turn, then take up three additional elements of

[18] The term patriarchy has been a useful political tool for teaching women about their subjugation, mobilizing against male resistance to change, and driving an opening wedge into a prevailing male-centered discourse on society and politics (e.g., Millett, 1977; Kuhn and Wolpe, 1978). Yet debate has raged over varieties of patriarchy and processes at work in women's subordination, given the ever-expanding range of historical-geographic knowledge being opened up by feminist research (see e.g., Moore, 1988; Ramanazoglu, 1989). We cannot do justice to this discussion and its now immense literature, but can only touch on a few salient issues necessary to confirm a degree of unity in the abstract concept of patriarchal gender relations.

gender relations: the family, the body, and the psyche. After that, we return to the interacting trajectories of gender, division of labor, and class.

Control over women's surplus labor, in the classic Marxian sense, is certainly a principal objective of male domination, but it is two further capacities of women, which are not included in the ordinary meaning of work – the making of new human beings and the creation of human bonds of affection and desire – which give gender relations their distinctiveness and their force in personal life. They relate, in part, to the unique capacities of the female body and are brought to bear overwhelmingly in the context of the family; yet neither the family nor the human body are sufficient to account for patriarchy and its historical force.

Gender, like any other system of relations in society, orders human interaction, creates positions and assigns people to them, and confers resources differentially. Gender relations have their material practices, institutional fabric, and forms of consciousness just as do class relations.[19] Gender structuring is a continual process, in which institutions, ideologies, and the system of relations itself are actively and recurrently constructed through everyday activity. It is, moreover, a process in which children grow up and develop as socially-defined men and women, rather than simply as nature-made males and females: hence the decisive turn to the concept of gender in place of sex. Indeed, the essential unity of men and women as creative, thinking beings means that gendering, desire, and consciousness have often been shaped in profoundly unnatural ways that have little to do with bodily and psychic attributes.

The first pillar of gender power is the control and appropriation of women's labor, which includes the immense mass of work maintaining the household (Oakley, 1974; Alexander, 1976); the constant labor of nurturing husbands, babies and relatives, as well as their own grooming and personal appearance for the pleasure of men (Ferguson, 1989); and work outside the household in activities as varied as commercial rice-growing in Africa, scavenging in Indonesia, and insurance underwriting in the United States (e.g., Boserup, 1970; Nash and Fernandez-Kelly, 1983; Pahl, 1984). Men, old people, and children, not to mention ruling classes, benefit from women's unremitting toil. But the crucial power relation for women, as women, is to men: men around the world have lived off of women's surplus labor as slaves, wives, and daughters, as well as employees. This is not to say that there are no pampered wives nor that poor husbands are not themselves exploited, only that male access to the products, benefits, and income generated by women's labor is extraordinarily pervasive.

Another pillar of patriarchy is the control over women's capacity to bear

---

[19] The critical observation that gender ideologies do not "accurately reflect" material power relations (e.g., Barrett, 1980; Moore, 1988) is not a valid reason to catapult "cultural representations" of gender into a position of causal superiority over material foundations of male-female relations; indeed, ideologies never simply reflect reality, much less reflect it accurately, and that is precisely why they are important social forces in their own right.

children, to breast-feed infants and to raise the young. Women are crucial economic assets who hold the key to future labor power and male heirs, as well as social assets for all the reasons parents and elders may value children (Meillassoux, 1981). The case for women's special powers and burdens of reproduction has to be carefully circumscribed to avoid naturalistic assumptions about their role in society: pregnancy and birth are biological functions, to be sure, but relatives, elders, and siblings are commonly involved in child-raising; wet-nurses, nannies, and servants may play the largest part in children's lives; children may be farmed out early on to day-care, apprenticeships, or boarding schools. Yet while the biological *mother's* role may be thus circumscribed, in the overwhelming number of instances it is still *women* who undertake the work of child-rearing.[20]

The third pillar of gender and patriarchy is human bonding, through what may broadly be called desire. Sexuality is the most dramatic facet of desire.[21] Sexuality is a constitutive element of gender, and its forms are always socially mediated, whether encouraged, channeled, or repressed. The dominant form has been strongly focused on genital sex in heterosexual relations, but this coexists in every known society with homosexuality, bisexuality, asexuality, androgyny, and so forth. Nonetheless, non-heterosexual forms have ordinarily been suppressed in patriarchal societies, often ruthlessly. Desire, moreover, goes beyond sexuality to include the social love, care, and bonding necessary to all human life. Ferguson (1989) argues that emotional ties and non-sexual desire have a place in personal security, parenting, and the formation of social groups, and that erotic bonding is only one form of uniting with a desired other. Social bonding plays an important role in gender formation, male power, and female resistance: hence the dangers for patriarchal order in unruly desires, comparable to the explosiveness of sexuality in Freud's view of psychic development and civilization.[22]

### The family, the body, and the mind

The primary trilogy of gender relations is woven into a larger fabric of patriarchal institutions and practices, or what Connell (1987) calls gender

---

[20] While some feminist writers, such as Connell and Moore, are inclined to abolish the distinction between child-rearing and other women's work, mothering remains a fulcrum on which balances women's special burden.

[21] The simple term "desire" seems preferable to borrowing the Freudian concept of "cathexis" as does Connell or the rather awkward "sex-affective energy" used by Ferguson. Sexuality receives by far the most attention in feminist literature, usually in an ongoing, sometimes rancorous debate with Freudian psychoanalysis.

[22] Ferguson takes this intriguing line of reasoning to a point where it virtually negates the troublesome aspects of sexuality altogether – including its bodily materiality, the struggle against a dominant heterosexuality, and conflictive desires for dominance and subordination, love and violence – to which the best insights of Freud and Foucault were directed (Rose, 1986).

orders. These can include everything from forms of the state to types of technologies that are male-dominated, but we shall confine ourselves here to three areas of long-standing concern: the family and household, the body and violence, and the mind and psychic life. Each is a structured reality independent of gender, but absolutely fundamental to the way patriarchal relations are developed and maintained.

The family is a central forum of male power and women's subordination, the principal forum within which sexuality and affection, the production of children, and women's labor have traditionally taken place (Barrett, 1980). Familial systems are held together through marriages, kinship networks, and systems of property which help determine the allocation of labor, personal rights and obligations, wealth transfers and income distribution, and so forth (Goody, 1976; Hareven, 1982). Families occur in the widest imaginable variety of forms, but have in most times and places been organized as methods of patriarchal control of women and children (Levi-Strauss, 1969; Stone, 1977). Patriarchy is not, therefore, bound to any one form of the family.

Family property is an important institution of patriarchy (Delphy, 1984); all property relations, therefore, are not class relations. Or, in the words of a United Nations report: "Women constitute half the world's population, perform nearly two-thirds of its work-hours, received one-tenth of the world's income and *own less than one hundredth of the world's property.*" (quoted in Cockburn, 1985, p. 7; italics added). It is not only family assets which have traditionally been the legal property of men, but also the woman herself and her children. This has often kept marriage and the family very close to slavery, certainly to master-servant relations (Lerner, 1986). Much of what Patterson (1982) says in his powerful study of slavery, in terms of degree of control and denial of name and identity could be applied equally to women. The husband's proprietary claims on sex and nurturing, and violent control of his prerogatives, are particularly highly charged. Marriage contracts, the legal heart of the family system, have thus been centrally concerned with property ownership, inheritance, and the wife's obligations to the husband.[23]

The importance of the family has led many feminists to identify the household as the site from which patriarchy extends outward into society (e.g., Barrett, 1980; Ferguson, 1989). This helps emphasize the importance of patriarchy, the productive nature of domestic labor and child-bearing, and the error of seeing families as bulwarks against exploitation, violence, and other outside dangers; it also has the virtue of depicting the family as an important set of social relations in its own right, not as a simple appendage of capitalism or other modes of production. Nonetheless, the family is not the elemental building block of society or a nature-given fact

---

[23] The relation of marriage to property, including traffic in wives, dowries, and brideprices, has long been a concern of anthropologists (Hirschon, 1984; Moore, 1988).

of social life. It arose with the emergence of gender relations and patriarchy and has been restructured many times (Lerner, 1986).[24]

Another site of gender relations is the body itself, as gay and lesbian theorists have made especially clear (Foucault, 1979; Fernback, 1981). The centrality of bodily forms for gender distinguish it from class or division of labor even though there are bodily forms and effects of all social relations: class may be sexualized, may appear in size differences, and may even kill. Patriarchal social relations are not inherent to sexual and reproductive differences, however, and cannot be justified by reference to biology.[25] Men and women are not differentiated so much by gifts of nature as by profound cultural processes of growing into and living "masculine" and "feminine" identities. These include an obsessive attention to appearance and dress, especially among young women aspiring to norms of beauty, but the effects of gendering cut much deeper. Gender identity works insidiously in ways of being that grow to be part of the physical presence of men and women, as characteristic forms of sexuality, aggression, bearing, movement, and so forth (Connell, 1987). Bodies can be substantially altered by gendered practices of eating, exercise, foot-binding, and the like: observe the dramatic effects of increased athleticism or widespread anorexia among contemporary American women, for example. The stress in contemporary culture is on an exaggerated masculinity and subordinate femininity, but intermediate and alternative forms such as "butch," "femme" or "androgynous" necessarily emerge from the real ambiguities of body and mind. Because biologically men and women are so nearly alike in so many attributes, a great deal of work goes into physical demarcations of gender and in every outward manifestation of difference (Connell, 1987).

[24] Similarly, the household is not a given dwelling and working unit, neatly bounding the family (Barrett, 1980), but like all social-spatial forms partly constitutive of what is considered the family-proper – particularly in the isolated household favored by the bourgeois nuclear family. Kinship networks regularly cut across household lines, as do wider networks of domestic labor and aid, as in immigrant neighborhoods in the United States (Stack, 1974), while several unrelated and unattached co-habitants may dwell inside the household.

[25] For critiques of sociobiology, see Gould, 1981; Rose et al., 1984. Most such differences – in genitals, breasts, average size – are obvious, and research has only reinforced the fact that men and women are more alike than different especially in mental capacities, the critical dimension of human speciation (Connell, 1987). The issue for social theory is what humanity has made of its very general endowments – hands, eyes, conceptual powers, genitals – none of which comes with a set of instructions for use. This guiding principle has served well for understanding in the face of reactionary beliefs in an unchangeable human nature, yet even historical materialism rests on essential claims about human attributes, albeit very general ones, such as laboring animals, social animals, thinking animals, and the like. Feminism brings to the fore our existence as sexual and childbearing animals, as beings with both a violent streak and a need for affection. Denying any of this flattens historical materialism into a cultural (even idealist) materialism, and eliminates certain problems of human existence that are better dealt with from the first. Human liberation is not built on denial but on transcending the past and its limits, overcoming whatever biological differences we have (must women catch up to men in size before men stop beating them? must all races interbreed before light ones stop demeaning darker ones?) and, in some cases, releasing basic human capacities, such as our delight in varied experience or love of hard work to a desired purpose.

One cannot adequately treat patriarchy and the body without also bringing in that fallen angel of desire, violence. Male violence and power directed at women's bodies, and ultimately to their minds, is pervasive in patriarchal societies, and feminists have widely documented the terror in the hidden abode of gender (Brownmiller, 1975; Daly, 1978). Violence is, of course, a general means of exercising power, and has been indispensable to ruling classes, national conquest, and racism (Patterson, 1982). Nonetheless, violence and its lesser cousins, harassment and degradation, are imbricated in men's power over women (and accordingly in their relation to other men) in a way that demonstrates the special importance of the body, desire, and personal (male) identity to patriarchal domination.

Gender relations also operate in an immediate and profound way in the realm of the mind and psychic life. Gendering is deeply implicated in personal identity, in one's own and others' expectations of what it means to be a functioning person. This follows from the reproductive, bodily, and sexual processes of gendering in childhood, long before work or property impinge directly on our lives. Class and division of labor also have massive effects on personal identity and on consciousness, but they are primarily economic relations of production and exploitation in a way gender is not.

Patriarchal domination is centrally about maintaining gendered identities, and masculine identity is both a manifestation of male power and a support for the exercise of power over women. To be a real man is to appear powerful, to be independent, instrumental, and competitive, among other things. In this masque, men must appear dominant over women to gain acceptance by their fellow men and to carve out personal space to exercise their autonomy. Femininity, on the other hand, denotes weakness and acquiescence to male initiatives and moods, self-effacing behavior in social situations, desire for attachment to a man, nurturing and affirmation of children and men, and so forth.

Feminists have turned repeatedly to theories in psychology and psychoanalysis for help in unraveling the mysteries of gender identities. There have been a range of explorations of a Freudian (e.g., Mitchell, 1975; Rose, 1986), semi-Freudian (Firestone, 1970; Irigaray, 1985) and counter-Freudian (Dinnerstein, 1976; Chodorow, 1978) nature concerning personality formation under patriarchy: why men are hostile to women, or more independent and competitive than women, why women identify closely with their mothers or develop more passive selves, and the like. Psychoanalytic approaches hold that the structuring of the mind in childhood, within the family, runs terribly deep and is, furthermore, a passionate, conflicted, and incomplete process. One can extend this idea to adult experience as well: lives are constructed of contradictory and incoherent elements, a person faces an endless stream of serious and often impossible demands, which can never be completely resolved, whose most terrifying and destructive elements need to be put out of conscious mind. The psychological component of gender consciousness speaks eloquently to the difficulty of casting out (or recasting) ideologies once they are literally incorporated in those who are the bearers of gender.

Admission of the unconscious, of desire and disgust in the human psyche, has the disturbing effect of asking social theory to face up to irrationality, passion, and violence in human affairs. Gender consciousness among women can mean personal identification of the oppressed with their oppressors, self-denigration, even passionate acquiescence in oppression. Male identity can equally be constructed on the obverse: disgust for the oppressed, desire for violent control, passionate disavowal of the effects of masculine behavior. Such deep insinuation of inhumanity within humanity points to elemental difficulties in rationalist visions of liberation.[26] None of this negates struggles for social justice, democracy, and socialism, but we stand a century further away from the Enlightenment than Marx and therefore can better measure the greater length of the long uphill push to achieve truly humane societies.


## Gender, Class, and Division of Labor

Although an appreciation of gender's impact on every sphere of social life must forever alter one's picture of history and of modern capitalism, we do not hold to the view that recognizing gender, race, or other relations of oppression makes class an obsolete category (e.g., Laclau and Mouffe, 1985; Cagan et al., 1986). Class theory is, certainly, not sufficient to handle all major forms of oppression and exploitation, but neither is patriarchy or racism. Feminism has itself run into the intersecting structures of power that divide women, giving them substantially different life chances. The cross-cutting effects of class, race, and imperialism have led many women to despair of retaining any unified sense of their plight at the hands of men (Ramazanoglu, 1989). The only way out of this impasse is to confront the intransitive and irreducible nature of each major structure of oppression in its own right, while realizing that gender, division of labor, and class are constructed simultaneously and reciprocally (Cain, 1986; Beneria and Roldan, 1987). Gender orders and gender formation differ markedly among societies, not least because of the different ways they have encountered class and the division of labor, and we should no more expect gender differences, patriarchal institutions, or women's consciousness to cleave reality along abstract theoretical lines than we can expect class alignments to crystallize easily against the countervailing structures of gender and division of labor.

To capture the mesh of gender and class, we must look for the way patriarchy is reinforced through capitalist class relations, the insistent gen-

---

[26] A point seen clearly by the Frankfurt School and writers dealing with racism and colonialism (e.g., Fanon, 1965), long before the current influence of Foucault and post-structuralism.

dering of capitalist class structure and stratification, and the indirect effects of capitalism on domestic and personal life, and vice versa. It is not possible to consider the integration of class and gender without crossing the terrain of the division of labor, however. Not surprisingly, feminist theory devotes an enormous amount of attention to women's work and, hence, to the division of labor. The sexual division of labor – between home and outside workplace, between husbands and wives, among occupations – is now recognized as having a history in its own right beyond patriarchy and capitalism, but here again the division of labor is never fully acknowledged as a structuring force. Division of labor creates differences that are part of an overall patriarchal and capitalist order and support the hierarchical arrangements of both, but cannot be entirely reduced to either. Patriarchal power depends heavily on the assignment of male and female work roles in every sphere. Capitalist exploitation gains both from a largely female domestic labor force and from a gender-divided work force facing a largely male employer class. At the same time, the division of labor evolves in ways heavily biased by the artifice of gender differences and inequalities.

While our chief concern here is with the way women's labor is harnessed by patriarchy, class, and specialization, we cannot entirely neglect the expressive side of the interplay of gender, class, and the division of labor. Life is not work alone; division of labor is not bereft of personal significance. Work specialities are crucially invested with meaning and social status, they touch deeply upon personal and social valuations of identity and worth. Division of labor imparts much of substance to prevailing notions of masculinity and femininity, and is altered, in turn, by those notions in the qualitative dimensions of its play of difference, hierarchy, and exploitation.

## Patriarchy and domestic labor

The family and domestic labor loom very large in studies of patriarchy. Feminists rightly object to the long-standing tendency, in both conservative and Marxist theory, to consign women and family to a sphere of reproduction created by, subordinate to, and strictly functional for, the greater domain of industrial capitalism (e.g., Parsons and Bales, 1956; Seccombe, 1974). Clearly, patriarchy and systems of domestic labor existed long before capitalism, and even as capitalism and industry have used women's domestic roles and altered the family, the influences do not run all one way – the force of family relations projects into the capitalist sphere, and there can be considerable friction between the two realms.

In many feminist studies, the shoe is simply shifted to the other foot, and the family becomes the principal site of patriarchal relations in society. This is manifestly so for theorists who privilege psychology, child-rearing, and sexuality over economics, (e.g., Firestone, 1970; Millett, 1977;

Chodorow, 1978), but it also crops up repeatedly among socialist-feminists who emphasize labor and direct exploitation of women by men. An early and extreme form of this dualism appears in a path-breaking essay by Delphy (1970), who posits class and gender as a two-class system based on twin modes of production, the capitalist and the domestic. Delphy argued that men pump out surplus labor from women in the realm of the family, through domestic work and services, in a way that is parallel (but not equivalent) to the exploitation of wage-labor by employers in the market. Women are not paid a market-determined wage, but get their keep.[27] She called the family/household nexus the *domestic mode of production*, in a challenge to the orthodox Marxist theory of an all-encompassing capitalism. While it was a salutary step to free gender from subordination to class, Delphy sunders patriarchy from capitalism, consigning it to the realm of the family alone.[28] Similarly, Barrett (1980) and Walby (1986) retreat, after long critical reviews of previous debates, to the domestic sphere as the principal locus of patriarchy. Barrett guards her flank by allowing that the family is largely an ideological construct, which would seem to wound its powers to act independently. Walby tries to soften the blow by adopting "patriarchal mode of production", then allowing that such a mode has no laws of development of its own, leaving one wondering why such a heavy-duty concept was wheeled into position without firing a shot.

Certainly, the family and domestic sphere are key sites of patriarchy and social life, but these formulations will not do. One cannot consign patriarchy entirely to one sphere of social life and capitalism to another. Patriarchy reaches well beyond the domestic sphere into the heart of capitalism and beyond, and when women enter the capitalist mode of production directly, crossing the boundary into wage-labor, they quickly discover the pervasiveness of patriarchy. Conversely, industrial capitalism reaches into the family, appropriating surplus labor and labor power, and reworking the contours of domestic work. We therefore follow Young (1981) in arguing for the gendered division of labor as a third force, cutting across patriarchy and capitalist class relations.

Virtually all attacks on the relation of capitalism and the familial realm fail to theorize the division of labor between domestic labor and industrial labor as an independent characteristic of modern societies.[29] It tends to be

[27] Matters are, of course, more complex than this, as some women have sources of outside income, others are subject to laws that render them virtual slaves to their husband-masters, and families extend far beyond this nuclear hub, but the basic point holds rather widely in capitalist societies.

[28] Delphy's theory was based on her work with French peasant families, and therefore applies more cogently to the agriculture-petty commodity production nexus than to families in general (cf. Friedmann, 1978). Delphy was also enough under the sway of Marxism to use the term "sex-class" in reference to men and women, thus retaining an overly economistic definition of gender relations, and failing to highlight the crucial differences between gender and class.

[29] Hence the kind of confusion manifest in Barrett (1980, p. 163): "The central point I am making here is that although the division of labour itself in capitalism is created by the economic require-

explained in its entirety by other things, such as capitalism, patriarchy, or simply "tradition" (e.g., Zaretsky, 1976; Edholm *et al.*, 1977). Barrett's argument (1980, p. 183), for example, boils down to the claim that "the relationship between domestic labor and female wage labor in capitalism has evolved through a process in which pre-capitalist distinctions have become entrenched into the capitalist relations of production". While the weight of tradition is not to be gainsaid, this is pure historicism; we need to investigate an ongoing process in which men and women, capitalists and workers, bend and shape the division of labor to their purposes – with the limitation that the division of labor, being tied to important productive and social functions, is by no means fully malleable.

The shape of the family as a site of patriarchy and exploitation is very much bound up with the social division of labor between domestic work and outside work. Women have throughout history been overwhelmingly consigned to child-rearing, household upkeep, and family nurturance. The origins of this may well be linked to the value of women's child-bearing capacities for maintaining the social band, as Meillasoux holds, but whatever the facts, two things are clear. No explanation based on a naturalistic theory of gender capacities and the domestic division of labor will suffice without considering gender struggle and the historical victories of men over women in a variety of areas of social life. And the historical process of assigning men and women to different tasks, whether in military conquest, agricultural production, government, or child-rearing, has been integral to the emergence of patriarchy and of class, as the brilliant explorations of Lerner (1986) reveal.[30]

The male monopoly on non-domestic labor has been particularly sharp where the bourgeois nuclear family has prevailed. While this form cannot be taken as universal – indeed, it is historically rather aberrant – it has been the touchstone for much of the feminist debate over domestic labor in contemporary Britain and the United States. This form is based on a shifting division of labor; it grew as capitalism sucked workers into workshops and factories. As productive activity was withdrawn, the familial wagons were drawn around the domestic sphere (Zaretsky, 1976). "In a sense the home became truly a private sphere only once production had left it. The constitution of home and work, the private and public as we know them, was in many ways a cultural artefact of the industrial revolution" (Cockburn, 1985, p. 37).

The feminization of household labor confines women to a relatively isolated sphere where they are more subject to the power of their husbands and fathers and less able to form bonds with other women. It renders

ments of capital accumulation, the *form* it takes incorporates ideological division [between men and women] to a considerable extent."

[30] It even extends to such little-theorized divisions as that between the moral wife and the prostitute. Yet even Lerner does not explicitly draw attention to the division of labor as a levering force in the marriage of class and patriarchy.

women dependent on the men who control family income and property and vulnerable to male dissipation of income on drinking or other pursuits (Hart, 1989), and it inculcates a powerful strain of nurturing that can disadvantage women in the competitive struggle with men in the larger world (Siltanen and Stanworth, 1984). The nurturing mother, then, is a joint product of patriarchy and the domestic division of labor.[31]

Domestic work is largely unpaid, and the place of this unpaid labor in an economy dominated by commodity production has long been a subject of dispute among feminists (Molyneaux, 1979). How can household work be so obviously useful for production and capitalism as a whole and yet without formal value in the system of commodity circulation and capitalist profit-making? The antinomy between value and use-value established by capitalist production for the market denies formal value to all production for direct use, whatever its gender (Rowbotham, 1973). This specifically capitalist coloration of the division of labor reinforces patriarchy by denigrating women's work in the dominant terms of capitalist society and giving them a strong incentive to marry so as to be attached to a wage-earner or capitalist income (Delphy, 1984). Furthermore, labor that has no "value" need not be paid a market price or even a living wage, with the result that domestic labor on the cheap has always been a wonderful way of lowering the cost of wage-labor, in the same way as labor recruited from peasantries (Burawoy, 1976).

The division of labor between the domestic and external (mostly capitalist) economies has been a shifting terrain. Women have been drawn into wage-work in larger and larger numbers over the course of the last two hundred years, but in episodic and uneven fashion; opportunities have come and gone as regional industries rose and declined, wars came and went, job categories expanded and shrank (Massey, 1984; Walby, 1986; Milkman, 1987). Capitalist piece work has been brought into the household at various times and places (Beneria and Roldan, 1987). Household labor has been drawn out into the capitalist spheres of food preparation and entertainment, while education and health care have become formalized beyond the confines of the family (see chapter 2). At the same time, women's domestic tasks have been altered by the introduction of household machinery, new fabrics, processed foods and the like (Miller, 1983).

All these encounters have transformed the family, making it anything but a stable center of patriarchal relations (Brenner and Ramas, 1984). Ferguson's (1989) model of this change in the United States is illustrative. She distinguishes father, husband, and public patriarchy, roughly corresponding to the 18th century mercantile epoch, 19th century indus-

---

[31] The relative weight given to patriarchy versus the division of labor affects the analysis of family dynamics. For example, in the debate over childhood psychology and personality formation, opposing positions have been staked out around the centrality of the subordination of women and children to the father (patriarchy) (Firestone, 1970) or the centrality of women's role in mothering (division of labor) (Dinnerstein, 1976; Chodorow, 1978).

trialization, and 20th century consumer capitalism.[32] Father patriarchy, characteristic of merchant and small manufacturer households, entails the power of the oldest male parent over an extended household that is society's basic productive unit; children are farmed out as apprentices and indentured servants; marriages are arranged; property passes to the eldest son; and the father is spiritual and moral leader of the family (Flandrin, 1979). Husband patriarchy applies to the Victorian bourgeois family and its lower class imitators. With the male working outside the home, the wife is left in command of children, servants, and household affairs. Women are constrained by sexual repression and sequestered in the home, but could foster a certain sentimental affection lacking in earlier hierarchical families and carve out a sphere of domestic autonomy and household skill (Ryan, 1981). Public patriarchy has, in turn, brought diverse households, involving singles, gay and lesbian couples, and complex step-families as kinship ties become attenuated, marriage arrangements more egalitarian, and sexual autonomy increases. Women have gone out to work in large numbers, and the dual income has replaced the family wage. Authoritarian control of children by parents has diminished. But along with some liberation from older forms of oppression, new difficulties have emerged in the form of double-time work, marital insecurity, male evasion of responsibility for children, female poverty, sexual objectification and performance demands, and childhood rebelliousness (Ehrenreich, 1983; Sidel, 1986).[33]

One should not claim that family patriarchy has been passively shaped by capitalism and the division of labor, however. For instance, patriarchal family property has blended in with capitalist property relations by serving as the original form of continuity in ownership and transmission of wealth across generations (Delphy, 1984). This was true, for example, of early merchant families operating without laws of incorporation, and family ties were also essential to relations of trust in long-distance trade (Pred, 1984). Family labor has been a means of primary accumulation in agriculture and manufacture, and the capitalization of women's labor was an important mechanism sustaining patriarchy in the transition from simple commodity production to capitalism (Middleton, 1983).

Conversely, patriarchal control of family property renders women and younger siblings propertyless, thereby serving as an initial source of proletarianization. The first large group of factory workers in America were the daughters of New England farmers without immediate prospects of marriage, and the practice of recruiting young rural women continues in rapidly industrializing places such as Malaysia (Ong, 1987). Unpropertied children have also been farmed out as servants, apprentices or farmhands; wives and children have left for the city and wage-labor to escape the

---

[32] Ferguson's model does not apply to most immigrant and non-white households, however.

[33] The first contemporary feminist critiques of women's station (e.g., DeBeauvoir, 1954; Friedan, 1963) were in many respects addressed to the last gasp of the bourgeois family, from which women were finally escaping in large numbers.

patriarch's heavy hand (Moore, 1988). The division of labor within the family has also affected the course of proletarianization and capitalist profitability. Women's domestic labor has traditionally underwritten the costs of reproducing the labor-power of the male wage-earner, and this continues even when women take in piece work or go out for part-time "supplementary" income (Seccombe, 1974; Gardiner, 1975). Indeed, the movement of men into certain kinds of jobs is often directly linked to the movement of women into others, and this duet is orchestrated through domestic relations, control of income and labor allocation within the family (Beneria and Roldan, 1987).

### Patriarchy and capitalist labor

By the mid-20th century in Britain and the United States women's entry into wage-labor was reaching levels comparable to men's (Cockburn, 1985). Women enter the capitalist labor force at a marked disadvantage, however, thanks to their placement in the domestic sphere of the social division of labor (Barrett, 1980; Beechey, 1987). They are constrained by the sheer exhaustion of working double duty, by interruptions for pregnancy, childbirth, and childcare, and by prevailing views that women's outside work is supplemental to the primary male wage (Hochschild, 1988). Women's self-identification with family and child-raising duties also keeps many from valuing wage-labor, career advancement, or entrepreneurial activity.

Women also carry the baggage of traditional divisions of labor into the labor market, as they are funneled into industries and occupations such as sewing, food preparation, or farm labor (Braverman, 1974; Alexander, 1976). White-collar jobs such as sales, teaching, and nursing, are still heavily typed for women, as they require nurturing attention. Even in work as secretaries or supervisors, women are generally expected to perform tasks of personal care, especially for male bosses. Education and training programs tend to reinforce the traditional biases. This sex-typing of jobs and career prospects splits the working class by gender from the outset of proletarianization (Hadjimichalis and Vaiou, 1990).[34]

This gendered capitalist division of labor is not simply a reflection of the domestic division of labor, however. It is very much a feature of the forces in the workplace and the labor market. Women are not channeled into certain jobs based only on their prior role and skills at home: an active process of sex-typing of occupations goes on within the sphere of capitalist production. The result is a massive skew of men and women toward particular industries and occupations, a systematic underpayment of women

[34] Even worse from the point of view of class solidarity is the widespread use of women in such wage-work as domestic service and putting-out systems, which leave them right back in the home, propping up patriarchal family relations and virtually unrecognizable as part of the working class proper (Beneria and Roldan, 1987).

who do the same or equivalent work as men, and consistent barriers to women's career advancement around the world (OECD, 1980). This gendering of occupations, of skills, and of wages is an extraordinarily rich process, vigorously contested by labor and capital, and by men and women workers, and fraught with meaning in terms of socially defined masculinity and femininity (Phillips, 1985).

The strategies of capitalist employers are one source of gender inequality. While economic calculation may move employers to accept a preexisting gender mix of skills, training, and experience as the most efficient way to match jobs and workers, capitalist practices are more actively biased than this. Job applicants are typed according to sexist (and racist) conceptions of qualities such as skill, work intensity, militancy, and time-discipline (Rubery, 1980; Wilkinson, 1981). Women are systematically employed on inferior terms in order to gain an extra measure of surplus value (women's wages are usually from half to two-thirds those of men for the same work, and their benefits are patchier). Capitalists also hire female workers because they are likely to be less unionized and less militant as a group than men. The gendered division of labor does not always serve capitalist interests, however: women could be brought into many male occupations at lower wages than men of comparable ability or trained to remedy skilled labor shortages; male fraternizing is commonly of a quality that does not encourage good work practices; and worker resistance can sometimes be broken by introducing women into all-male work groups. Yet patriarchal practices persist despite the supposed gender-blindness of capitalist profit calculations, and there can be considerable tension between the two systems of power and exploitation.

Patriarchy at work extends to male employees, as Hartmann (1979) and her followers have shown. Hartmann showed that 19th century craft unions resisted women's entry into manufacturing. Subsequent studies have confirmed that male workers persistently collude with employers in creating and sustaining gendered divisions of labor, to protect wage advantages and hard-won skills, as well as out of love of male fraternization and fear of feminine inroads on their masculinity (Cavendish, 1982; Cockburn, 1983; Walby, 1986). Along with their interests at work, men have wanted to keep women at home to serve them and to comply with a domestic vision of women in the social order (Milkman, 1987).[35]

The gendered division of labor within industry is a dynamic process. Skills and jobs are often defined and redefined according to their bearers: women's skills are repeatedly downgraded by employers, male workers,

---

[35] Clear evidence of the society-wide front against women's fair entry into wage labor in Britain and the United States is provided by the experience in each of the World Wars, when huge numbers of new female workers were recruited into the widest possible range of jobs, only to be summarily hustled back into their homes after each war, with an enormous loss of accumulated skill and potential labor power. Fears of recession and mass unemployment among returning soldiers figured in these mass redomestications of women workers, but cannot explain everyone's enthusiasm for the project.

and even women themselves, while men's work is inflated in value (Phillips and Taylor, 1980). As women take over formerly male preserves, new job distinctions are created which shift the basis of male advantage (Humphrey, 1987). Beyond hierarchical sex-typing, the real substance of work is structured in favor of men, who dominate technological know-how such that, even in the face of dramatic technical change, jobs are continuously restructured to keep the most advantageous bits in the male domain. One thread running from the industrial revolution to the present is male command over the machine, through repair, machining, and engineering design skills (Cockburn, 1985); another is the roles of supervision and management that go to men and the greater degree of surveillance that befalls women (Cavendish, 1982; Nelson, 1984).[36] A third is a virtual monopoly of the elite professions: men are doctors, professors, engineers, scientists; women are nurses, teachers, lab technicians (Larson, 1977). In this way, men have used leadership positions of the technical and hierarchical divisions of labor to reproduce patriarchal power and to lever themselves into the most elevated zones of the working and middle classes.

Yet even the combined influence of family position, capitalist maneuvering, and male worker reaction cannot fully explain the gendered division of labor in the capitalist workplace. As Milkman shows, sex-typing of occupations is remarkably unpredictable as new sectors and job categories open up; emergent divisions of labor continually pose the problem of gendering of work anew. Furthermore, "once sex-typing takes root in an industry or occupation, it is extremely difficult to dislodge" unless there is significant restructuring of an industry and its technology, or a major crisis such as war (Milkman, 1987, p. 3). This view can be reconciled with Cockburn's subtle tracking of technical change, job restructuring and gender struggle, as long as we recognize that the men and women, employers and employees, involved in the war of position in the workplace are maneuvering in and around a division of labor with its own dynamics and rigidities. The division of labor, in short, functions as both pivot and barrier to changing sex roles in the working class.[37]

Regardless of the obstacles thrown in their way, women have never been content to go quietly back to the family. Women have been pushing their way into the capitalist labor force over the whole sweep of industrial history (Tentler, 1979; Kessler-Harris, 1982). After a brief post-war reversal, women again flooded onto the labor markets of the advanced capitalist countries, and, as they grew more confident, educated, and militant, began to challenge gendered divisions of labor and sexist practices at

---

[36] The differential gendering of occupations across industries also rests on the varying economic, technical, and social histories of each industry (by country) (Milkman, 1987; Storper and Walker, 1989).

[37] Milkman, too, cannot bring herself to posit the division of labor formally as a social structure independent of patriarchy and class, appealing instead to "historically specific analysis" and employer power.

work (e.g., Remick, 1984). Capitalist labor demand grew vigorously at the same time, drawing on women as a major untapped labor reserve; indeed, the availability of women at low wages encouraged the expansion of certain "service sectors" in which they have been concentrated (Crompton and Jones, 1984). In the Third World, women have also worked their way out of the confines of the family and the countryside, as well as being drawn into cities, factories, and office work by rapid industrialization (Ong, 1987; Beneria and Roldan, 1987).

Even if the gendered division of wage labor breaks down, there remain larger fields to conquer: women in capitalist society stand in a subordinate relation to men as capitalists and governors, as well as husbands and fathers. Capitalists are overwhelmingly (white) men (Eisenstein, 1979).[38] That is, class is married to patriarchy at the top of the social pyramid, and has been in all known class societies (Lerner, 1986). This marriage goes beyond capital to include the critical powers entrusted to the state, such as the military and police, regulation of commerce, property and marriage, and the provision of major social benefits such as education, childcare, and health services (Wilson, 1977). Men overwhelmingly control the key positions in the state apparatus (McIntosh, 1978; Eisenstein, 1984; Moore, 1988).

In short, it is necessary to abandon the idea that patriarchy is lodged principally in the family. Historically, the subordination of women is probably as much a class and state process as a familial one, and has had a cumulative character across a range of histories.[39] There is no single locus of patriarchy, either in the family or in capitalism (Connell, 1987; Clawson, 1989). An additional, powerful force in gender differentiation and domination is, as we have shown, the division of labor – even as patriarchy helps reshape the division of labor itself.

### Stratification and gender

The stratification of capitalist societies cannot be assessed without clearly recognizing the overlap of class with gender and a gendered division of labor. The same muddle we observed with regard to class and division of labor arises in the sociologists' efforts to draw class maps in the face of a recalcitrant gender ordering of society, because gender cannot be reduced

---

[38] It is doubtful, however, that capitalism could not survive without patriarchy, as Eisenstein claims.

[39] The earliest urban civilizations of the Near East appear to have developed the slave mode of production out of military conquest and by incorporating subjugated women into the households of the conquering rulers (Lerner, 1986; Silverblatt, 1987). Ancient Greece incorporated women into several class and familial forms: as ruling-class bearers of property, as (chiefly domestic) slaves, and as independent peasant-wives (Wood, 1988). In medieval Europe different forms of family obtained across classes and places, but all had a strong sense of hierarchical obligation consistent with feudal relations of duty and military service, and emphasized kinship alliances over conjoint households in defining families (Stone, 1977; Flandrin, 1979).

to class (Garnsey, 1982). The traditional practice has been to assign women (and children) to their husbands' class position, a method which presumes a classic bourgeois family with the wife working in the home (Goldthorpe, 1983; Wright, 1985). Where many women enter into wage-labor, this assumption produces absurdities, since many women's jobs are quite different from their husbands' and husbands are sometimes the dependent spouse (Walby, 1986; Marshall et al., 1988). A possible way out is to treat everyone as individuals, acknowledging that the employed woman is working class in her own right (Stanworth, 1984);[40] if her position differs markedly from her husband's, one speaks of cross-class families (Britten and Heath, 1983). By the same logic, of course, house-wives, children, and other dependents have no direct class position.

This is not a way to read women (spouses) out of the working class, but rather to acknowledge women's material conditions in all their speci-ficity.[41] Wright (1989b) thus appeals to a concept of *mediated* class posi-tions, which recognizes the constraints of the family on women and dependents, whether they work outside or not. In fact, wives have a strong associative class membership via their husbands (or vice versa), which colors class formation and class consciousness. It may blur class, as with wives of capitalist and professional men who work in proletarian jobs as secretaries or nurses, but develop little working-class attachment. Con-versely, it can strengthen class, as when wives are the backbone of success-ful working-class struggles such as the great sit-down strike in Flint, Michigan, that launched the United Auto Workers union.

Stratification studies must also consider the gendered division of labor in the industrial realm. Women disproportionately occupy the lower rungs of the hierarchy of the division of labor and the elevation of certain occupa-tions above the mass of wage-labor is very much tied up with their monopolization by men. The new middle class enclaves in the professions are based not only on skill and long education, but specifically on a fierce exclusion of women and the young, combined with grueling and exploi-tive apprenticeships for their own novices (Ehrenreich, 1989). This has propelled doctors, lawyers, and architects into closer identification with the bourgeoisie, while pushing the mass of female clericals, saleswomen, teachers, or nurses downward. Trapped between the class fortresses of the male professions and the bastions of male labor in the "true" manual working class, women have often been caught in a kind of limbo between

---

[40] This class affiliation continues during absences from the labor force, which are commonplace for all workers in capitalism (Harrison and Bluestone, 1988). Of course, long-term unemployment attenuates the class relation, but this is not confined to women. A more interesting association is the one between women and part-time or "temp" work, which gives them a quite different relation to boss and job than men (Applebaum, 1987; Christopherson, 1988).

[41] The same need for specificity applies with children and the elderly. Children are frequently absorbed directly into their parents' class (and gender) roles, but they are also prime candidates for mobility across class or gender lines.

the new middle class and the working class, filling up the nebulous category of the "lower middle class".

Although the gender schism usually weakens the working class, women can be a profound force for change. Their double oppression as workers and as women sharpens certain classic demands for wage increases and unionization, and adds such novel interests as wage-equalization by gender (comparable worth), flexible and part-time work for mothers (and fathers), and maternity leave (Christopherson, 1988, 1989; Christopherson and Storper, 1988). Furthermore, the disproportionate presence of women in divisions of labor not traditionally associated with male trades and industrial unions, such as teaching, social welfare, hotels, health care, and sales, provides a base for extending the sway of working class organization beyond the dwindling pool of manual and factory labor (Walker *et al.*, 1990).

*Gendered divisions of identity*

In joining class, patriarchy, and division of labor we cannot forget the personal elements of desire, the body, and identity inherent to gender relations. This important dimension is not taken up by dual systems theorists of a decade ago or other recent economistic approaches (e.g., Walby, 1986). It will not do to lodge capitalist class relations at the economic level while fobbing off patriarchy to the ideological level, as Mitchell (1975) and Barrett (1980) do. Class and patriarchy operate at all levels, from the labor process to consciousness. Indelible patterns of masculinity are created and reinforced through the apparatus and exercise of class (and state) power. The military and warfare are obvious training grounds for violent masculinity; but the distancing, self-control, and instrumentality of command positions in business and other kinds of hard masculinity are nurtured in such venues as elite schools and athletic fields.

Ruling class masculinity is defined in relation to subordinate femininity and a dominant heterosexuality, as well as to the instruments of power, such as weaponry. Wives, servants, secretaries, female production workers, skilled employees, welfare mothers are all sucked into the vortex of this overarching masculinity, which demands deference, attention, and sexual favors, and metes out harassment and bars independence. Even the most favored of women, in class terms, still live daily with the debilitating effects of sexism. Homosexual and working-class men face this dominant masculinity in other complex ways, through imitation and distancing in both directions: working class "lads" brashly putting down middle class "earholes" at school, workers picking up military values through youthful enlistment, businessmen valuing proletarian football virtues, boys engaging in bullying and gay-bashing, etc. (Fernbach, 1981; Connell, 1987).

Division of labor also enters into the cauldron of masculinity and gender-formation. A sex-typing process takes place around work specialization that involves an expressive stereotyping of men, women, and jobs in ways that polarize characteristics which would otherwise overlap. Such qualities become part of gender formation and identity, just as do the more familiar genderings of household labor, and are invested with intense emotions. Typical polarities used to differentiate masculine and feminine work are:

strong / weak
heavy / light
dirty / clean
dangerous / safe
hard / soft
interesting / boring
mobile / immobile
rational / emotional
professional / unprofessional

These ideological markers cannot, by their nature, be applied consistently but are, rather, used opportunistically to signify superiority and inferiority as conditions change. Thus, according to Cockburn (1985, p. 235):

> [I]n engineering masculine ideology made use of a hard-soft dichotomy to appropriate tough, physical engineering work for masculinity and thus ran into a problem when it came to evaluating its 'opposite', cerebral, professional engineering. . . . The ideology coped with this contradiction by calling into play an alternative dichotomy, associating masculinity with rationality, intellect, femininity with the irrational and with the body – incidentally turning an almost complete conceptual somersault in the process.

Similarly, dirt, danger, and discomfort are often used by men as criteria of superiority where no other elite markers are available, but where such work can be exchanged for more attractive jobs, these qualities will become stigmatized and associated with females (or racial minorities).

Domination and exploitation can therefore be as much about the *qualitative* distribution of work, valuation, and reward as about simple *quantitative* amounts of surplus labor. Women not only do paid work for lower pay and a great deal of work for no pay, they are also saddled with a great deal of low status and unpleasant work. Ascriptions of low value are, of course, based in dominant ideologies shaped by men (capitalists, professionals, Anglos, etc.), which means women can rarely win: as an area of work opens up to more women, it becomes devalued. Of course, there are cases where men, by virtue of their supposed masculinity, are led to dangerous and dirty work like lambs to slaughter; but men have a virtual lock on the most prestigious and fulfilling jobs at the top of the occupational and class hierarchy.

## Conclusion

We have taken pains to show the need to consider class, gender, and the division of labor as distinct, if wholly intertwined, structures of social oppression, exploitation, and difference that cut particularly deep into the body politic. A good deal of conceptual clarification follows if we sort out abstract categories before tackling more concrete and taxonomic investigations. Many of the problems that have dogged class analysis and feminist theory reflect attempts to make single structures and abstract concepts carry more analytical weight than they can possibly bear. We have shifted some of that load onto a new pillar, the division of labor, and shown how it intersects with class and gender. While the division of labor has always been present as a kind of room divider in the house of capitalism or patriarchy, it has never been treated as a bearing wall. We think this move helps to solidify the framework of social scientific thinking about fundamental matters of social structuration.

The political implications of acknowledging the division of labor as a force cutting across class and gender, as of admitting gender into the heart of the discourse on class, may be discomfiting at first. It means recognizing the degree to which the working class and the female gender are not simple unities waiting to break free from the chains of their oppression; it also means acknowledging the material sources of the often frustrating lack of class and gender consciousness despite tragic degradation.

For the socialist left, it is particularly hard to admit that patriarchy and the gendered division of labor affect the course of class formation and class struggle under capitalism. Roughly speaking, gender divides and weakens the working class, whereas it unites the (male) ruling class, and even provides some solidarity between male employers and male workers. One must conclude that, overall, it has helped maintain the capitalist order. Where husbands have gone out to work and wives have stayed home, women have been distanced from class by their own lack of wage-work experience, by the exclusionary solidarities of men that grow up around the workplace, and by direct exploitation at the hands of their husbands. Where both men and women go out to work, they do so largely in different industries and occupations, around which, again, form relatively exclusionary cultures. Patriarchy also became embedded in the skill and administrative hierarchies of the division of labor.

Worse for the course of class struggle, the instruments of working class organization have largely failed to overcome gendered occupational division. Women were long excluded from trade unions or forced into separate auxiliaries; even where they entered unions in large numbers, union leadership has remained largely a male preserve as has the leadership of working class parties. Cockburn (1985, p. 40) observes: "Men were not misled in perceiving women as a weapon in employers' hands by which their own wages could be kept down. Where they were misled was in their response.

Instead of helping women to acquire skills and to organize their strength, they weakened women (and in the long run the entire working class) by continuing to exploit women domestically and by helping employers exploit them in secondary labor markets." Even the best (male) labor radicals easily fall prey to an image of a (white) male working class in defining the organizing needs for a renewed labor movement after the political and economic onslaught of the 1980s (Frank, 1989). The need for a new vision of labor organizing across genders, meeting diverse demands, and yet unifying a larger working class than ever, is a political task of the first order.

For feminists, it is equally difficult to face the fact that class and the division of labor have debilitating effects on the solidarity of women. Women are divided by workplace, by occupation, by trade unions, by professional boundaries, and so forth. Women who work in clothing factories on the Mexican border, as clerks in San Francisco, or as lawyers in Los Angeles will often have little in common. The women's movement still faces the vast schism between wage workers and housewives, many of whom regard feminism as a threat to their dignity. Women attached more or less strongly to their capitalist and working-class families split along class lines. Add race and national differences, and it is easy to understand the strain of fragmentation and hesitancy appearing in feminist writing today. None of this is to deny, of course, the kind of bonding that occurs through shared work, even in domestic networks of mutual aid, and forms a pediment on which most larger solidarities are built (Ferguson, 1989).

The inability of all women to see plainly a broad unity of gender condition and interest is hardly surprising, considering the world's complexity. Yet it leads some feminists, such as Ramazanoglu (1989), to move away from the difficulties of analyzing the unities behind a divided reality and toward an idealist solution: that women can achieve unity only in the act of creating their own liberation. This is quite the same end-run around difficulties in the theory and practice of class that we encountered from Przeworski. While it is certainly true that organization, mobilization, and radicalization are all means of forging unity within a class or a gender, such efforts canot possibily succeed without some structural basis in a shared condition. Indeed, as the often sad history of working-class politics shows, even with a structural basis solidarity can only be imperfect or momentary.

Nonetheless, there is some political comfort to be found in the fact of cross-cutting relations of difference and domination: the failures of working class struggles and women's liberation movements to bring unity and social transformation are not a product either of false identification of oppressions of class and patriarchy, as conservatives maintain, or of weakness of political will, as the ultra-left keeps on about. Real schisms and oppositions lie at the heart of the enormous collectivities constituting the oppressed due to the cross-cutting influences of class, gender, and division of labor, making unity a difficult, fragile thing.

There is another side to the matter. What fragments in one direction can unify in another. Competing structures may have a galvanizing effect, as in the case of nationalism and nationalist class struggles, which have been some of the fiercest of this century. Gender and division of labor may also act to solidify class formation: coal miners are frequently the most class-militant of workers, yet this has rested in large part on severe isolation from other groups and extreme masculine identity (Cooke, 1983). Conversely, women separated into lowly occupations under heavy male oppression are often driven to class militancy (Cho, 1988). The issue, then, is not whether or not classes are fragmented, but whether the fragments provide a basis for weaker or stronger formation of class-oriented institutions, practices, and insights; the same applies to gender (Giddens, 1980).

A final conclusion that flows from our analysis is that the real elements of difference and of oppression in the division of labor need to be recognized and grappled with for what they are, and not forever soldered onto patriarchy and capitalism. The prerogatives of mental labor or the cleaner work environment of office workers compared to garbage collectors are not merely the products of capital or patriarchy (or race), although they are certainly compounded and twisted by the latter into peculiar and inordinately oppressive forms. The division of labor presents humanity with a particular set of problems to solve, now and in the future; on this Marx had the correct inclinations in his early writings, if not the right answer.

Just as 20th century communist countries did not overcome patriarchy (Molyneux, 1990), they have not solved the division of labor. The ascension of a new ruling class in such systems has, of course, been very much a political process tied up with the Leninist parties, Stalinist dictatorships, and concentration of property in the state (Kornai, 1986) and is by no means simply a matter of the rise of a new class of intellectuals, managers, or bureaucrats (Szelenyi, 1986). In fact, socialist revolutions have often made the difficulties of economic development worse by scaring off the middle class, as in Cuba and Nicaragua.[42] Nonetheless, the latter are a pivotal element of the class structure under socialism. A post-Stalinist socialism, or any new order for the capitalist countries, must confront the powers and prerogatives of skilled and managerial workers which will remain as a potential source of inequality and exploitation. These privileges trigger struggles across the division of labor, and even now are deeply insinuated in the class struggles in the Soviet Union over *perestroika* and democratization, and in those shaking China over the private accumulation of fortunes under forms of market socialism and the prerogatives of the intellectuals. We shall return to the problems which the division of labor creates for economic planning and socialism in the last chapter.

---

[42] Both because of the antipathy of the revolutionaries to all previous elites and because of the craven refusal of many skilled workers to remain and share the difficulties of building a new society, instead of running off to make their fortunes under capitalism.

# 2

# The Brave New World of the Service Economy: the Expanding Division of Labor

The concept of "services" has entered the language with too little critical examination. The notion that the advanced economies had entered a new era in which the services sector and service occupations had replaced manufacturing and manual labor as the engines of economic growth became widespread by the late 1960s (Fuchs, 1968). A resounding declaration of this position was Daniel Bell's *The Coming of the Post-Industrial Society*. Bell's more fulsome claims have been amply criticized and the theory of the service economy itself has come in for careful scrutiny (e.g., Kumar, 1978; Gershuny, 1978). Although recent literature has contributed important theoretical clarifications, the lack of a systematic analytic framework continues to dog the field. As a consequence, disparate phenomena are haphazardly loaded onto a single overburdened concept, services. For those who want to see them, services are everywhere – "transportation and distribution services" such as airlines and wholesalers, "business services" such as legal counsel and equipment leasing, "consumer services" from hotels and fast food chains, "personal services" such as medical care and education, "repair services" by auto mechanics, "financial services" from insurance companies, and so on. Indeed, the only things not counted as services in most presentations are manufacturing and resource extraction.[1]

This promiscuous labeling simply does not hold up. Few of these putative services are indecipherable in the language of capitalist industrialism; but the service literature cuts so wide a swath across the whole of economic theory, and has borrowed and bowdlerized so many concepts that clearing out the conceptual Augean Stables is a truly Herculean task. Our purpose is to bring a little order out of this chaos, while laying out an alternative vision of the new social economy and its trajectory. In our view, the division of labor provides an irreducible axis for analyzing the theory of services. What is called the transition to a service economy is best characterized as a widening and deepening of the social and technical divisions of labor, part of a more general process of industrial evolution and capitalist development.

---

[1] For reviews of classification systems, see Delauney and Gadrey, 1987; Elfring, 1988.

The idea of transition to a service economy is a crude attempt to capture the momentous transformations in modern capitalism. This was the express purpose of Fisher (1939) and Clark's (1940) early efforts to distinguish among primary (resource-extraction), secondary (manufacturing), and tertiary (service) activities, with nations marching forward through stages of development from agriculture to industrialization to service domination (Petit, 1986). These gropings were followed by more expansive theories of post-war liberalism which introduced the idea of a new industrial or post-industrial economy (Galbraith, 1967; Touraine, 1971; Bell, 1973). For these writers, a transition out of classic industrialism had been marked by advances in management, science, and education, the growth of the technocratic elite, and the emergence of knowledge as the axial principle of society.[2] These liberal pronouncements have been echoed by later writers (e.g., Stanback *et al.*, 1983; Nusbaumer, 1987) but recent research into services has chiefly been occupied with economic strategy in an era of manufacturing decline in much of Europe and North America (e.g., Riddle, 1986; Cohen and Zysman, 1987; Elfring, 1988). A repeated claim in the service literature is that the economies of the advanced capitalist nations have gone through a structural change to the point where services constitute over 50 percent of gross domestic product – 67 percent in the United States.[3]

Our purpose is to use a classical and systematic set of categories to frame the issues of development and to clarify what is new and profound in the modern industrial economy and what is not. To be sure, the evolving, revolutionary process of capitalist industrialization has carried the world forward into a new era – there is no point in treating the late 20th century as a simple quantitative extension of 19th century industrialism. Nonetheless, we insist on certain indelible economic continuities with the past, continuities which the theory of services denies. In the course of the discussion, we shall come across some elements which others have seized upon as *the* explanatory factor behind current developments, such as the information, financial, or retail revolutions, but it will become apparent that none of them carries the wide-ranging explanatory power of such classic concepts as industrialization, the division of labor, and capitalism.

In the usual discourse the shift toward a service economy involves five leading elements: the development of service sectors, i.e., certain kinds of output; the multiplication of service occupations and informational services, i.e., certain types of labor and management inputs; distributive services which mediate between production and consumption; and, finally, personal and social services received by consumers. These correspond very roughly to a more classic set of categories: production, labor process,

---

[2] Similar ideas could be found across the political spectrum, from Habermas (1971) to Drucker (1968). For critiques see Ross, 1974 and Kumar, 1978.
[3] For comparisons of data among advanced capitalist countries, see also Singelman, 1978; Gershuny and Miles, 1983; Petit, 1986; Ochel and Wegner, 1987.

circulation, and consumption.[4] We use the latter because they cut more deeply.

We shall first take up production, in terms of the industrial base and its outputs. Industrial products are forever expanding in number and changing in form, and the sectoral division of labor has broadened along with this. The outputs of industry can come as either goods or labor-services, but the bulk remains as goods though there have been significant changes in the substance and content of industrial products. A great many sectors are mislabeled as service industries because of confusion over the diverse forms taken by the products of human labor.[5]

Production requires inputs, the most important of which is human labor, so we turn next to the labor process. Labor is defined by Marx (1863) as human activity that transforms materials into useful products with the help of tools. This stripped-down version of labor has been overthrown in the course of capitalist development and the labor process has been broadened from the work of the individual craft worker to modern social labor, which includes, among other things, applied technical knowledge, highly specialized operations, and the management of collective projects. The division of labor in complex processes has vastly augmented the numbers of workers engaged in indirect productive activities – wrongly labeled service labor – leaving proportionately fewer workers engaged in the direct, hands-on tasks of transforming materials into useful forms.

Following that, we take up circulation, or the flows of output from production to consumption, from seller to buyer. Circulation consists of three parallel movements of commodities, money, and property rights. The division of labor between production and circulation is fundamental to all capitalist economies, and circulatory activities have been much elaborated by their own divisions of labor, but cannot be sundered from the industrial base by relabeling them either service outputs or service labor. Circulation remains a vital category of economic analysis in its own right (Marx, 1893).

Lastly, we move to final consumption, both personal and social. Service theorists put great emphasis on a shift from industrial production to the satisfaction of consumer needs. Closer analysis, however, suggests that the social division of labor has, if anything, been shifting toward the consumption of industrial goods, the industrialization of everyday life, and capitalist domination of the consuming public. Still, the level of popular consumption has risen dramatically over the last century, including health care and education as well as consumer durables and leisure pursuits. The shift from capitalist to state provision of consumption goods and labor-services has also been marked. Together, these developments pose more

[4] The best of the service writers, such as Delauney and Gadrey (1987, ch. 7), come very close to using these categories, in effect, but cannot bring themselves to adopt such a classical approach, perhaps for fear it will annul their efforts to break with economic orthodoxy, left and right.
[5] As well as because of diverse forms of provision, capitalist, state, or non-profit/non-governmental organization.

starkly than ever the problem of harnessing society's immense industrial powers to serve human welfare.

The service thesis is linked to a political theory of social change. Behind the nitty-gritty of classifying service outputs and labor inputs lies a sweeping modernization theory, in which the natural forces of the economy lead the way, capitalist social relations are an afterthought, and ugly notions of class struggle and social rupture are banished from consideration.[6] In response, we close with an explication of the continuing capitalist character of the new industrial economy. Some may charge us with granting too much to the growing division of labor and industrialization as processes in themselves. Yet we are convinced that many fundamental aspects of industrial economies cut across different modes of production, and that this in no way denies the role of capitalism in driving industrialization, nor the way capitalist society warps industrialism. Marx and Engels were right in theorizing economic history in terms of both relations of production (capitalism) and forces of production (industrialism), including the unities and the tensions, the blending and the contradictions between the class system and the productive base.

## Products: What Comes Out

We begin with industry and its output, or what comes out of the production system, because uses of the term services often rest on the nature of the product of labor. It is widely held that the old-fashioned good is passing from the scene, along with its gritty world of factories and machinery, to be replaced by the service, a term that implies personalized labor, immateriality, information content. As the following elucidation of categories makes clear, this is not the case.

### Goods and labor-services

The service theorists create confusion from the outset by failing to distinguish carefully between goods and labor-services. The distinction cannot rest on the fact that only one renders a service to the consumer – all useful goods provide a service in the sense of a benefit, or filling a need; that is their use-value. All useful labor renders a service.[7] A valid distinction

---

[6] The Fisher-Clark stages theory of history is bereft of any relations of production at all. In the more sophisticated versions of post-industrialism that followed, social actors can be found, but disturbing political struggles over economic development are replaced by a social transformation involving the gradual euthanasia of the old capitalist and working classes in favor of a new technocracy or intelligentsia.

[7] The following definition of services, then, is really a definition of useful labor: "Services are economic activities that provide time, place and form utility while bringing about a change in or for the recipient" (Riddle, 1986, p. 12; cf. Hill, 1977). Gershuny and Miles (1983) see the common usefulness of goods and labor-services, but think it applies only to final products, even though intermediate goods are also useful to producers (cf. Marx, 1863).

between goods and services can therefore only lie in the material form taken by the product of labor.

A *good* is a material object, such as a ship or can of beer; in its simplest form, it is tangible and mobile. The products of mining, agriculture, and forestry (so-called primary production), of manufacturing, construction, and utilities (so-called secondary production) almost entirely take the form of goods. In contrast, a labor-service works directly on or for the consumer but does not take the form of a discrete product; typical would be theater productions, manicures, and housecleaning. A labor-service is thus normally irreproducible and involves a unique transaction between producer and consumer (cf. Hill, 1977). This does not mean, however, that labor-services are immaterial (cf. Delauney and Gadrey, 1987).[8]

This is not always an easy distinction. The difference is not that between standardized mass production and craft labor, even if the latter is unique and irreproducible. The crux of the matter is the discrete product and its alienability from a particular user. A haircut is a tangible product of labor. If it were a toupee, it would unquestionably be a good, but it is generally considered a labor-service because it is part of the wearer, making it personal, unique, and irreproducible. Conversely, a sculpture is a unique act of labor, often for a particular buyer or site, but because it is a discrete object (and reproducible through casting) it is ordinarily considered a good.[9]

The production of goods is the mainstay of industrialism today, as in the past. The output of the basic goods-producing sectors continues to rise in every country, barring depression. Nonetheless, manufacturing's *share* of economic activity has declined since mid-century, prompting speculation that it will be superceded by services. The absolute quantity of labor-services has also risen over time. Employment in labor-services has generally grown even faster because productivity is lower than in industrial goods production, which is easier to rationalize and mechanize. Where labor-services go to final consumers they are termed consumer services: haircuts, domestic help, home nursing, education, and so forth. Where they are sold to businesses they are known as business services: consulting, audits, legal counsel, and the like. However, a great many things are lumped under these headings which do not belong there, such as consumer

---

[8] There is no validity to the idea held by some Marxists that only material goods can be commodities and objects of productive labor (Miller, 1984). Nor should labor services as *outputs* be confused with labor *inputs*, a common error of service literature (e.g., Hill, 1977; Petit, 1986; Ochel and Wegner, 1987).

[9] Adam Smith (1776) defines a service as that which disappears in the moment of consumption, confusing the materiality of the good with its useful life. Many solid, tangible goods disappear in the moment of consumption, e.g., hamburgers and BIC razors. Some, such as bars of soap and tubes of toothpaste, are consumed bit by bit. Some are hardly diminished at all by consumption, e.g., radios. But note that the useful life of most labor-services is also normally not extinguished in the moment of consumption, like the memory of a great play or information learned from a good teacher.

finance, restaurants and hotels, which really represent goods production or circulation activities.

## Joint products

Many labor processes have more than one issue; these are known as joint products. Nonetheless, there is a good deal of confusion about production that joins goods and labor-services. Restaurants and other food outlets, for example, are ordinarily considered part of the consumer service sector, yet the meal is a good: produced in the kitchen and delivered to the consumer as a tangible, discrete entity. Indeed, the meal may simply be emptied from a can and reheated, and no one imagines that canned foods are anything other than industrial goods. Because attendance by a kind of hired servant, the waiter, is considered part of the pleasure of dining out, this labor-service is often provided jointly with the meal. Such is not the case at MacDonald's, where service has been all but eliminated in the interest of mass production.

Similarly, hotels principally offer shelter, a useful aspect of a type of good, i.e., buildings. Only fine hotels also offer personal labor-services. Having someone take your money and clean up after you does not count as service; it is sales and maintenance work. Restaurants and hotels should therefore be counted partly under manufacturing and partly under retailing, with a residual of labor-services. Conversely, some true labor-services, such as concerts or consultancies, provide decorative goods, such as a souvenir program or annual report along with their services, in recognition of the difficulty consumers have in conceptualizing their intangible purchase (Riddle, 1986).

## Structures and infrastructure

Houses, dams, office buildings, bridges, etc. are large and relatively immobile goods. The first problem they raise is construction *in situ*, for which a specialized branch of industry has arisen. In rapidly industrializing and urbanizing areas, construction can reach 20% of employment (Kuznets, 1966). Some writers include construction among the service sectors even though its output is unmistakably goods (e.g., Clark, 1940; Riddle, 1986). Once erected, however, structures can be difficult to classify because of varying forms of ownership, use, and circulation.

Many smaller structures such as houses and septic tanks are purchased and used individually, but larger ones such as sewage works cannot easily be divided for private ownership and consumption. The output of these elements of the physical infrastructure are frequently counted as services, even though they are material goods (e.g., Illeris, 1989). This is particularly the case for utilities such as water, gas, and electricity, whose outputs can be used and sold as ordinary alienable commodities, though their enormous generation, storage, and distribution systems mean they must be shared by large numbers of users. Utilities have been a mainstay of urbanization and industrialization since the 19th century (Teaford, 1984).

Another staple of infrastructure since the outset of industrialization has been transportation, which reached its peak employment in the 19th century but continues to expand in output, extent and speed (Gallman and Weiss, 1970; Vance, 1986). Transportation is almost invariably listed among the service sectors. The transport network involves both immobile structures, such as highways and airports, and mobile machines such as cars, locomotives, and ships. The latter are produced as goods, but transport systems are used collectively and their output is the transfer of other goods and people from one place to another. Goods transfer is a geographical necessity if something is to be of use to its consumer (Marx, 1863); indeed, it is no more than an extension of the transfer function within all industry (Walker, 1989b). Personal transport, on the other hand, involves the use of a collective good (e.g., an airplane) for an individual purpose. This raises a question of the form of circulation, which we shall return to, but the only labor-service involved is provided by porters or flight attendants.

Some forms of infrastructure have true labor services as their main output. Fire departments, security services, armies, and professional sports teams are examples. Those collective-use products are frequently provided by the state because of the heavy fixed investments involved, the social character of distribution, and because they meet basic needs, so they may appear under the "government" heading in lists of service activities. The ownership of these industries is a political matter, however, not a natural consequence of their physical features. Some are provided commercially by private utilities, carriers, security firms, and the like; others such as police and armies, are almost never provided privately because the state monopolizes armed force and control of the national territory.[10]

### Information and substance in products

Confusion over the changing physical nature of goods has led some observers to see services where none exist. The *Economist*, in calling services "everything you cannot drop on your foot," manifests the antiquated notion of goods derived from the mechanical age. This view fails to see that computer software, which consists of electronic signals on a tape or disk, can be every bit as much a material good as a chair. A customized program written for one customer would be a labor-service, but a packaged program such as Lotus 1-2-3 served up on the shelf at ComputerLand

[10] The state covers so many disparate functions that it transcends the debate over industrial and service economies, or even modes of production. We cannot go deeper into the matter in this book, however.

is unquestionably a good. Even a custom program written on my machine and transferred to yours by diskette is a good: it has a discrete, tangible, and fungible form (which the carrier disk helps to emphasize), unlike a true labor-service. Yet software is generally misclassified as a business service. The real distinction here is between things that are easily seen and grasped, and those that are not.

There is a coordinate but broader problem of seeing written, informational products as goods rather than services. A legal brief or an environmental impact statement is simply intellectual craft applied to paper, as a chair is woodcraft applied to lumber. As long as the brief remains in the lawyer's head it may be used as a labor-service, such as advice-giving, but once it is on paper and takes a material form, it is potentially useful to anyone. Hence many technical consultants produce goods rather than direct, personal advice to their (usually business) consumers.

Some service theorists think we have entered an age of information and communication, leaving the world of industrial goods behind (e.g., Porat, 1977; Jonscher, 1983; Howells, 1988; Hepworth, 1990).[11] The information explosion in the contemporary economy is readily apparent, but information is not a free-floating ether; it must be pinned down. Information can be either part of industrial products or, as we shall see, part of their production, circulation, or consumption. We consider its role in outputs here.

All goods communicate symbolic information, if only in their appearance: chairs carry the utterly conventional meaning that they are things to be sat on – though in the case of a throne, a great deal more is implied. In other words, human beings speak through their objects and actions as well as through their throats (Miller, 1987). Only certain goods, however, have as their principal use-value an ability to store, transfer, and interpret information. Similarly, not every labor-service is informational: some have relatively mute purposes, such as trimming a hedge, while others are meant to impart wisdom, as in engineering consultancies.[12] The electronic age has unquestionably increased the ability to package greater amounts of information in more sophisticated products, and industrial output has become more information-rich, but there is no convincing evidence for a large, distinct information-producing sector in the economy as a fourth sector beyond agriculture, manufacturing, and services (Miles and Gershuny, 1986; Lyon, 1988).

A handful of sectors loosely designated as "the media" sell goods and labor-services particularly laden with information, and are for this reason commonly classified as services (e.g., Illeris, 1989). These include such

---

[11] For a review see Lyon (1988). Daniel Bell (1980) is now an advocate of the information society, as is Castells (1990).

[12] A bookmaker writing a tip-sheet produces a good; one who whispers advice in your ear performs a labor-service. The information content may be the same in both cases.

activities as printing and publishing (of books, magazines, and newspapers), film-making, and recording which were in the past considered branches of manufacturing. The shift in label cannot be justified, because the media remain industries producing chiefly goods that store, transmit, and impart information, whether in the form of music on records, stories on film, or news programs on tape. These forms of storage and transmission raise additional problems of circulation, to which we shall return.

A group of business labor-services is also information-heavy. They fall under three main headings: software, technical consultancies, and data banks (Howells, 1988). Software is a necessary adjunct to information-processing by machine, including communications systems; technical consultancies sell specialized dollops of advice in areas such as product engineering, marketing, and finance; data banks offer massive amounts of unprocessed information for use by those with particular expertise, armed with computers, usually in media companies, management, or financial investing.

Communication networks, such as telephone and telegraph systems, have almost a pure information-handling function, with no additional output, as a rule. They act as a special type of infrastructure, which is used collectively and often state-run, though private systems are becoming more common. For business consumers, communication has always been an integral part of production (and circulation) (Pred, 1973). Within the fabric of production communications acts in a slightly different way: a division of labor requires that information pass between workers and workplaces to coordinate social labor processes. Intermediate goods cannot be silently passed along the production chain in hope that workers at the next step, on the shop floor or at the next factory, will understand what is to be done, so there is communication, in the form of speech, *kanban* boards, packing labels, etc. Similarly, distributors and final consumers need information about the goods they are handling or buying. Of course, communication networks are also utilized for purposes of final consumption, as in personal phone calls, but these are secondary to business uses.

In sum, while the informational aspect of goods is increasing, the use-value (information content) of a commodity should not obscure the materiality of physical goods, thereby leaping over the tangible labor involved in their production. We are still talking about production and industries. We do not wish to ignore the transition to a more electronic or informational age; we do want to emphasize that such a transition still lies within the bosom of industrial capitalism.

*Intermediate output: producer goods and services.*

Not all products are meant for final consumption. The general division of labor in modern capitalism creates differentiation and specialization across immensely complex production systems. Economic growth rests squarely on the multiplication of the division of labor represented by a deepening web of intermediate inputs and outputs, as Young (1929) and his followers have insisted (Storper and Walker, 1989). Although it is possible to carve the economy into sectors with discrete commodity outputs (whether goods or labor-services), they must ultimately be joined together into systems involving intermediate as well as final outputs. Intermediate outputs serve as materials and means of production for other sectors, and goods and labor-services sold from one business to another can ultimately contribute to other goods or other services of any kind. A steel bar can ultimately appear in diverse forms: as part of a car, a tin can filled with processed food, or a bank building, and tracing intermediate commodity and labor flows is the staple of modern input-output analysis.

Traditionally, the most important intermediate outputs were capital goods (i.e., machinery), but producer services have been the most important intermediate outputs and the fastest growing segment of the service economy in recent years (Stanback *et al.*, 1983; Ochel and Wegner, 1987; Moulaert, 1989).[13] The growth of producer services does not alter the fact that final output remains largely in goods, as with an engineering consultant who helps design a factory to produce diskdrives (Gershuny, 1978; Gershuny and Miles, 1983). The expansion of business services does indicate that the social division of labor in the production of all outputs is steadily enlarging.

All goods require labor-service inputs and, conversely, all labor-services are produced with the use of goods (Gershuny, 1978). Ultimately, all production requires both labor and goods (materials and machines): it is production of commodities by means of commodities, as Sraffa (1960) so aptly put it. The regress is infinite, and as we shift from the realm of outputs to that of inputs into production, we arrive eventually at the labor process. It is possible with the right assumptions in the end to reduce all productive inputs to quantities of labor (Walker, 1988). Individual jobs almost never have a discrete product, so all work becomes service work, if we are to take Gershuny literally – an interesting route by which to rediscover the labor theory of value.

### The persistence of industrial production

The dominance of goods production in the modern industrial economy is simply not in question (Blades, 1987). There has been no large or sweeping shift toward a different sort of output, direct labor-services; instead, many classic industries, such as electric power, construction, food processing, and transportation have been swept into the capacious bin of the service sector, falsely enlarging its size in national accounts.

---

[13] Greenfield (1966) appears to have been the first to make this distinction.

At the same time, the social division of labor has been changing. First, new sectors have arisen whose output is atypical of earlier industries: electronic devices and plastics embody new technological principles and are materially different in ways that can be profoundly dislocating. With the shift toward electronic miniaturization and information storage, especially, the tangible material substratum diminishes radically while the meaningful content soars. Second, the division of labor has not only widened, but deepened as the layers of intermediate inputs have multiplied. Mostly this involves unfinished goods or components, but there has been a notable growth in business services, which are both intangible and closely related to more sophisticated production technologies, systems of management, and circulation functions. These are considered in depth below, where we shall see that distinguishing between goods and labor-services as the final outputs of production, which is central to many discussions of services, does not take one very far in analyzing the economy, its division of labor, and its developmental tendencies.

An important tendency in industrialization is that intermediate outputs are increasingly directed toward expanding areas of labor and away from traditional manufacturing. Compare the shift from agriculture to industry: the agrarian economy was once the principal source of industrial materials and the chief user of manufactured goods, but the manufacturing economy soon became its own largest market and source of inputs (Page and Walker, 1991). Today, steel and concrete are going the way of wheat and corn, which – while still much in demand and produced in large volume – occupy only a small fraction of the modern labor force. Agriculture is no longer the heartland of the economy, and neither are traditional heavy industries; electronics, software, and optics have moved into this position (Forester, 1980; Freeman, 1984). The leading edge of product development is increasingly directed away from manufacturing altogether: the largest use of computers is not in running steel furnaces or metal-cutting machines but in accounting, design, and retailing.

Because the industrial base is continually replenished with new products, there is no discernible tendency for goods production to fade away. True, manufacturing has been shrinking slightly as a share of national output and employment in the advanced countries; even if we include business services, output and employment have not grown in recent years (Petit, 1986). But this trend is complicated by the overall slackening of economic growth in the last two decades, relative to the quarter century after the Second World War, and by the uneven performance of national economies: after all, the two top-performing industrial countries in the world today, Japan and West Germany, are also those with the largest manufacturing components (Cohen and Zysman, 1987). Nonetheless, overall, there has been a shift from manufacturing toward other parts of the economy, in particular toward circulation and social consumption activities.

## The Labor Process: What Goes In

A rather different tack on the service economy is taken by those who stress service occupations, or inputs, over service sectors, or outputs. Stanback and his colleagues observe that one must not only talk about "what we produce" but "how we produce", and Gershuny argues that occupational data offer a better entry to the problem of services because service occupations add up to almost 50% of all jobs, while service outputs amount to only slightly more than 10% of final production.[14] What he has hit upon is the dilemma of complex labor, in which workers perform only pieces of larger processes. From this angle services are *labor inputs*, and the variety of such inputs is dictated by the division of labor in large production systems. We are not speaking here so much of the *social* division of labor among major industrial sectors, such as steel and autos, but of the technical division of labor within commodity systems. (This distinction is by no means hard and fast, however, since internal technical divisions can be externalized to become independent outputs in the social division of labor). The rise of service occupations involves more than shifting specializations: the nature of the work itself is felt to be qualitatively different in terms of material people work with (information), the place of work (offices), and the kind of labor involved (mental), in contrast to the classic blue-collar factory work. Many writers who emphasize service occupations, such as Nusbaumer (1987), are most concerned to stress the increased role of knowledge and mental work in the modern economy.

Marx's notion of the labor process is still germane here, for what is really at issue is the nature of the labor process in modern industrialism. The service theorists are grappling with something for which they have no name: the development of social labor. On one side is the increasingly social nature of labor processes, which leads to massive confusion about the output of individuals versus collective work. On the other side lies the evolution of the social powers of production, which generates further confusion about the relation of direct and indirect labor. Marx (1973, p. 705) had a surprisingly good idea of what was at issue here:

> In this transformation [to large-scale industry], it is neither the direct human labour performed, nor the time worked, but rather the appropriation of the worker's own general productive power, understanding of nature and mastery over it by virtue of his presence as a social body – it is, in a word, the development of the social individual which appears as the great foundation-stone of production and of wealth.

Nonetheless, Marx's treatment of the ways in which the productivity of social labor were increased – in terms of cooperation, the division of labor,

---

[14] The idea of service occupations is closely allied to the debate over the service class, to which we have already referred in chapter 1.

mechanization, and industry's appropriation of science – could not anticipate the level of development of the productive forces reached by the late 20th century. We must clarify the ways in which the indirect component of work has grown with the mammoth expansion in the technical division of labor in every labor process, whether in the sphere of production or of circulation.

Gershuny's simple reduction to service work writes virtually everyone and every kind of work out of traditional industry and into the service economy. This conflates three different aspects of the division of labor. The first is the *extended* division of labor in all types of production, that is, necessary work performed before, around, and after the direct labor process. The second is the *hierarchical* division of labor associated with every complex labor process, what is usually termed the work of management. The third is the *mental* division of labor, which parcels out the application of knowledge and the manipulation of materials.

### The extended division of labor: the temporal and spatial dimensions of production

To deal with the modern division of labor, it is necessary to break with the fiction that everything happens in an instant and begin to grapple with *extended labor processes*, that is, work that takes place before and after products are actually formed by direct, hands-on labor, as well as work complementary to the immediate labor process. Extended labor, from R & D to janitorial work to automobile repair, is regularly mislabeled as services, especially where it is sold in a commodity form, when it is, in fact, nothing more than work that can be separated in time and space from the core of direct labor.

Pre-production labor usually includes research on materials, processing techniques, and product uses as well as development work on process engineering, prototype design, trial production runs, and product testing. This should not be confused with the early steps of a sequential process; for example, the initial clearing of a field for planting or the breeding of experimental livestock are pre-production labor but each year's plowing and calving are part of the ordinary cycle of agricultural labor. The dividing line between short developmental runs and regular production can be imprecise, and one should not assume that design ends at the moment production is regularized. Product design and process engineering were long the province of skilled workers or capitalists themselves, while research was an esoteric specialty of outside scholars (Forty, 1986). In the 20th century, research and development and engineering departments grew to be major divisions at the large corporations (Hounshell and Smith, 1988).[15]

[15] General scientific research is harder to classify than ordinary R&D. It is a highly socialized activity that rarely pertains directly to only one product or labor process, and must be adapted and

Auxiliary labor comprises those tasks which take place on a regular basis during actual production, but which back up or complement direct labor without ever coming in contact with materials or products. Auxiliary work has expanded in step with the growing technical division of labor within the workplace. Three general types may be discerned: tasks that feed into and out of the production line, such as the work of the stockboy or quality checker; routine repair and maintenance of machines and buildings; and gathering and exchanging information: instructions, feedback on performance, and interaction among workers. There is a strong spatial element here because workers have to communicate across factories, between departments, and even between far-flung workplaces.

Post-production labor is work applied to, or an adjunct to, a good after the immediate production process. Such labor begins within the factory with inspection, labeling, and packaging; continues with transport (delivery) of a good away from its point of origin; and includes the additional work required in wholesaling and retailing, such as minor assembly and clean-up of furniture or automobiles. Even after sale to the consumer there can be further work involved in delivery, installation, adjustment, and instruction for use – especially with equipment sold to businesses. The product is useless without such labor, so it cannot be considered only part of the sales effort.

Finally, repair and maintenance can continue for years after a product has been put into use. This work is a regular part of the litany of the service economy, but such labor merely restores goods to proper order, as with car repairs, road work, or house painting. In other words, the labor process has been extended because the original good has a long life and is cheaper to maintain than to throw away or remake. The need for maintenance and repair has grown along with the growing stock of fixed capital, infrastructure, and consumer durables, and therefore represents an enlargement of the world of industrial goods and their further penetration into personal life. No personal service is given except to soothe the customer's nerves.

### The hierarchical division of labor: coordination and direction by management

Management stands as a category apart in the social division of labor, irreducible to the labor it organizes and commands. The task of coordinating and directing complex labor processes is a necessary part of all modern industrial activity and has grown enormously over the last two

enhanced for practical industrial use. Most scientific research enters into the stream of available knowledge and cannot easily be commodified, yet basic knowledge about physical systems underlies specific productive activity – much as urban infrastructure lays the basis for city growth. In some cases, however, the products of scientific research are sold directly as marketable commodities (patents or licensing rights) where they may be classed as informational goods or labor-services.

hundred years (Melman, 1951; Pollard, 1968; Chandler, 1977).[16] The useful results of managerial labor are such things as improved marketing strategies, more effective accounting systems, or new investments. Because the product of managerial labor is rarely embodied directly in a good, conventional service theory commonly includes management as a service.

Every working group requires some oversight and trouble-shooting, whether its product is a symphony or an organic solvent. As the scale and complexity of the labor process increases, an element of specialized technical competence enters into these tasks. This is true whether the ultimate product is a good or labor-service (e.g., limousine companies need managers), or whether the activity involves circulation or government (department stores and fire departments need managers, too).

In the wider economy, work units of diverse kinds are commonly linked within large corporate entities, state agencies, or inter-firm networks, and this calls forth additional layers of administrative hierarchy. Management's purposes include reducing the cost and time of circulation, coordinating far-flung production processes, and linking pre- and post-production labor inputs for more effective product development and sale. Information handling – gathering, processing, and using knowledge about the many activities to be coordinated – is a basic element of managerial labor (Arrow, 1985). While corporate management has received most of the attention and credit for directing and regulating modern industry, other forms of management exist, including commercial capitalists who weave together the functions of distribution and production (see chapter 3).

Capitalist management encompasses another dimension, as well: the extraction of surplus value. Social labor is also in the service of capital; and while this does not make for a service economy, it means that there is an ongoing dialectic between the hierarchical division of labor and class. The control and disposition of the surplus, or capital investment, is, along with the planning of future action, the role of the very highest echelons of management, and the labor of disposing of capital the finest flowering of the capitalist division of labor (Herman, 1981).

The division of labor *within* management has become increasingly elaborate, as well. Separate functions such as accounting, finance, and purchasing have grown up within every large industrial firm, and this proliferation has brought forth an army of clerical workers to carry out the labor of information gathering, storage, and communication (Crompton and Jones, 1984). Labor control has evolved from individual supervisors to large personnel departments staffed with psychologists and doctors

---

[16] The ratio of administrative to production workers in US manufacturing has increased across the board and, while the level varies systematically across industries, average growth rates are roughly the same everywhere. Melman (1951) also found that other common explanations for managerial growth at the level of the firm – such as bureaucratic padding, sales effort, mechanization, and number of technicians – were not significant, although this finding is subject to critical comparisons over time and place (Bowles *et al.*, 1983; Dore, 1987).

(Edwards, 1979). Finally, management has become so sophisticated that it demands an immense array of new inputs and their forms of auxiliary labor, such as data processing, advice on international law, and money-market manipulation (Lambooy, 1988).[17] Those who perform these functions frequently break off into separate firms selling business services (Stanback, 1979), particularly a new segment called, variously, management, informational (IT), or high-tech consultancy (Moulaert *et al.*, 1989). Such firms come from any of several directions, such as accounting, finance, and data processing. They help corporate management plan, design, and implement new approaches in technology, organization, and labor relations, and are vital to industrial restructuring. They can also function as a kind of new merchant capitalist, bringing buyers together with suppliers of new IT equipment, software packages, etc. It is no wonder that management consultancy does not look like goods production. It stands at a very far remove of indirect labor, at the pinnacle of the capitalist hierarchy and the leading edge of industrial change.

### The mental division of labor and the growing sophistication of technology

The technical division of labor has also been deepened in terms of knowledge content and technological sophistication. Mental aspects of production have come to be emphasized and rationalized as the more purely mechanical, or manual, acts of labor once were. Marx had it right in seeing that the heart of the first industrial revolution was the ability to transform the human element in production, before widespread mechanization was possible (von Tunzelman, 1978). But Taylorism took that sort of rationalization of manual labor about as far as it could go. Already in the late 19th century (and Marx had an inkling of this, certainly), industrialists had begun to look to materials science, chemistry, and mechanical engineering to change product capabilities and the labor process in ways that went beyond the immediate acts of direct labor; this has continued with the contributions of synthetic chemistry, electricity and electronics, and bio-engineering (Freeman, 1982). In the early 20th century, both Fordist assembly lines and corporate management made huge strides in orchestrating complex labor processes and sets of labor processes, and this continues in the Japanese achievements in mass production – which is not simply Taylorism writ large, as we shall see in chapter 4. This growing technological and managerial sophistication depends on applying human intelligence to the mysteries of nature and of social labor itself. And, in the late 20th century, the focus has shifted even more directly to the rational-

---

[17] Some of the labor counted as "management" is, in fact, technical labor used directly for production, such as engineering services, or circulation labor used in the sales effort, such as advertising, or for handling finances, as in accounting. The lines cannot be too sharply drawn, and it is difficult if not impossible to label these bits precisely in the national accounts.

ization and appropriation of knowledge and the powers of the human mind itself.

Neither the extended nor the hierarchical division of labor adequately captures these changes in the technical division of labor. We need a third term, the *mental division of labor*, to grasp them. The increased emphasis on the conceptual dimension of human labor means more overall time, and more specialists, devoted to design and planning, communication and cooperation, evaluation and self-regulation, teaching and learning and the like in all labor processes. These tasks cut right across the extended and hierarchical divisions of labor. A successful product design team, for example, does not just produce drawings, but devotes considerable time to conceiving the new product, discussing ways to improve it, doing conceptual tests and evaluations, learning from their mistakes, seeking outside information, and so forth. Such mental labor is not confined to skilled designers. It is also required by the production worker, who must think about how the cloth will act in the sewing machine, how to adjust to breaking thread, and exchanging ideas with fellow workers. Where labor processes grow large and complex enough, some of this mental work breaks off into discrete jobs within and between work units.

Just who gets to specialize in mental labor is very much a political question affected by class, gender, and previous position within the division of labor, as noted in the last chapter. We do not join those liberals, such as Bell and Galbraith, who celebrate an elite of think-workers as the key to modern industrialism (cf. Ross, 1974; Gershuny, 1978; Kumar, 1978). Capitalists continually try to appropriate the powers of social labor, including the knowledge in every worker's head, and in this struggle many of the craft workers' arts have been annihilated (including some which exceed in sophistication those of advanced industrialism), as Marx showed. On the other hand, it is by no means management and its allies among the technocrats who unquestionably win the struggle over mental labor, leading to an unbridgeable gulf between "conception and execution" of the sort depicted by Braverman (1974) and his followers (Manwaring and Wood, 1984; Adler, 1985).

Behind the mental division of labor lies the social knowledge contained in all labor processes, which has risen inexorably as part of the development of the forces of production. This social knowledge lies behind more capable machines, esoteric materials, better products, and improved work organization (Florida, 1991). The service theorists have attempted to capture this dynamic of an increasingly sophisticated industrialism and its attendant mental division of labor. But they have it wrong in two fundamental ways. First, they seize upon some aspects of the general advances in knowledge and technique to the exclusion of all others: one favorite is the contribution of science to industry, or the science-based industry thesis (e.g., Bell, 1973); a second is rising worker education levels, or the human capital thesis (e.g., Becker, 1964); a third is the contribution of electronics, or computer revolution thesis (e.g., Nora and Minc, 1980; Perez, 1985); and a fourth is the explosion in information, or the information-intensity thesis (e.g., Porat, 1977; Jonscher, 1983).[18] This is not, of course, to deny the far-reaching effects of the "information technology" revolution (Forester, 1987; Castells, 1990; Freeman, 1991). Second, service theorists are idealists who ascribe industrial progress to disembodied forces or highly rarified social actors: scientific knowledge developed by a white-coated priesthood, education as book-learning without work experience, information flowing from data bank to data bank, computers humming along in silent witness over robotic factories. But while industrial production is indeed "the power of knowledge objectified", as Marx (1973, p. 706) said, that does not mean ideas are magically translated into effects, without the practical mastery of machines, materials and products, and of social interaction, which can only develop with long periods of experimentation, activity, and learning (Storper and Walker, 1989). A growing mental division of labor does not signify the triumph of ideas over practice.

We have no wish to deny the great advances in labor processes to which the service literature points (Lyon, 1988). With the developing forces of production, the mental division of labor has unquestionably broadened: there are more jobs involved with information-handling, teaching, computing, engineering and the like. Along with this, the materials and objects of labor have become more intangible, as workers spend more time transforming data, manipulating ideas, or organizing social labor than in hewing wood, drawing water, or bending metal. This is the kernel of truth in the idea of "service occupations". Stanback et al. (1983, p. 1) come close to the truth when they speak of "an economy characterized by an increased sophistication in terms of what it produces and how it carries out production". But they, too, persist in trying to analyze this in terms of services, particularly the growth of intermediate business services, and their grasp of technical change and complex divisions of labor remains quite thin.

## Conclusion: the growth of indirect labor in advanced industrialism

Capitalist industrialization has been synonymous with increasing the productivity of labor, or relative surplus value, and with increasing the turnover time of capital, or lowering the time (and cost) of circulation (Harvey, 1982). This has had a number of effects, among which the transformation of the technical division of labor *within* social labor processes is one of the most important. In the press to revolutionize the labor process, the technical division of labor has not only increased, but its center of gravity has

[18] Machlup (1962) and his followers (Porat, 1977; Rubin and Huber, 1986) have tried to argue that an overwhelming number of workers – perhaps one-third in the USA – are occupied in some way with handling information. But their categorization of information sectors and workers is so overbroad these estimates need to be taken with a grain of salt. Rubin and Huber, for example, count musical instruments as information machinery.

shifted from *direct* to *indirect* labor. The hands-on work of processing, assembling, and moving materials has diminished relative to the work of regulating, administering, organizing, and improving production systems. Indirect labor is at the crux of the service-labor phenomenon, as Gershuny (1978, p. 58) recognizes in defining service occupations by their "distance from material production" not by their final output. Yet he persists in writing of "service work", so his insights have been largely misunderstood.

This shift in the division of labor has affected everything about work, workers, and workplaces; it has moved work from the factory floor to the office, put more people to shuffling paper and fewer cutting metal, emphasized packaging and shipping, built more distant hierarchies of command and coordination, and so forth. These changes come wrapped up in new social garb, not only the switch from overalls to suits, but from foul factories to clean offices, wage packets to salary checks, and tyrannical foremen to bureaucratic rules. Under capitalism or socialism, these revolutions in everyday work and life have had a tremendous impact on everyone's experience and perceptions of industrialism. No wonder the notion of a service economy has caught on so widely.

We must avoid the sort of tunnel vision that looks at either the shop floor or office desk, but never both at once. Limited vision has plagued Marxist accounts as well as post-industrial ones, by discounting indirect labor as unproductive, but this is true only if we make no allowance for the necessary elements of time and space (extended labor), coordination and direction (managerial labor), or knowledge (mental labor). It is absolutely crucial to break with the view that an individual worker, standing at a work station from 8:00 to 5:00, is the essence of industrial labor. Labor is collective, continuous, and has many parts that must mesh like the gears of a good machine.

For instance, consider the work of organizing production. This helps coordinate the disparate parts of complex labor processes, and directing, monitoring, and regulating production processes increases the performance of all types of labor. Effective forms of organization have permitted a degree of industrial productivity far greater than would have been possible through market integration of minuscule work units. Similarly, the labor of thinking about a problem beforehand, of working out better ways of designing products, building machines, or doing a job is a way to make subsequent labor – one's own or that of others – more productive and more useful. The janitor, the computer operator, the receptionist, the repair person keep the work flowing from minute to minute, day to day. Work in training sessions or committee meetings teaches people new skills eases group interaction, or selects a collective course of action.

Attendant upon this elaboration of the division of labor, and particularly of indirect labor, is the unfinished nature of most pieces of any complex labor process. As noted earlier, those working somewhere in the pipeline have a hard time seeing the tangible results of their labors. Moreover, when the work is concerned with insubstantial material and involved with manipulating information, or manipulating people and organizations, it is even easier to confuse indirect labor with final labor services. This confusion is further compounded by *externalization* of business services, where certain kinds of labor inputs are purchased from independent firms.[19]

Everything said here with manufacturing in mind applies equally to labor processes in the realm of circulation, which we consider next. Handling money, selling groceries, and advertising have their quirks, but they are still labor with definite results: accounts are cleared, stocks moved, and advertisements appear on television. And the drive to increase productivity applies just as forcefully in this sphere. Indeed, the elaboration of complex labor processes and the growing sophistication of work technology extends into every nook and cranny of the industrial system, and, in further iterations, to labor processes within labor processes, such as the specialities arising within management or in R&D labs.

The growth of indirect labor in the 20th century may well indicate a revolution in social production as profound as earlier shifts from workshop to factory or from mechanical to electromechanical machinery. Certainly, the embellishment of the division of labor in the late 20th century appears to be part of an ongoing industrial revolution (Walker, 1991), but the concept of service occupations does not go very far in elucidating this process.

## Circulation: What Goes Around

Almost all circulation activities are given the service label, simply by virtue of not being manufacturing, yet the service economy literature lacks an overarching concept to explain links among such things as banking, retailing, and insurance. Instead, these functions are called distributive services, integrative services, intermediate services, and the like. In fact, circulation is a fundamental complement to production in industrial capitalism which stands on its own as an economic category and has only a tangential relation to the idea of service. The key to circulation is not the form of the product, the type of labor input, or the relation to the consumer, but rather the ceaseless flow of industrial commodities and money.

Circulation begins with the exchange of the commodity between buyer and seller; once this exchange becomes generalized, a veritable storm of goods (and labor-services) moves in and out of the pores of society.

---

[19] Of course, some pieces of indirect social labor may become embedded in discrete goods such as software packages, books, or specialty semiconductors. This takes us back to the situation of a growing *social* instead of *technical* division of labor. Such goods are likely to have a high information content and to be design-specific, as purpose-built machines have always been. This tends to increase the emphasis in modern production systems on product innovation over process efficiency.

Circulation also includes a counterflow of monetary payments, and a background movement of information, property titles, and income. The service theorists' usual conception of exchange, drawn from neo-classical economics, sees a linear movement from producers to final consumers, conceived of as individuals, households, or governments. In fact, there is a massive intermediate flux wholly within the realm of production, which is a virtual extension of production in time and space. Moreover, circulation is not just an exchange or interweaving between buyers and sellers: it involves the return of profits on outlays of invested capital, and the upward spiral known as the accumulation of capital.

### Commodities: products that circulate

In looking at circulation we immediately encounter the social relations of commodity exchange. In capitalist economies, products circulate principally as commodities, i.e., as private property exchanged for money through market transactions.[20] Commodities can be either goods or labor-services, and they are both useful products and embodiments of value, a measure of the abstract (socially necessary) labor time used in production. Circulation therefore encompasses flows of values as well as of use-values. Capitalism is deeply inscribed with the value-form of the products of human labor, the generalization of money as means of purchase, and the restless search for value and surplus value (in the form of money). Where Bell believes the service economy is increasingly "a game among persons", we agree with Marx that the capitalist economy remains critically a game among commodities.[21]

### The distribution of commodities: wholesaling and retailing

Retail and wholesale trade are ordinarily counted among the service sectors, usually under the heading of distributive services. Trade is largely concerned with goods, however; because the labor expended in handling

---

[20] Not all goods and labor-services circulate as commodities, because much of the division of labor is strung together through non-market forms of integration; for example, goods and labor-services can be transferred within firms, without any price formation, change in ownership, or exchange of money (see chapter 3). Some objects of nature which are not produced goods are also exchanged and hence take a commodity form and have exchange values, but they do not embody labor value.

[21] On the other hand, commodity production and circulation exist as a subordinate form, in many precapitalist modes of production and even within socialist economies. Many of the elements of circulation described here extend beyond capitalism, particularly the distribution of goods, but also money (even if used only in consumer and international transactions) and means of transferring possession of goods (even where personal usufruct rights and not capitalist property). Nevertheless, the difference between capitalism and Soviet socialism is actually greater in the realm of circulation than in production, where engineers, line workers, and janitors have a strong affinity across modes of production. These matters are treated further in chapters 5 and 6.

and selling commodities does not appear separately from the product, it is mistaken for labor-services as a final output.[22] Wholesale and retail sales usually involve joint labor-services to the buyer, such as information about the good or instruction for its use (Stanback et al., 1983). But most of what goes on in wholesaling and retailing is akin to post-production labor, i.e., it simply extends production across time and space toward the point of consumption.[23] Still, the trading function is predominant, so wholesaling and retailing must be counted as circulation. Insofar as sales-labor promotes the product, or merely transfers it from one owner to another, nothing has been added to the product's use-value – no production has taken place. Such labor is wholly that of circulation.

Wholesaling consists of shipping, warehousing (storage), and distribution to retailers, as well as the strategically vital mediation function of the merchant or commercial capitalist. Wholesale merchants buy and sell to other businesses, weaving together the disparate fabric of production (cf. Riddle, 1986). Some, like Benetton, go so far as to create their own retail chains and directly organizing production through subcontracting networks; others are no more than commodity brokerages, setting up trades but leaving the handling of goods to other wholesalers and shippers. Retailing, by contrast, is concerned with sales to final consumers in the wide public. Large retailers, such as Sears, usually move upstream into wholesaling and subcontracting, while industrial corporations frequently move downstream into shipping and direct sales to business; occasionally industrialists sell directly to consumers, as with the Levi-Strauss-owned clothing chain called The Gap.

The web of commercial trade has grown right along with commodity production and exchange, and long predates the industrial revolution (Braudel, 1979); indeed, wholesaling and merchant capital dominated the early economies of the United States and Britain (Dobb, 1947; Porter and Livesay, 1971). Retailing arose as a separate function by the end of the 1800s and probably enjoyed its golden age with the dominance of the giant department stores in the early to mid-20th century (Fraser, 1981). The huge size of the wholesale and retail sectors today is a direct result of high industrial productivity and the mass of goods in circulation, but productivity improvements in the sphere of circulation have held the growth of employment in check, particularly in wholesaling. Nonetheless, the capitalist sales effort continues to go up more than proportionately to the mass of goods to be sold, as we shall see farther on.

---

[22] Labor-services are performed directly for the consumer, so they do not require as much mediation of their exchange.

[23] This is why some authors put wholesaling and transport under distribution services and retailing under consumer services, or by itself. No one, to our knowledge, has sorted out the different elements within the wholesale and retail sectors.

*The circulation of value: money, credit and capital in the financial system*

Financial services are always included in the litany of elements in the service economy.[24] Yet in the first instance financial activities involve neither a good nor a labor service with a consumable use-value, but the handling of money. The origins and functions of money are completely untheorized in the service literature.

The money system develops as part of constructing the institutions of market exchange (Harvey, 1982). Money serves as the medium of exchange and the measure of value, and long antedates both capitalism and industrialization. It also serves as the store of (labor) value accrued through commodity production, which gives the bearer command over a portion of the social product. Finally, it functions as capital to be invested in profit-making activities. It is hard to discern any new functions for money in the contemporary economy that would justify the idea of a new world of services, but the forms of money, methods of handling it, and means of extending its powers have certainly been enhanced.

The heart of the financial sector is the banking system, which arose as a means of issuing and transferring money, storing savings, and guaranteeing value (Studinski and Kroos, 1952). Money as hard currency and payment at the moment of purchase long ago proved too limiting for the widening compass of commodity exchange, so credit-money arose to facilitate payment over time and space, bridging gaps between sales, between production and sale, and between paychecks. Commercial banks have the power to create that credit-money, in the form of checks and credit cards. With credit comes interest, and a segment of surplus value enters directly into the financial sphere of circulation. The first great lenders in England and North America were merchant-capitalists, followed by banks. In the 19th century an additional method arose for channeling savings into large capital investments: securities, or stocks and bonds, with their specialized institutions, such as the stock exchanges, brokerages and investment banks. Mortgage banks and savings institutions were added to the system early in the 20th century followed by consumer credit branches of major industrial corporations. Behind the banks and capital markets stand sovereign states, with their powers to mint currency, create paper money, regulate financial institutions, control the rate of credit creation, and settle international accounts (Harvey, 1982; Ingham, 1984).

The last twenty years have witnessed an explosive growth of financial activity and credit (debt), led by developments in London and New York which has transformed the way business is conducted (Thrift and Leyshon, 1988). Money has radically changed form; with the growth of plastic cards

---

[24] The US Census category is FIRE (finance, insurance, and real estate). This mixes basic finance with interrupted circulation and the circulation of property rights, which we consider under separate headings.

and electronic funds transfer and storage, money can be nothing more than magnetic imprints on a hard disk and money-handling can be done entirely by computers. Meanwhile, funds can be held in sophisticated portfolios and easily traded for both present and future. Credit has been extended and made more liquid through new forms of bonds and the development of a whole series of secondary markets for long-term, fixed-interest instruments. New players, such as venture capitalists and quasi-bank brokerage houses, have appeared on the scene, and banks are effectively acting as much as intermediaries for securities as sources of loans. Along with these changes, money and capital have attained new levels of international and intersectoral mobility (Warf, 1989).

Financial activities have thus grown apace in the leading centers of world capitalism.[25] They require a substantial portion of social labor to speed money on its way, as the division of labor widens and deepens in the financial sphere, and intermediaries proliferate. At the same time, some enterprising banks and brokers are selling "financial services" in a commodity form – essentially direct advice (a labor service) about one's financial affairs – further amplifying the social division of labor (Thrift, 1987).[26] Nonetheless, the growth of financial activities bespeaks the growing volume of commodities, value, and capital in circulation, i.e., the expansion of the industrial system, not its transcendance.

### The circulation of income and savings

Income flows creep into accounts of the service economy in three ways (Petit, 1986). First, international accounts often label flows of repatriated wages and profits, plus government spending abroad, as services. Second, housing rents are sometimes counted as consumer services. Third, outputs of service sectors are sometimes measured by income (usually wages expended) where output is difficult to quantify. But income should not be confused with outputs, including labor-services. Income flows are money made from the sale of labor-power, the ownership of property, or the tax power of the state. The state may, in turn, distribute some of its share as social income to retirees, the sick, or other dependents.[27] These income flows bear no necessary relation to the value of output, or to the contribution of different inputs to production, as held by the marginal productivity theory of neoclassical economics, but are determined by social struggle over distributive shares – which is itself constrained and enabled according

[25] As Thrift (1987) points out, however, a goodly portion of what is labeled finance capital is actually commercial capital, and ought to be counted under the heading of merchant/wholesaling.
[26] Such so-called "financial services" as packaged loans or brokered securities are nothing more than the traditional function of banks and investment houses dressed up in new clothes, however.
[27] Under Soviet socialism, the state also tithes the enterprises for social payments, but the critical difference from capitalism has been state control over investment funds and the elimination of private claimants on rent, interest, and profits. This takes place in a less monetized economy in which state claims may be directly on a proportion of output to be transferred to other enterprises.

to the workers' position in the division of labor (Storper and Walker, 1989). Income cannot, therefore, be used as a surrogate measure of anything else.

Insurance and pension funds, usually included under financial services, should be treated separately. They are special financial instruments created to handle stores of value set aside for a rainy day, i.e., they are savings or interrupted circulation. Large sums of money are diverted for the purpose of increasing (or maintaining) production, income, or consumption levels in the future. Bank term accounts are the simplest form of savings. Insurance is a highly controlled form of saving aimed specifically at reducing losses from death, injury, fire, theft, and other plans gone awry, in business or personal life. Pension funds, a newer form of savings, threaten to overshadow all other players in the financial arena and may be held in banks or invested in securities and real estate.[28]

All such forms of saving would disrupt the circulation of money and commodities were it not for the activities of financial institutions, which put the pools of withdrawn money to use. Interruptions are an unavoidable part of the circulation of money, commodities, and capital in a goods-producing economy and the development of institutions to overcome them is as old as merchant capitalism. Their present growth marks progress toward the perfection of capitalist circulation, not beyond it (Harvey, 1982).

### Shadow circulation: the flux of deeds, titles and claims

All commodities in a capitalist economy are private property. This holds equally for the productive equipment in private firms, consumer durables enjoyed by individuals, and infrastructure owned by the state. As commodities are exchanged, so are the titles to them, either implicitly or as formal legal documents. So titles circulate in the same direction as commodities and opposite to the flow of money, a "shadow circulation" normally lumped in with finance in accounts of service sectors.

The primary form of shadow circulation is the brokerage of deeds to real property, such as buildings and machines. Because land and buildings cannot easily be carried away, paper titles are needed to complete the exchange and establish ownership. Corporate stock is another form of paper title, representing part ownership of the firm's real assets, which may be sold or traded. Under capitalism property has a second nature; it grants the right to income generated by something, as well as the possession of the thing itself. A second form of shadow circulation therefore arises to deal with financial claims, such as bonds, mortgages, and stocks.

---

[28] Inventories are also a form of interrupted circulation, or savings, though in commodity rather than monetary form. They are always counted as part of the assets or capital outlays of industrial firms.

These instruments are issued in the process of credit-creation, but continue to be traded in secondary markets, often at a discount price to allow for risk. A tertiary form of shadow circulation embraces futures and options. These are prospective purchases, which may be traded many times before the actual transfer of property takes place. They may be considered a kind of delayed circulation, which tries to reduce the insecurity of the future.

Shadow circulation appeared in the normal course of capitalist industrialization to facilitate commerce, investment, and profit making. It is, of course, highly subject to the speculative temptations of investors, as revealed again by the global stock market crash of 1987. It has no role at all to play in the direct process of production, nor in the consumption of use-values; it serves no discernible service function except to grease the wheels of circulation, a quintessentially capitalist endeavor.[29]

### Temporary circulation: rental and leasing

Leasing services are yet another commonly cited growth sector of the service economy, but they are part of the normal circulation of commodities. Leasing represents a temporary separation of ownership and use, in which one person pays for the use of a good owned by another for a fixed period. It is an incomplete, or temporary, form of exchange. The principal reasons for leasing are difficulties in taking possession, lower initial expense, and short-term need.

While renting buildings is a long-standing practice, leasing equipment such as computers, drilling rigs, and vehicles is a relatively recent phenomenon, leading to the mistaken view that something new is afoot. Thus business leasing is counted under business services, yet no labor-service is involved. Nor is leasing properly to be counted as part of consumer services. For example, the ticket to a movie theater is a form of leasehold on a seat for the duration of a film. Given the high cost of several reels of film, watching the movie in a theater is an economical alternative; so the theater owner leases the movie from the studio and, in turn, rents seats to viewers. The result is not a labor-service as a play would be, but the consumption of a good. Hotel rooms, also a major category of consumer services, are another example of rentals. A hotel is simply a large building; for a visitor, it is more economical to rent a piece of that building for the night than to purchase it outright individually.

Workers, too, can be rented for a short term by one company from another, the temporary employment agency. The practice of subcontracting labor (the gang-boss system) is as old as capitalism, of course, but has burgeoned of late, particularly in the area of office work, both clerical and

---

[29] But not exclusively capitalist, of course: personal property in housing and commodity futures are surely compatible with socialist ownership of the means of production and a leading role for central planning.

professional. Temp agencies are commonly included among business services. The renting of labor has certainly had an effect on labor-management relations, but not on the emergence of a service economy.

Communications and transportation normally involve considerable temporary circulation, with the rental of electronic networks and seats on large machines. The telephone network lies in wait, fully staffed and maintained, for the next caller to use the system for a few minutes. Most people rent a share of the available voice-carrying capacity, although some large firms now have their own microwave relay systems. The only labor-service involved in a phone call is getting help from the operator. The same applies to air or rail transport, where planes fly and trains run by schedule, and passengers or freight forwarders merely rent space.

### The circulation of information

Information must be passed from person to person in the process of circulation to weave together the multitude of production units and consumers throughout the wide social division of labor. A vast circulation of information thus accompanies the flow of commodities, money, and capital (Hepworth, 1990). Handling such information is a necessary part of the work of circulation, and hence of the social division of labor; it is not a labor-service unless sold in a commodity form. Innovation in this sphere is quite ancient and includes double-entry bookkeeping, newspapers, and the telegraph (Pred, 1973); today new technologies have so increased the volume, speed, and quality of data that the circulation of information has outraced the volume and pace of commodities and money, but the information is still largely a representation of their movements. As a part of circulation in general, the expansion of information and of the labor devoted to it is a straightforward indicator of industrial growth, not of a shift to a service economy.

From the earliest times, traders have borne vital information on conditions of demand and supply. Today, industrialists, wholesalers, and retailers are constantly seeking information on trends in demand, and are busy putting out quantities of information (much of it of dubious quality) designed to tempt buyers. Advertising, market research, and sales forces all generate and distribute information in the service of commodity circulation. On the input side, masses of information must be collected and assessed about material flows into production, and inventories on hand.

In one sense, money, as the measure of all things, is the chief bearer of information in the commodity system. Prices are essential data for buyers and sellers. Outlays of money tell the buyer the costs of particular actions; incoming revenues inform sellers of their success or failure. Thus better and faster feedback about consumer purchases is an important element of modern-day retailing. Where capital is invested, the flux of profits tells of the health of the enterprise – with more capital in circulation, the demand for better information about conditions of production, realization, and turnover time grows by leaps and bounds. Not surprisingly, financial data are the single largest item in the present-day flow of business information (Hamelink, 1983).

Communications companies increasingly lease access to their networks for the circulation of information between buyers and sellers, a practice commonly mislabeled telecommunications services. There are true communications labor-services on offer, designed to deal with the increasing sophistication of data handling and the greater complexity of transactions within and among firms. Specialized informational technology can be tailored to fit the needs of any firm needing faster communications, better information, and improved analytic capabilities. These closely overlap other areas of information services such as data banks and consultancies.

## The distinctive role of circulation

Activities pertaining to the handling and movement of commodities, money, property, and information take up a sizeable share of the social division of labor. Their employment share has not grown markedly in recent years, due in large part to automation in warehousing, retailing and finance – but the run up of financial incomes in the 1980s made circulation appear larger than ever.[30]

If it is incorrect to call circulation activities services, and disconnect them from manufacturing and the industrial base, the service literature does point up some deficiencies in the traditional handling of economic categories. Most apparent is the simple need for a coherent category, circulation, which can contain several related segments of the division of labor. This has been quite lacking in conventional economic discourse. In addition, the labor diverted to circulation activities must be counted as indirectly productive. Marxian economics has had an unfortunate tendency to assert the priority of production over circulation without adequately recognizing the relative autonomy and absolute importance of circulation (Allen, 1988; Britton, 1989). This tendency may be traced to the debate over productive versus unproductive labor which goes back to the 18th century (Hunt, 1979; Delaunay and Gadrey, 1987; Elfring, 1988). The Physiocrats and Adam Smith argued that production and improvements in production – whether in agriculture, manufacture, or modern industry – were central to national development, against the mercantilist idea that trade and specie were the key to wealth. Marx, who followed Smith in this regard, held that no value is added by mere exchange of commodities or money handling (Miller, 1984). Marx recognized, how-

---

[30] There is an extensive debate in the service literature about productivity trends in service sectors, of which a large proportion fall in the circulation sphere (e.g., Riddle, 1986; Petit, 1986; Elfring, 1988). It is indicative of the specific nature of circulation labor that its output and productivity are difficult to measure (Mark, 1982; Nusbaumer, 1987).

ever, that labor expended in circulation is absolutely necessary to the completion of exchange and the accumulation of capital, and his reference to circulation work as "necessary but unproductive" labor hangs uneasily in mid-air.[31]

As we have seen, circulation consists of more than acts of exchange or money counting. The labor of circulation involves such absolutely vital functions as geographic distribution, ensuring access to consumers, mobilizing investment funds, extending credit, and reducing uncertainty. Circulation also integrates complex labor systems straddling many privately owned firms, and in this sense comes very close to administrative labor within the firm. Even as he confuses circulation labor with other services, Nusbaumer (1987, p. 52) is on the right track when he extols such work as "lubricating ... the various channels of economic activity". A similar insight (and confusion) lurks in the statement that "agriculture, mining and manufacturing are the bricks of economic development. The mortar that binds them together is the service industry" (Shelp et al., 1984, p. 1).

In the broadest view, development of the forces of production in the sphere of circulation can serve as a lever to capitalist growth (Harvey, 1982). Better circulation of capital – in terms of fuller realization of value (sales), more rapid turnover, and lower costs of circulation – accelerates the accumulation process. This formula brings one very close to the role of circulation in the development of the forces of production over space and time. Not only do better commodity flows enhance direct labor productivity by linking workplaces, expanding markets, and lowering input costs – this formulation refers only to the simple rate of surplus value – faster circulation contributes to the development of the forces of production by generating larger masses of investable funds, more rapid investment in new technologies, more practical experience in production, more funds for R&D, etc. In this dynamic sense, labor expended in circulation is comparable to indirect labor in production, and is productive in conjunction with a developing industrial base. Furthermore, circulation sectors are increasingly purchasing inputs from other firms, making for self-referential expansion of some circulation activities (Riddle, 1986). In other words, the interpenetration of circulation and production in a complex social division of labor elides any hard and fast line between productive and unproductive labor, and makes the hasty dismissal of circulation functions a grave error.

The pitfalls of undervaluing circulation show up most clearly in debates over national and regional development. Given the possible extent of

---

[31] The classic political economists were also making a case for the productive classes against the rich and their attendants who live off the wealth of others – except that Marx's definition of the parasitic classes included the capitalists whom Smith was defending. In contrast, the great defender of the idle rich, Thomas Malthus, was the first to support the idea of consumer services as productive enterprise (Elfring, 1988). Marx was particularly emphatic that mere ownership of property, while it gave control over the labor and surplus of others, was not a contribution to production, but rather a social arrangement that could be replaced.

spatial divisions of labor, massive concentrations of specialized circulation activities can be found in such centers as London, Rotterdam, or San Francisco (Massey, 1984). Commercial and financial capital have a life of their own in such places and do not function as appendages to industry (Thrift, 1987; Allen, 1988). Indeed, class structures and international empires can be built on these foundations, and they can last for centuries, as the debate over the peculiarities of English capitalism has revealed (Ingham, 1984; Anderson, 1987). While it is still possible to argue that a predominance of commercial capital stunts industrialization, it is nonetheless true that the wealth of the English or Dutch nations rests substantially on the vital role they play in world commerce and finance, some of it indirectly productive, some of it useless gambling with other people's money, but all of it highly remunerative.

## Personal Consumption: What Results?

Consumption has so far been a silent partner to our discussion of production and circulation, as is too often the case in left analysis. Personal consumption has been continually expanded by the wealth of goods pouring forth from an ever more productive industrial system, as well as by new areas of consumer (labor-)services. As the mass of consumption has grown, the division of labor has shifted more toward work done in support of the immense social process of selling and using industrial products, expending free time, and disposing of higher incomes. Along with this, the nature of consumption, the relation of producer to consumer, and the possibilities for human welfare and personal satisfaction have been significantly altered.

It is important to take consumption seriously as a moment in the process of production and the expenditure of human labor as a whole. The dialectics of capitalist production and consumption have, if anything, intensified over the last century as the mass of industrial goods has risen, consumption has taken more and more human energy, and the links between producers and consumers have grown tighter. The best theorists of a consumer service society point in this direction (Delauney and Gadrey, 1987; Murray, 1988), but others naively portray the revolution in consumption as a new era of liberation for the masses of consumers (e.g., Bell, 1973). While it is not satisfactory to simply assert that industrialism and capitalism have priority over consumerism, the service writers' portrayal of a consumer-led and people-serving society begs a number of critical questions about the use-values, or final services, rendered by the modern economy.

In this section, therefore, we not only highlight changes in the division of labor but also consider how industrialization serves the consumer. That is, we take up the fourth meaning of the term "services": the service rendered by any product or labor by virtue of its use-value. We investigate what is consumed, how it is sold, and what it does for people. We shall be concerned broadly with the relationship between industrial production and final consumption, the social relations of personal consumption, and the kind of services rendered by industrial capitalism, giving closer scrutiny to the true consumer services, household labor, the sales effort, and the infrastructure of private consumption.

### Personal labor-services and the servant sectors

A considerable portion of final consumption is in the form of personal labor-services, rather than goods. We passed quickly over this in the opening section of the chapter in order to establish the priority of goods production in the industrial economy. True consumer services are made up of activities done directly for or to the consumer, such as sports and concerts, barbering and beauty treatments, and recreational travel. (Health care and education are similar, but warrant separate consideration, so we hold them aside for the moment.) While the absolute amount of income, consumer time, and social labor devoted to such activities has undoubtedly increased over time, there is no evidence for a significant shift toward an economy of hair salons and mud-baths. On the contrary, personal services have grown little over the course of this century as a proportion of the division of labor and output (Stanback et al., 1983).

Early writers on the service economy argued that consumer tastes had shifted to favor personal services over goods (Fuchs, 1968; Bell, 1973). The underlying belief was that with greater affluence, people develop more sophisticated needs that can only be met through personal attention. This drew its inspiration from the theory of a "hierarchy of needs", an idea with no more basis in history or anthropology than the trivial notion that as people gain more income they eat fewer potatoes and more steak. Human needs and the capacity to fulfill them develop in dialectical fashion; there is not a pre-existing list of wants. The materially impoverished do not simply fulfill basic needs while the rich attend to all the finer aspirations; this trivializes human culture and elevates the activities of the wealthy, who are by no means necessarily in the vanguard of civilization (Bourdieu, 1977; Walker and Greenberg, 1982).

Indeed, the wealthy have always enjoyed the benefits of personal services, either by hiring domestic servants or being coddled by ranks of workers hired by hotels, restaurants, and the like to serve an elevated clientele. Where class differences are great and the mass of the population is impoverished, as in India or Brazil, such servant relations are pervasive and add a special twist to the overall class structure. In the United States and Britain, the golden age of personal servitude in the capitalist epoch was the 19th century, when domestic servants were almost as numerous as industrial workers (Marx, 1863). In the 20th century, the mass of workers broke free to establish their own households and partake of consumer services of their own, such as baseball games and nights out dancing.

An enormous superstructure of luxury consumption still remains, of course, which would collapse without the value generated by modern industry and high labor productivity. Luxury consumption is not confined to big houses, deluxe cars, and ski lodges. The well-appointed penthouse offices of executives and their corporate jets are feathered with both luxury goods and luxury labor-services: the servants of the rich reappear as pilots, fashion designers, and architects. The river of surplus value in modern economies is so large that millions of luxury service workers can receive paychecks whether or not they make any contribution to capitalist production or other useful purposes. Whole sectors of consumer services, and of goods production and circulation, remain in the thrall of luxury demand, a twist on the modern division of labor on which the service theorists are mute.

For the common run of consumers, on the other hand, one element of personal services which has been expanding rapidly in recent years is recreation and travel. This was well begun by the mid-19th century, with the appearance of holiday resorts, dance halls, and popular sports, and has grown in participants and expenditures ever since (Fraser, 1981; Pred, 1981). The division of labor diverted to supporting such leisure pursuits has also grown precipitously: e.g., athletes, sports managements, hotel staffs, amusement park employees, etc. At the same time, leisure activity has been intensely commodified and industrialized, so the amount of personal labor-service involved has in fact stayed relatively small. The common person is not waited on hand and foot but takes a great many pleasures from the personal use of industrial goods, generating what Gershuny (1978) has called a "self-service" economy. On the input side, every hobby, sport, or mode of travel has become heavily laden with industrial goods: hiking boots, electronic games, power tools, and the like. Indeed, what was once done by consumers themselves at home, or bought as a labor service, is now often virtually replaced by consumption of a good. On the output side, popular recreations from sports to music have been professionalized and commodified, seized upon as business ventures and raised to sublime heights of money-making. The range of travel now possible takes millions of people across the globe in search of new experiences and pleasant environs, equipped with bundles of industrial goods from luggage to cameras to beach chairs. Tourism itself has become big business for tour bus companies, innkeepers, restaurateurs and souvenir-makers who receive the golden hordes. In between step the agents of circulation, the travel companies, who not only arrange travel on request but package entire holidays as a single commodity (Urry, 1990).

The implications of these free-time activities are ambiguous. Without question, many, many people are enjoying themselves in new ways and broadening their experiences of the world; one should not dismiss industrial tourism or professional sports fans with the gestures of cultural elitism so common to intellectuals of both left and right (cf. Bourdieu, 1979).[32] Even the most industrialized recreations can have much to offer, such as a more musical ear, a broader palate, a greater concern for environmental conservation, even a budding internationalism. At the same time, travel can be engineered to keep people from experiencing almost anything new, while entertainment too often means being passively glued to a glowing box and evading real world responsibilities.

### Commodification, consumption and household labor

If the 20th century has been characterized by the mass consumption of goods rather than personal service, the principal reason is relative cost. A good and labor-service can be close substitutes in the provision of a use-value, or satisfaction of a given need: concert or record, vacuum cleaner or maid, French laundry or home washer-drier. With the growing productivity of labor, industrial goods have achieved a tremendous cost advantage over personal labor-services (Gershuny, 1978). Capitalist penetration of the household has also come through the appearance of unprecedented products and the stimulation of needs that did not exist before.[33] The most important types of goods to be mass-produced and mass-consumed in this century have been consumer durables: automobiles and houses, and the appliances and accessories that go with them. The home has been overwhelmingly the prime locus of consumption, the household the principal consuming unit. This sort of consumerism has been aggressively privatized, especially in the United States.

The substitution of consumer goods for personal services is part of a larger process of moving production and labor out of the household into the wider economy. The steady incursion of commodities into realms previously dominated by household labor, from clothing to gardening (Tryon, 1917) has meant a shift in the social division of labor from the realm of domestic patriarchy to the capacious House of Capital. And the conquest of household production continues with a vengeance: eating out in restaurants, for example, is a vigorously expanding mode of consumption (Elfring, 1988). Central to this phenomenon is the spread of fast-food outlets, each a small factory employing highly rationalized and standardized methods that replace the traditional handicraft work of food prepara-

---

[32] A good time need not be sublime, a taste of the world need not be refined. Moreover, such popular entertainments as football and jazz have lured the upper classes. Nor is direct participation always better than observation of those with more skill: the football fan is no less engaged than the symphony goer.

[33] The lure of the modern and the fashionable plays a role here, as we see once more in the way Moscovites have been flocking for their first taste of a Big Mac, and the propaganda barrage of advertising helps things along.

tion in home kitchens and small restaurants, much as industrial canning did in the last century (Luxenberg, 1985). It is quite absurd to refer to this as "the industrialization of services" (Levitt, 1976), without comment on the problems this raises for the concept of service itself.

The household nonetheless retains its character as a principal site of social labor – providing from one-third to two-thirds of formal output (Hawrylyshyn, 1976) – a dimension of the social division of labor overlooked by service theorists. An important aspect of domestic labor is work associated with consumption of goods: shopping, preparing, mending, and so forth. Consumption is not a passive act of contemplation, but requires time and effort: one employs a cookbook, stove, and store-bought vegetables to prepare a meal; uses a vacuum cleaner, detergent, and mop to clean the floors. That is, externally-purchased commodities are intermediate goods for use by household labor. It might be more appropriate, then, to speak of the *correlative labor* of consumption.[34] This gives the concept of the self-service economy a further spin.[35]

Beyond the household, a considerable amount of additional social labor and fixed capital broadly support the domain of personal consumption. Harvey (1982) calls the sunk capital involved "the consumption fund", but we prefer to speak of the infrastructure of personal consumption. It consists of both goods and labor-services, in an elaborate division of labor comparable to that of industrial production and circulation.

Indeed, the infrastructure of consumption is in many ways the mirror image of the infrastructure of industrial production, and often shares the same institutions and goods. Hotels on the Costa del Sol, fishing piers, sporting palaces, and the whole paraphernalia underlying true consumer services fall under this heading. Transportation and communication systems are widely used by private citizens for personal ends; the financial system exists, in part, to handle personal savings, insure personal belongings, and extend credit for consumer purchases, particularly for major consumer durables, which require a host of cleaning, maintenance, and

---

[34] We have previously referred to post-production labor, or work that takes place after a commodity is finished in order to make it serviceable. Much household labor amounts to the same thing and here again there is a choice of whether to do one's own work or purchase the same result, as in the assembly of furniture or car repairs.

[35] Women's work dominates household labor, and, in light of this, Gershuny's concept of the self-service economy is suspiciously genderless. The story of self-service commodities is a contested field of men and women's changing roles. In the 19th century United States and Britain this meant primarily expelling women from agricultural and industrial production and confining them to the household, where they were to become experts in consumption and the management of privatized families (Ryan, 1981). From the turn of the century, household labor has been increasingly mechanized, with the help of factory-made consumer durables, but this did not lighten women's work because of the concomitant loss of servants to wage-labor and of daughters to schools (Miller, 1983). Women have re-entered the capitalist labor force in large numbers in this century, creating so-called two-earner households, which must turn to outside commodities (or to domestic servants, often from destitute immigrant groups) to replace the lost output of women's domestic labor (Silver, 1987).

repair services (Petit, 1986). Indeed, the entirety of the urban system of roads, utilities, phone lines, buildings, and the like can be seen, from one angle, as a gigantic support system for household consumption (Walker, 1981; Harvey, 1985b; Belec et al., 1987). Even public spaces such as parks and city streets are used primarily for individual ends rather than collective purposes. Behind all this lie vast divisions of labor creating, supporting, and managing the social space of private consumption and personal life, such as the work of the sewage plant engineer, grounds-keeper, garbage collector, or urban planner. Many of these functions are, of course, provided by the state.

### The sales apparatus: promotion of industrial goods

One reason for the broad availability of industrial goods in capitalist economies has been the sales apparatus, a huge part of the social division of labor devoted to promoting consumption. The urgency pushing the capitalist to realize the value in commodities – to enlarge the market – has turned the sales effort into a cutting edge of industrial revolution and service theorists are correct in calling attention to the amount of economic activity bound up in all this. But they are mistaken in failing to see that the sales apparatus evolved as the right arm of industry, as well as to service consumers and their needs. There is a subtle but essential difference be-tween a tightening bond of production and consumption on one hand, and the ability of final consumers – the mass of the populace – to turn industry into their servant. These are, indeed, "the stakes of the service society", as Delauney and Gadrey (1987) put it.[36]

The history of selling, and the burgeoning division of labor diverted to the sales effort, follows closely the awakening powers of industrialization from the earliest years (Fraser, 1981; Forty, 1986). It has brought ever-larger stores, chains of stores (multiples), and now diversified groups of chains under monolithic ownership (Gardner and Sheppard, 1989). A closely allied device is the franchise system, as in car dealerships, brewery-owned pubs, and gasoline stations, and today's burgeoning fast-food out-lets (Luxenberg, 1985). Armies of sales workers have been hired to staff the new acreage of retail sales, and to entice customers with a feeling of personal service.

The art (and labor) of product presentation have been essential to this effort. Large stores, catalogues, and shopping centers offer variety in a

---

[36] The manifest failure of the Soviet model to "deliver the goods" has had its own pernicious effects of too little availability of some goods and different kinds of obsession with ferreting out scarce items. In that case, however, there is little division of labor around the sales effort. One lesson of Soviet industrialism's failure to serve consumers is an appreciation of the social-technical problem of creating the means to distribute consumer goods and meet consumer demands with some reliability, flexibility, and innovation (Elson, 1988). Capitalist economies have done this with staggering thoroughness.

convenient array. Packaging, labeling, and display entice purchasers, brand names catch the customer's eye and heart. Fashion changes orchestrate the rise and fall of commodities within a common line, as in clothes or automobiles. Advertising has exploded, even becoming a kind of popular art form but a form of deception as well as delight, including forays into basic tricks of psychology and group dynamics (Ewen, 1976; Fox, 1984). Concern with design to meet and attract consumer demand is as old as industry, but today, "design" has become the watchword of product, store, and shopping mall success, along with sophisticated targeting of subpopulations, and presenting them with complete "lifestyles". The "Holy Grail of the designer [is] the perpetual attempt to achieve that magic fit between production and consumption, to measure and pre-empt people's deepest needs and desires perfectly" (Gardner and Sheppard, 1989, p. 74).[37]

Rationalization of selling has reached new heights in the "retail revolution" of recent years as computer stock control and electronic point-of-sale systems have both lowered the cost of inventories and labor on hand, and drawn a tighter rein between sales and orders, allowing closer targeting of markets and tracking of fashion changes. Retailers are less and less merely offering shelf space for goods, and more and more actively mediating the producer-consumer relation. At the same time, advances in batch production and the ability to switch more quickly between products (at least between cosmetic variants), have lowered response time, leading some enthusiasts to claim greater consumer (or retailer) control over industry and its output (e.g., Murray, 1988). But that is quite dubious: if Ford offered the customer any color car as long as it were black, then Benetton offers any color pullover as long as it has a bright stripe across the middle.

The coin of producer-consumer intimacy has a darker side, the invasion of the home, the mind, and the passions by the purveyors of products playing on the unfilled spaces of personal desire. The sales apparatus enters the intimate relations, psychic life, and symbolic language of the consuming public, implanting the seeds of consumerism deep within the soil of personal expression and social rebellion (Jameson, 1983; Baudrilliard, 1988). One has only to consider the crude but successful promotions of the tobacco companies, which have knowingly put millions of consumers in their graves.[38] Even the promise of flexible production has a

[37] Retail stores also sought good locations along well-traveled thoroughfares or junctions, creating the main or high street of the 19th century and strip development of the 20th century. Much of this was fortuitous, but there is a parallel history of planned developments such as arcades and galleries, zoning ordinances, and, after World War II, enormous single-developer shopping centers or malls – the original department store idea blown up to gigantic scale. In these venues, the circulation of commodities and the circulation of people mingle, commerce becomes a spectacle in itself, and shopping becomes a form of entertainment (Harvey, 1989).

[38] Nevertheless, the excesses of full-blown consumerism in capitalist societies spring from a deep geology of social life, not just from the manipulations of retail promoters. For example, class inequality has long stimulated those above to express their position through tasteful consumption

Janus face, suggesting a consumerism emptied of all content except the style of the moment (Harvey, 1989). The sales effort has created a genie that is very hard to squeeze back into the bottle, making it difficult to be sanguine about the arrival of the consumer-service society.

Nonetheless, condemnation of the inauthenticity of consumer culture, long a staple of elitist critics of modernism, has usually been based on romantic notions of pre-industrial or post-capitalist society (Miller, 1987). Consumption, after all, represents the social appropriation of the useful products of the mighty industrial system, and is a necessary part of modern life. No proposition based on a past golden age, in which good taste and good craftsmanship were the norm, can stand close inspection; rather, history shows a mixed picture of parallel development and erosion of consumer skills, with a slow accretion of popular knowledge about both the growing range of things to buy and the traders' swelling bag of tricks. Retail capitalists know the balance between serving and stimulating consumer demand can be a delicate one, and the increasingly sophisticated sales effort speaks to the greater sophistication of the modern consumer, rich and poor. Consumption can be an important expression of personal choice, even playfulness, whether in clothing, house furnishing, or sports cars. Mass consumption has a strongly democratic element, and can be closely tied to cultural expressions of youth, minorities, or women, thanks to their rising disposable income and independence (Hebdige, 1979; Willis, 1990). There is, in short, an ongoing struggle between capitalist and consumer over the appropriation of use-values from the industrial system, which has posed an intellectual and moral dilemma since the early 19th century (Williams, 1982), and which has by no means been resolved in favor of a consumer-service society.

### The search for consumer sovereignty

Without question, the wealth of the industrial economies – especially the capitalist ones – supports an unprecedented level of consumption. Consumption is chiefly of goods, which are what these economies are best at producing, although true consumer services have grown in the realms of leisure, recreation, and luxuries. Along with these goods and labor-services come a sales apparatus, household work, and a social infrastructure which represent major additional divisions of labor in capitalist societies. One cannot make a leap from these conditions, however, to the neo-classical idea of consumer sovereignty in which the consumer proposes while the producer disposes. This is extremely dubious, whether we are speaking of

and those below to try to catch up by conspicuous display and imitation (Veblen, 1899; Bourdieu, 1979; Levine, 1988). Sellers of commodities are able to play on fantasies of instant gratification, of obliterating class and wealth differences; given the yawning chasm of human expression across the gender gap, it is easy for advertisers to play on unsatisfied sexuality and use women's bodies to sell everything.

capitalist or Soviet modes of production, especially if we add the gender dimension.

To exalt an economy of consumption is to be blind to the two-faced character of modern development (Miller, 1987). On the one hand, industrialization heralds the potential for human satisfaction and liberation: the proliferation of useful products ought to be accompanied by a corresponding wealth of personal care and well-being; the shift from household to industrial production ought to free personal life from drudgery and household labor. But reality often belies this hope. The industrial system is good at putting out things, but does not guarantee personal fulfillment. Neither capitalist nor Soviet socialist relations of production have harnessed industrialism's powers satisfactorily in these terms. Capitalist industrialism renders disservice in cascading advertisements and mad fashion changes. Soviet industrialism has played its own infernal tricks on the hapless consumer though persistent shortages and poor quality.

A democracy of consumers will not come from disposable income alone. Those who control production still command the heights of the industrial system, not the common people whose welfare is supposed to be the end result of all the productive power unleashed by industrialism. The capitalist's interest in serving the customer is to sell the commodity, make a profit, accumulate capital: production for money's sake. Soviet central planners and communist parties have had other interests in mind such as rapid industrialization, military power, class prerogatives, and subsidizing of basic goods. The better sales effort and distributional apparatus of capitalism are manifest, but even as the friendly hand of consumerism has been extended, the other hand has kept a tight hold on the leash that links commodity production and realization.

## Social Consumption and the Infrastructure of Modern Life: Results as Preconditions

Three of the most striking areas of expansion in the division of labor in the 20th century are health, education, and general government. These are almost always included among the service sectors, and if that definition is accepted, their huge amount of social labor carries the day for the thesis of transition to a service economy. Together, education and health represent more than half of all employment under the broad category of consumer services and virtually all of its growth – even though the growth of employment in these sectors, very rapid after World War II, has tapered off or fallen recently (Stanback et al., 1983). Early service theorists lumped health and education under consumer services, but recent writers usually place them under the heading of "social services". Government services are normally kept to the side, largely as an afterthought.

Health and education mostly take the form of labor-services directed at individual bodies and minds. Yet each involves huge numbers of workers, elaborate internal divisions of labor, a great deal of machinery, and enormous workplaces that can exceed in size the largest factory complexes. Both engage some of the most advanced knowledge and sophisticated labor processes known. The state is similarly elaborate and engorged, but seemingly produces no output of its own. How, then, are we to understand these vital sectors of social labor? A full answer would take us very far outside the ambit of a theory of services, so we shall make only a few observations germane to the division of labor and the results of the industrial system.

Our starting point is that the outputs of these immense systems of labor are not so much consumed as ends in themselves, but as essential means of social life for all workers and consumers. One must be educated, healthy, and governed in order to work or to consume; in this sense, health care, education, and the state act as general societal infrastructure much as the urban built environment does, even though they are not primarily things or industrial goods.[39] Modern industry prompts the improvement of people's lives which is realized in better health care and longer education, but the politics of social welfare and human development cannot be reduced to an equation of capitalist or socialist reproduction, as in traditional Marxist thought. And while service theorists have equated improvements in health and education with a new society, the liberating elements in both often remain trapped within structures of production, organization and control that undermine their full promise.

### Health care: industrial cures

Health care is peculiarly important because physical well-being is a precondition for virtually all other activity. The output of the health care system is primarily a labor-service – direct action on the body of the patient – but behind this stands an immense industry with a complex division of labor like any other, despite the singularity of its project. The gulf separating midwife, herbalist and country doctor from the modern hospital is as great in approach, methods, and organization as anything else the industrial revolution has produced. The vast increase in labor and resources devoted to health care is indicative of the rising wealth of industrial economies, while the division of labor brought to bear on medical problems speaks to the sophistication of modern health care delivery. Yet health care illustrates with special poignancy the ambiguous legacy of consumer services delivered by contemporary capitalist and socialist economies.

---

[39] The category of social infrastructure could be much expanded, of course, and is chiefly a political and historical definition. Housing, for example, can be treated as social infrastructure, as it was for years under British social democracy, or it can be completely privatized as in the United States or Thatcher's Britain.

The division of labor in health care has proceeded along several lines. On one side are drugs, goods produced by the huge pharmaceutical industry, many sold directly to consumers, bypassing formal health care altogether. Here, as elsewhere, consumers can (or must) take their health care in the form of goods. On the other side are the highly sophisticated industrial goods machines of diagnosis and treatment, from X-rays to surgical knives. In between stand the medical workers, doctors, nurses, orderlies, receptionists, and so on, dealing with patients' needs. And behind them lies a depth of social labor: lab technicians, drug sales representatives, university researchers, etc. The whole apparatus is brought together chiefly in and around large hospitals and clinics, under a centralized and rationalizing administration. The patient must come to the health care "facility" to be processed efficiently along with many others. In its typical modern form, health services are individually consumed but collectively produced. This lends a further dimension to the term social infrastructure.

Despite its many achievements, the health care industry's industrial technique, professional aggrandizement, and institutional power have been subject to sharp criticism (e.g., Illich, 1976; Brown, 1979; Starr, 1982). Modern medicine has offered a kind of care that is deeply colored by deemphasis on kindness toward patients, bias toward drugs and surgery, focus on treatment over prevention, and almost wholesale ignorance of nutrition and environmental hazards.[40] The industrial model of health care is not limited to capitalist countries, as the Cuban approach demonstrates, nor is it simply the result of capitalist directives, as the National Health Service in Britain shows. Nonetheless, the distinctively capitalist origins of certain aspects of modern medicine, such as drug advertising, the financing of health care, and the private administration of hospitals, cannot be overlooked (Navarro, 1976; Doyal, 1979; Himmelstein and Woolhandler, 1990). Health care thus shows a face other than that of consumer service: one composed of producer sovereignty, the factory system, and industrial cures. This has too often come in place of generalized social attention to good health through good food, good shelter, good work, and good social care in the lives of all people. If this is the best example of a true service sector we have, it shows how far we are from real consumer service and sovereignty over fundamental areas of social welfare.

### Education and the production of labor power

Education, the other great growth segment of social services of the mid-20th century, is primarily taken as a labor-service from a specialized group of workers called teachers, and is the quintessential information-bearing,

---

[40] The displacement of women from the center of health care has come in for particular criticism by feminist historians (e.g., Gordon, 1976; Ehrenreich and English, 1979).

immaterial type of output.[41] It is, like health care, individually consumed and collectively produced. Education generally takes place in large class-rooms backed by a gigantic infrastructure of schools and colleges, research arms, libraries, administration, marketing and procurement departments, and the like. Education is not something easily industrialized and material-ized as goods, and remains quite labor intensive despite substantial inputs of industrial goods such as textbooks, paper, and pencils – and the now-ubiquitous personal computer. Yet schools and universities have been criticized for impersonal, factory-like processing of students and a peni-tentiary-like atmosphere of control. Certainly the checkered results of modern schooling make pronouncements about a people-serving consumer service ring a bit hollow.

In good part, the ambiguous character of schools reflects the purpose of education under industrialism and class societies, which is not principally to serve as an end in itself but as an input to industry and a means of social differentiation. The principal goal of modern education is to prepare people for work. Gaining knowledge and work discipline is a positive recipe for the skilled elite, but education is decidedly truncated for the mass of working class kids, who usually face a future of mind-numbing jobs and authoritarian workplaces (Bowles and Gintis, 1976). Hence, working class youth often opt out of what appears to them as a pointless exercise (Willis, 1977; Wolpe, 1988).[42]

Nonetheless, the significance of a rising level of mass education cannot be gainsaid, and is indicative of the developing powers of social labor, for which large numbers of people require technical literacy and general sophistication (Adler, 1985). Though not yet a knowledge-centered pro-cess, industrialism is nonetheless a knowledge-intensive one (Florida, 1991). Moreover, education is essential for the development of individual talents and the elevation of a democratic citizenry, for liberation beyond narrow economic concerns. In all these respects, mass education has been a major concern of a wide public drawn from every class.

The last great period of expansion in the US education system came at the end of World War II, with the spread of universal secondary education and mass entry to colleges for the first time, at least for men; British higher education expanded through the Labour governments of the 1960s. Education stopped growing in both countries thereafter (Petit, 1986). Indeed, with the assault on the schools by the conservative governments of the 1980s, one is hard pressed to see how the hopes of the service theorists for a well-educated post-industrial society will be realized (Block et al., 1987).

---

[41] This definition ignores on-the-job learning, personal study, and all other forms of gaining knowledge.

[42] Of course, education does allow some working class kids to climb up the social ladder, but recruitment to the ranks of business and the professions is highly skewed towards the children of the upper classes by reason of family advantage and personal networks (Bowles and Gintis, 1976; Domhoff, 1970, 1974).

*State services*

Government activities are frequently included among social services, or lumped under "government services", one of the service theorists' more egregious acts of casual categorization (e.g., Stanbach *et al.*, 1983; Illeris, 1989). Clearly, state labor, from social work to road repair, renders definite labor-services or goods. To a large extent, however, these are just relabeled categories. The state is more accurately seen as a provider of certain goods and services which could be delivered by capitalist enterprise or other non-governmental organizations. State provision is common for collective goods such as transport, communications, and utilities,[43] and social labor services such as health and education, which have been pre-empted from the capitalist sector by political demands for egalitarian provision. All such government services have been in retreat in the 1980s, and hopes that state intervention would fundamentally ameliorate inequalities in housing, health, and communications have faded.

Social security and other income-maintenance programs, another huge slice of government spending, represent income flows based on special non-property claims to the social surplus, mostly to supplement wage-income of the poor during sickness, unemployment, and retirement, and for care of children. They have no direct relation to use-values or useful labor in the economy, and should be eliminated from any accounting of service outputs. But they do have a big impact on access to goods and labor services of all kinds, and are highly contested politically.

The really distinctive output of government, as Max Weber (1978) put it, is organized violence. Among the largest segments of most state budgets are the police and military, by which governments supposedly provide a public good called national peace and security. The inputs to this effort are a mix of goods (armaments) and labor-services (by uniformed men and women). While some policing is necessary to social order, and could be categorized under the infrastructure of social production and consumption, the heinous quality of most military activity certainly does not qualify it for inclusion in a people-serving economy. It has principally to do with securing class power domestically and internationally (Miliband, 1969). In short, the state is the principal *consumer*, as well as a primary producer, of military services.

Finally, there is legal and civil administration, which maintains the state and manages government functions. This bears a loose relation to the general augmentation of economic management (and, to a lesser extent, circulation) in the private sector in capitalist economies, and a more direct one in Soviet systems. It includes such specialized labor as that of judges, inspectors, standard-setters, attorneys, and civil administrators, as well as

---

[43] The capitalist state may also run certain nationalized industries or help coordinate and plan industrial output, thus entering fully into the production sectors of the economy. The Soviet state went much farther, of course, and also illustrates the dangers of confusing state provision or consumption of goods and labor-services with the transition to a consumer service society.

thousands of subaltern office workers. Their labor contributes very indirectly to the sustained operation of the industrial economy.

In short, large portions of the social division of labor may be under the wing of the state (Meszaros, 1987) – though the state is more than a creature of the division of labor. In many cases, there is a political option between private and public provision of the same good or labor-service, and to classify anything as a service on the basis of ownership amounts to ignoring every other quality of that sector, and is theoretically untenable. In other instances, such as the military, judiciary, and foreign policy, the essence of the state as an entity apart from civil society is at issue, and we move out of the realm of economy and industry – and out of the service debate – altogether. Growth of state activity, while significant, bears little relation to a shift from an industrial to a service economy.

## Surplus Value: What is Extracted

Industrialization has revolutionized human capacities to transform nature over the last 200 years, and industrial society is a necessary but not sufficient title for the setting of this earth-shaking process. We require some concept of social order that provides a human face and human actors, for no machine ever raised itself from the dark soil. Even the most compelling critics of Bell's idea of post-industrial society, such as Kumar, fall back on an equally suspect notion of industrial society, *sui generis* (cf. Kerr, 1983). We hold, to the contrary, that capitalism's logic of accumulation and ruling class power has been the principal (though not exclusive) force behind the industrial revolution of the last two centuries, and remains so (Brenner, 1986). This is not to deny the remarkable impact of industrialization on modern social life; indeed, our whole discourse here is an extended argument for the causal powers of the division of labor as part of ordinary processes of production. We take issue with those Marxists who exaggerate the effects of capitalism by ascribing all money, commodification, mental-manual divisions, and the like to capitalism, *per se*, and who thereby perpetrate a romantic view of the abolition of the division of labor and of industrialism by means of a strictly political revolution (see discussion in chapter 6). Nonetheless, certain Marxian insights retain their sting, including the effects of capitalism on the shape of the division of labor and the condition of different classes.

### Pursuit of value and surplus value: the social drive behind an expanding division of labor

The pursuit of surplus value drives capitalists to seek ways of raising labor productivity (the search for relative surplus value), and this is compounded by the competition among capitalists (the search for surplus profit). Marx saw this expressed most clearly in the division of labor within the work-shop and in mechanization of the factory, where the rationalizing powers of the capitalist held sway. He zeroed in on the connection between capitalist goals and the transformation of direct production because he was aware of the revolution in human history effected by advances in manufacturing and the factory system in the 18th and 19th centuries, and saw that the logic of surplus value had pushed capitalists to move from the simple extraction of absolute surplus value to forcing revolutions in all methods of working.

It is similarly incumbent upon us at the end of the 20th century to take a hard look at how the developed world got to its present enormous wealth. Here again the service literature is either silent or adopts a Schumpeterian view of technical advance as an exogenous product of scientific progress (cf. Storper and Walker, 1989). In fact, the capitalist logic of economic gain continues to dominate industrialism. As a result, the present-day productivity of labor in all goods production, from agriculture to manu-facturing to utilities, is astronomically above all previous standards. The growth of the division of labor along novel lines, generally mislabeled services, continues the long-term process of raising the system's productive power, particularly the expanding technical division of labor in complex labor processes of the sort we have discussed here.

There has also been a vast proliferation of new goods and labor services in pursuit of profit. Any product of labor that becomes a commodity is a bearer of value. Capital can and will commodify anything, Marx noted, but he was insufficiently attentive to what is usually called product in-novation. Schumpeter (1939) put product innovation on a par with pro-cess innovation in the profit-seeking efforts of capitalists. Indeed, Marxian value theory provides a compelling reason for capitalists to introduce new commodities: they provide a new source of value and surplus value for the system, as well as offering competitive advantages over firms making obsolete products. Contrast this with the traditional Smithian view that the division of labor is a creature of an exogenous demand (Stigler, 1951).

The capitalist does not care about the material substance of the new commodities. Industrialization has transformed the kinds of materials em-ployed, as in the case of wood giving way to fiberglass or copper to plastic tubing. And, as noted earlier, the material substrata of goods has become increasingly insubstantial, involving paper, electrical circuitry, etc. The useful content of output has also been changing, with a growing element of knowledge or information; we must not cling to an outdated image of industrial production tied to big, heavy objects alone.

These changes have been closely associated with another shift, in which indirect labor and circulation labor have become increasingly commodified. Examples include technical consultancies to department store manage-ment or use of communications by sales forces. This eagerly propelled, self-referential development of new activities is captured, in a rough and tumble way, by Gadrey's (1987) "double-dynamic of services". In other words, raw segments of complex labor processes and circulation continually break off to become part of the industrial "base", i.e., to become commodity outputs that embody value and can be sold for profit. But the use of labor to produce commodities and profits continues as before, despite the metamorphosis of what is produced and how.

The divisions of labor explored in this chapter are not given to us as natural or necessary constraints; yet the service literature implicitly accepts every service activity as a normal part of modern economies. In fact, the division of labor has been pushed and pulled in a variety of ways by the play of social forces, including gender, race, and state politics. The luxury element of capitalist class consumption, for example, alters the makeup of resource allocation and output. But the implications of capital for the configuration of the industrial system go further than this, and it is worth noting that the excesses of the drive to accumulate stretch the social division of labor in peculiar ways – among them, exaggerating certain arenas associated with the expansion of the service economy, like finance and management.

Product proliferation, for instance, is more than a technical requirement of an amplified division of labor because capitalists will try to create a commodity out of anything and everything. Hence the feverish multiplication of new goods and labor services. Knowledge is for sale, in the form of patents, information services, expert systems, and data banks (Howells, 1988). Life itself is now patentable, ownable, and saleable thanks to genetic engineering (Kenney, 1986). This self-referential process may go too far, as capital presses for new outlets in areas ever more removed from direct production. For example, producer services are bloated with brilliant schemes ("Theory Z") for reforming management which lead nowhere.[44] The fierce debate over services and national development, mentioned previously, is not only about different forms of useful or productive labor, but also about truly useless and parasitic forms of money-making at the top echelons of economies such as the United States and Britain – countries in which the ability to manufacture goods efficiently and competitively has dropped off markedly (Cohen and Zysman, 1987; Florida and Kenney, 1990).

Financial speculation is another area in which capitalist excess is notable. Money is, after all, the condensation of all the social powers to produce and command commodities (Harvey, 1985a). Because of this power, the temptation is always present to seize upon money itself as the source of value, the goose that lays the golden egg.[45] In the casino-like

---

[44] One must, of course, allow for some failure in any process of innovation, but this cannot justify wholesale diversion of resources into games capitalists and young professionals play upon the backs of the working class.

[45] Two aspects of capitalist social relations augment this propensity. One is that property rights in means of production, or access to capital, is itself a means to the end of breeding money by means

atmosphere of the 1980s, money bred money, and the number of high-flying players in the game grew very large, from junk bond kings down to the hired schools of bright young sharks swimming in the rich waters of high finance (Strange, 1986; Sweezy and Magdoff, 1987, 1989). Almost none of this financial activity has been remotely necessary or productive. The bare cupboard of this sort of service economics lies clearly revealed in the gleaming skyscrapers and post-modern palaces for the rich (Davis, 1985).

A third area of overdevelopment in the capitalist division of labor is in the sales apparatus (Baran and Sweezy, 1966). The fevered pressure of competition and capital accumulation forcing capitalists to create new demand, has produced a fantastically engorged machinery of image-mongering, fashion change, and stimulation of desire from computer-enhanced graphics for commercials to those post-modern cathedrals of shopping, the megamalls (Ewen, 1988; Harvey, 1989). More grievous than the seduction of consumers may be the sheer waste of social labor (and capital) involved in blanket advertising and luxurious accoutrements for elite buyers.

However, it is not necessary to argue that all capitalist excesses, as reflected in the division of labor, are beneficial to capital. Indeed, the hard laws of international competition have recently proven that Anglo-American managerial hierarchies have been excessively swollen with un-necessary middle managers, now dispersed in droves under the pressure and example of Japanese manufacturing (Bowles, *et al.*, 1983; Harrison and Bluestone, 1988). And, as the debacles of London's big bang, Tokyo's stock slide of 1990, the US savings and loan collapse, or the junk bond im-plosion illustrate, capital itself sometimes pays a huge price for indulging in speculative dreams.

### The extraction of surplus value: conditions of the different classes

Capitalists are the principal beneficiaries of industrial production. A huge proportion of surplus value goes straight into the hands of the owners of property, whose ownership grants them the right to appropriate interest, profits, dividends, rents and so forth; but ownership itself is not produc-tive. Part of this surplus goes to personal fortunes and corporate wealth; part returns as capital investment for further rounds of production. A large dollop is diverted into the pockets of those close to the capitalist class or with strong market power owing to their favored position in the division of labor, such as managers and professionals, even if their labor is wholly necessary and useful. Finally, a major portion of surplus passes through

of money, for they yield the fruits of someone else's labor. The other is that the investment which all capitalists are forced to make in order to accumulate is by its very nature speculative because one cannot predict the future (Harvey, 1982). The speculative and rentier aspects of capital are never far in the background.

the hands of government, for various purposes, via the tax system. An exploiting class is not a *service* class in any meaningful sense, despite the promiscuous use of the term in the sociology of industrialism (see chapter 1).

Marx was at some pains to distinguish the production of surplus value from its appropriation by the owners of the means of production. This rested not only on a moral condemnation of the rights of property, raw exploitation, and class inequality, but on an historical understanding of the priority of production (human labor) in the transformation of the modern world. It is in this double light that we should see his stern refusal to designate acts of exchange, capital investment, or the work of the capitalist as productive of surplus value. Even though we have dropped the classic duality of productive and unproductive labor here, we must keep the essence of the Marxian position in mind: the necessity for real work and the presence of real exploitation, and a refusal to let the powerful and the parasitic take credit for their own good fortunes or for the work behind industrial wealth.

What of the workers themselves? Industrialism is not neutral with regard to conditions of work, rewards, and assignment of jobs. Many advocates of service theory present a halcyon vision of labor in a new society, where work has become more pleasant, creative, and independent, because service jobs and workers are more skilled and professionalized than those in manufacturing. For example, Bertrand and Noyelle (1988) argue that information technology has brought an upgrading of skills in banks and insurance companies because it creates new products, new levels of expertise, and closer customer assistance. This is an unduly salutary view of what is actually "a rolling process of deskilling and reskilling" (Baran, 1987, p. 58), with quite mixed results in terms of worker welfare. Even with net reskilling, there can be considerable job loss, oppressive use of machines to intensify work, blocked career paths, and low wages (Hartmann *et al.*, 1986). Indeed, the bulk of non-manufacturing jobs are low-paying, monotonous and dead end, regardless of implicit skill levels (Freedman, 1976; Rumberger, 1981; Harrison and Bluestone, 1988).

The shift to so-called service work does, nonetheless, alert us to the shortcomings of the classic Marxian critique of the degradation of work, which rests heavily on the capitalists' use of the division of labor to parcel and reduce labor to the lowest common denominator (Braverman, 1974). In fact, human labor, which remains the indelible core of production, cannot be expunged by the capitalists' machinations, it can only be moved around within complex divisions of labor. Deskilling is repeatedly offset by the creation of new skills in other areas which the capitalists can never completely command; the general technical competence required for social labor processes has increased, and with it the educational levels of the labor force; the locus of skill has shifted from the individual craft to the capacities of the collective worker, including a range of skills outside direct labor; and worker learning and participation in problem-solving is still essential to improvements in productivity. Of course, the division of labor, machinery and knowledge can be used as weapons by employers against workers, but there is no strictly technical logic to the degradation of all industrial work, nor is the division of labor simply putty in the capitalists' hands (Walker, 1989b).

Without question, any shifts in the social division of labor toward the kinds of work associated with the service economy are fraught with potentially dire consequences for the working class. In a dynamic situation, we can expect an ongoing class struggle as new terrains of production open up (Storper and Walker, 1989). The social causes of the degradation of work and of workers show up clearly in the pattern of job creation in the service domain. Making jobs that are low-wage, part-time, and temporary (and non-union) amounts to a massive redistribution of income from labor to capital, and is a result of the political offensive against the working class in the 1980s (Harrison and Bluestone, 1988).[46] At the same time, a select group of skilled workers has prospered in every sector from electronics to banking (Walker et al., 1990). Finally, there is the extraordinary coincidence between divisions of labor and divisions by gender and race with women and people of color herded into the bad jobs and white men filling most of the good ones (Hartmann et al., 1986). This suggests that the degradation of work is part of the degradation of the workers in an unequal society and will remain so for as long as the double purpose of labor is to be the bearer of surplus value as well as the provider of useful goods and services.[47]

### The social character of the division of labor

Oddly enough, most social scientists never inquire as to the origins of an expanding division of labor, merely taking it as given or perhaps crediting scientific and technical advance (Garnsey, 1981). In the Marxian theory of capital, on the other hand, the division of labor is expressly motivated by the pursuit of gain. In our view, this capitalist character of the burgeoning division of labor has not been transcended, but extended and deepened. Not only are the great majority of so-called services still the classic activities of an industrial economy, these activities are still aimed at an ever

---

[46] Elfring and Kloosterman (1989) tie bad jobs entirely to growth of the service sector (93% of all low-paid jobs in 80s in Netherlands), poor productivity of service labor, the push of labor supply, and part-time work – yet, they hardly mention the capitalist attack on unions and the working class.

[47] The sharply gendered new division of labor, the growing number of bad jobs, and the double exploitation of women's labor at home and at the office, put the feminization of the labor force at the heart of the question of the service economy. Scholars have barely begun to grasp this nettle. Reorganization of the division of labor on the scale that has occurred in 20th century capitalist countries has blown the lid off the secret abodes of patriarchy and provided an important lever for social criticism and change today.

more profound development of the powers of labor to generate, circulate, and realize surplus value. As long as this is the case, one cannot treat work in modern society simply in terms of its usefulness in an industrial system, as does Gershuny (1978). An ineluctable tension will continue between the promise of more knowledgeable, more useful, and less arduous labor which the service thesis discerns, and the reality of unnecessarily stupefying work, repressive bosses, and poverty wages; of financial shenanigans, bloated corporate hierarchies, and corrosive advertising; of skyscraping monuments to executive egos, golden parachutes, and armies of corporate lawyers – in short, of the blatant disservice to humanity that follows from gross inequality, unchecked power over capital, and the pursuit of accumulation for its own sake.

## Conclusion: The Struggle for a Truly Social Economy

This chapter has sought to lay out the key axes of the theory that the service economy is the new dominant in modern life. We have looked from every possible angle – service sectors, service labor, distributive services, consumer services, and service society – and inspected every major thesis of service theory, from the growth of information and knowledge as productive forces to the movement of work from workshop to office. As a unifying concept for understanding the new social economy, the term services is sorely wanting. What coherent abstraction underpins the panoply of ideas swept into this capacious vat? The service concept forces salient economic issues under a single heading only by an immense corruption of familiar categories such as transportation, retailing, and money; by a libertine eclecticism that obliterates the differences between production and circulation, inputs and outputs; and by a conservative masking of fundamental issues of class, gender and exploitation.

One by one, we have stripped the service label from the major categories of activity in the production system, revealing the familiar face of industrialism. That is, the economy can still be characterized in classical terms as a system dominated by industrial production, whose outputs come mostly as tangible goods: goods circulating primarily as commodities, accompanied by the circulation of money; produced by human beings with the help of machines, technical knowledge, and rational organization; and consumed by the various classes of society in proportion to the distribution of social income.[48] Most important from our perspective, the proliferation of so-called service sectors and service occupations can be explained in terms of burgeoning social and technical divisions of labor throughout the industrial system.

[48] "Far from being service economies, OECD economies remain firmly anchored in the production of goods" (Blades, 1987, p. 166).

We have repeatedly noted the historical roots of many features of the industrial system, so as to correct the tendency to confuse the contemporary with the new. In quantitative terms, a lot of people in the advanced capitalist countries have been doing something besides direct labor in manufacturing for a long time. In the United States employment in manufacturing, narrowly defined, has never been higher than 27% of the workforce (Lebergott, 1966) and in manufacturing plus utilities never higher than 36% (Singelmann, 1978). In 1880, the manufacturing workforce was only 19% of employment, while services, conventionally defined, were 25% of the total (Lebergott, 1966; Gallman and Weiss, 1970). In other words, the United States has long been a "service-intensive" economy and did not follow a path from agriculture to manufacturing to services nor, in recent years, did Japan or the Asian Four Tigers (Singelmann, 1978; Riddle, 1986).[49] It would be more accurate to speak of embellishment of the industrial division of labor than of transition to a service economy.

Nonetheless, there has been continuous upheaval in the shape of industrial capitalism, and this is indicated by the raw data on the growth of certain sectors and kinds of employment. Change ought to come as no surprise given the tumultuous dynamism of capitalism and industrialism. The marriage of these two primal giants gave birth to the mythic juggernaut of eternal disorientation known as *modernity* (Marx and Engels, 1848; Berman, 1982). The lesson of modernism is one bemoaned by every generation of the aged and forgotten by every generation of youth who consider themselves the very embodiment of the modern (or the postmodern) (Jameson, 1989; Harvey, 1989).

Without question, long-term trends of industrialization have had a profound impact on what goes into and out of the economic system. Commodities are produced, exchanged, and consumed in ever-larger numbers, to be sure, but also in some new and exotic forms; commodities are advancing further into everyday life than earlier forms of consumption; monetary flows and financial operations have become fabulous in scale and rococo in intricacy; methods of producing and handling commodities have become increasingly extended and sophisticated over time and space; the sales effort has grown more elaborate and fanciful; systems of industrial management have improved markedly.

Capitalist industrialization has, above all, expanded and reordered the division of labor in society. It is this, more than any change in the form of products or their use-values, that is at the heart of the debate over the service economy. The locus of social labor has shifted over the last century not only from agriculture to manufacture but from production to circulation, and from direct labor to indirect labor, including technical and

---

[49] Over all OECD nations the average of service employment has risen from 24 percent in 1870 to 58 percent in 1979 (Britton, 1990). As Britton points out, one should expect wide variance in national development paths with respect to service expansion, given the wide range of activities included under that heading.

managerial work. Thus the *indirect economy* has increasingly become the focus of industrialization – i.e., of new products, automation, and creative organization – feeding on itself as an engine of growth. Behind all these changes lies a growing sophistication of labor across the whole range of modern activities. Industrialism has made a series of leaps that have carried us forward to new levels in the development of the forces of production.

At the same time, there has been a deepening integration of complex production systems and labor processes bound together by webs of finance, management, technical knowledge, subcontracting, and close relations among buyers and sellers of intermediate commodities; simultaneously, there has been a tightening embrace of consumption within the ambit of production. The social division of labor and industrialization have rendered production more thoroughly "social" in ways that both Marx and Durkheim were grappling with. Labor processes today can only function with large-scale coordination and a knowledge base utterly beyond the mastery of any individual. Production systems demand an intricate and far-ranging tissue of circulation, which can only be established and maintained through immense social undertakings and constant social attention. Producers can no longer hurl goods upon the market without relating to consumer wants, while the consuming public cannot possibly survive without the never-ending inputs of industry. One cannot, therefore, think of work as an individual pursuit or consumption as an individual satisfaction. They are irreducibly social production and social consumption on a scale, and to a depth, never before seen (cf. Delauney and Gadrey, 1987).

Over the course of the 20th century there have been repeated declarations of the end of capitalism by the managerial revolution, the scientific revolution, the corporate revolution, the welfare state, and social democracy (cf. Ross, 1974); now we hear similar declarations about the service society. There is an irony to the ideology of services, which speaks to the real hopes and fears of those who live with capitalism: the industrial system *ought* to serve common people, but the system, its powers, and the workers who embody the forces of production are also busily serving capital. The struggle to liberate the powers of humankind from class domination remains very much uncompleted. However, we hold no brief for a socialist model that abolishes capitalism without addressing the ills of patriarchy, racism, and the difficulties of harnessing industrialism and the division of labor. While the Soviet model has stood as a distinctive mode of production, it has nonetheless been a form of industrialism commensurable with industrial capitalism in a variety of ways, including pervasive sexism and racism. A debate which tries to reduce all the essential characteristics of the new social economies to either capitalism or socialism will founder in Cold War rhetoric. Nuclear power plants, credit money, and industrial management have certain irreducible properties and problems that cut across the societies in which they are embedded. Similarly, the elaborate social division of labor needed to produce semiconductors, distribute goods to consumers, or administer large industrial undertakings appears under all modern systems. Neither the immense swirl of commodities across vast divisions of labor nor the material base of high popular consumption can magically be done away with (Miller, 1987). The struggle for a human-serving industrialism must be extended beyond the failures of communist societies to other, as yet unclear, models of economic relations.

# 3

# The Expanding Horizons of Industrial Organization

After years in the backwaters, industrial organization has moved to the forefront in discussions of the changing landscape of the new social economy. There has been a worldwide upheaval in business practices, and industrial organization theory is scrambling to catch up. This chapter tries to consolidate theoretical advances on several fronts and to frame a more comprehensive model of industrial organization than now exists. To begin, we insist that the analytics of organization start from the problem of production and the division of labor. Then we expand the palate of organizational forms well beyond the conventional dualism of markets and firms to include the many varieties of workplaces and inter-firm networks, and further still into the realm of territories and states. Finally, we push industrial organization into the theory of capitalist dynamics, where it enters debates on technical advance, uneven development, and the changing face of private power.

Industrial organization theory has been hobbled from the outset by the dominance of the neoclassical paradigm in economics. In the basic model of Marshall (1890), the economy consists of many small firms operating in well-formed markets and guided by clear price signals. Industry is efficiently organized by market exchange, and that's the end of it. Of course, the existence of large corporations in the 20th century has long stood as a direct challenge to this idealized model of perfect competition, but the great bulk of critical thinking has been directed to the monopoly power of the big companies, as in the theories of imperfect, monopolistic, and oligopolistic competition of the 1930s (Robinson, 1932; Chamberlin, 1933) or the studies of barriers to entry in the 1950s (Bain, 1956). Such approaches implicitly accept the Marshallian universe of small firms as the ideal condition for economic efficiency, without regard to the real foundations of industrial production or the dynamics of capital accumulation.

The left has not done much better. Marx's theory of centralization offered an explanation for the greater survival of large firms over small, but chiefly as a function of financial mass and stability, not development of the productive forces (Marris, 1979; Mueller, 1986). As a result, Marxist economics has remained largely within a discourse of monopoly capitalism that emphasizes unfair competition of large against small, alliances of corporations with the state, and tendencies to stagnation (Steindl, 1952; Baran and Sweezy, 1966; O'Connor, 1973). The attention to capitalist dynamics at the core of Marx's theory was lost in this account. Among non-Marxists on the left, Veblen's (1898, 1909) prescient critique of Marshallian economics offered great promise, but the institutionalist school that followed him has never been able to generate more than a set of sociological caveats on neoclassical theory, lacking a systematic alternative to it (Hodgson, 1988).[1]

Perhaps the most significant contribution to the field of industrial organization has come from students of the internal management of modern corporations. Veblen (1921) was the first economist to praise the engineers for overcoming the limits of the market, but the study of modern management dates principally from the 1930s: Berle and Means (1932) were the first to document the rising power of managers over financial capitalists; Barnard (1938) began to think about administrative behavior; Burnham (1941) declared the coming of a "managerial revolution"; and Schumpeter (1942) incorporated the large firm into his thinking about technical innovation and capitalist development.[2]

The single most important contributor to industrial organization theory has been Chandler (1962, 1977), a business historian who documented at great length the achievements of corporate managers in operating large production systems, integrating forward into marketing and diversifying through multi-division structures. For Chandler, the search for monopoly profits was a false step abandoned by progressive management early in this century, and the concentration of capital in large firms is purely a function of their productive capabilities.[3] Chandler's paean to modern management has definite limits, however, as do the US companies on which it was based. To see this, it is necessary to move outside the boundaries of the self-contained firm.

In the last decade, the transaction cost theory of Williamson has come to dominate the field of industrial organization. Williamson (1975, 1980) sought to bring Chandler's insights back into the Marshallian fold, explaining the resort to internal management (hierarchy) by reference to the failure of markets as a means of economic integration, but keeping markets as a lively counterpoint to hierarchy in the set of choices available to

---

[1] Keynesian theory never took up industrial organization to any degree. The left Keynesian Joan Robinson (1956) later regretted her wrong turn into imperfect competition rather than the theory of capital accumulation, but did not return to the question of organization.

[2] This literature broke with the prevailing concern of the previous thirty years with the dominance of finance capital, as in the writings of Lenin, Hilferding, and Brandeis. Berle and Means' work led, in turn, to an endless – and often sterile – debate over who controls the corporations, which we will review later on. Simon's (1947, 1957) follow-up to Barnard led into the dead-end of behaviorism (Cyert and March, 1963). Veblen and Schumpeter's romance with the technocracy was picked up later by Galbraith (1967).

[3] Chandler and Marx's theories were effectively merged by Hymer (1972) in his imaginative vision of the multinational corporations of the post-war era (cf. Stopford and Wells, 1972).

entrepreneurs seeking to minimize their costs of doing business. Recent events have conspired against the tidy reconciliation achieved by transaction cost theory, however.

First came the unexpected success of industrial districts made up of networks of small- to medium-sized firms in the Third Italy and Silicon Valley, allowing the theory of flexible specialization to rise in direct challenge to the whole shared narrative of Marxian, monopoly, and managerialist schools about the inexorable growth and domination of large firms (Piore and Sabel, 1984).[4] Second, Japan's ascent to the top of the international competitive heap was achieved with an industrial system organized around combinations of giant firms, huge industrial groups, and layered subcontracting networks that do not readily fit Anglo-American or European experience (Dore, 1987).[5] Indeed, the evident success of organizational models beyond the Anglo-American pale, especially in Germany and Korea, has opened a veritable Pandora's box (Katzenstein, 1989; Amsden, 1989). Finally, the management structures of the Anglo-American multinationals themselves have been undergoing profound readjustment toward less hierarchical forms than the classic model described by Chandler (Martinez and Jarillo, 1989).

In short, all hell has broken loose and there is simply no overall theory of industrial organization that can contain the outburst. The whole problem needs to be rethought from the ground up, beginning with the elemental fact of the expanding division of labor – the thing that requires organizing in the first place. The discussion must then escape the noose of the Marshallian trade-off between markets and firms, and see the whole broad horizon of organizational form. Further, the dynamic visions of Chandler and Marx need to be joined so we can grasp the interactive development of organizational technology and capitalist social relations. Lastly, the whole discussion must be placed in a comparative framework of uneven development within the globe-straddling advance of capitalist industrialization.

## The Organizational Problem: Division and Integration of Labor

Industrial organization theory begins with neoclassical presumptions: the fundamental problem of economic life is allocation of scarce resources, exchange is the basis of action, businesses are rational actors seeking

---

[4] Flexible specialization theory provoked a salutary revival of interest in the social division of labor (external economies) as a source of productivity (Storper, 1989), the forgotten work of Marshall (1921) on industrial districts, and the long-neglected challenge to Marshallian partial equilibrium and marginal cost theory by Young (1929). Scott (1988a,b) has attempted to put all this into the formalizations and trade-offs of Williamson's system.

[5] Desperate attempts have been made to shoehorn Japanese industrial organization into the Williamson model (Aoki, 1984), the flexible specialization mold (Friedman, 1988), and the Chandler model (Amsden, 1989). But it will not be bent, spindled, or mutilated into such convenient slots.

efficient solutions, and the essence of production is cost minimization. The nub of Williamsonian organization theory is *transactions*, which are nothing more than exchanges by another name, and the goal of organization is to minimize transaction costs, which is static optimization theory applied to a new realm.[6] But these are not the purposes of capitalist economic life. On the contrary: one must first produce in order to have something (a use-value) to exchange; one must meet certain standards of competition and the law of value in order to remain viable in the market; and one must generate a surplus and continually seek competitive advantage in order to keep pace in the race to accumulate capital – the real goal of the capitalist economy (Marx, 1863).

## *The division of labor and its mirror image*

The organizational problem in the industrial economy begins with production rather than exchange. Bringing together labor, materials and machinery poses an elemental organizational problem at the heart of the labor process (as we shall see in the next chapter); but it is the process of social labor that creates the organizational problem as usually understood, i.e., the puzzle of effectively piecing together the complex divisions of labor inherent in all modern production. The mirror image of the division of labor is, therefore, the *integration of labor*.

Integration begins with the direct labor of producing a commodity, which normally consists of several steps. Two kinds of production chains are ordinarily distinguished: sequential processing and component-assembly systems. A simple integrative act would be to bring the various steps of either under a single roof, as in a factory, or under single ownership, as in a multi-plant firm. This has traditionally been called vertical integration. In contrast, horizontal integration is the bringing together of several different commodity lines (each with its own production chain) under one firm's control. The standard vertical/horizontal dualism is too simple to encompass the real complexity of integrated production schemes, however. To begin with, actual production systems cut across a wide matrix of divided labor. Input-output analysis has demonstrated that the economy can be divided into hundreds of thousands of production cells, and each of those cells takes inputs from many places and sends outputs in many directions; in a highly developed division of labor, everything ultimately connects to everything else. Production systems overlap, sharing inputs from common sources and sending outputs from one process to many others. Steel goes into drill bits, automobile frames, and ball-bearings; ball-bearings are used in bicycles, automobiles, and textile machinery; automobiles may, in turn,

---

[6] The basic transaction cost calculus is one of efficiency and cost minimization to which are added extreme self-interested behavior ("opportunism") and limits of perfect knowledge ("bounded rationality") which can cause markets to fail in certain circumstances. This extends the insights of Simon and behavioral sociology from the large firm to the market.

be used by pizza parlors, gardeners, and airline companies. This creates systems of nested and branched integration.

The degree of overlapping integration makes industrial organization appear impenetrable. It means there is an arbitrary element in all practical systems and in the boundaries between factories, firms, industries, and the like (Auerbach, 1988). Nevertheless, there are good technical and economic reasons why some connections matter more than others; the parts of an airplane are by no means arbitrary. It is therefore possible to discern certain shapes or root and branch connections, which may be called production systems. Production systems provide a partial framework around which the organizational fabric of the economy may be wrapped; hence Chandler's (1977) cogent argument that large corporations and their management systems evolved to meet certain technical problems peculiar to large-scale or far-flung production. But the argument cannot be taken too far: in technical terms alone, the intertwining nature of production systems still precludes any tidy way of slicing things up and putting them back together. One cannot reduce the immense practical difficulties of organization to mathematical puzzles around input-output tables.[7] The social organization of production is not to be gainsaid.

Direct labor is not the only organizational problem, moreover. As we have seen, modern industry includes correlate branches of indirect labor, what we call extended processes of production, hierarchical management systems, and mental divisions in handling specialized knowledge (see chapter 2). Production must be complemented by the divisions of labor in the sphere of circulation as well, including distribution, communication, and finance. Production systems in the full sense are thus far more than sequences of physical manipulation of materials; they are also sequences of research-development-manufacture, purchasing-manufacturing-shipping, planning-financing-education, command-response-effect, advertising-distribution-sales and the like, each of these involving feedback and interaction in a way that belies any simple notion of either sequential or parallel acts of labor. We do not yet have an adequate conception of integration to handle these intersecting pathways in a completely coherent way, in part because the various bits of the puzzle have long been consigned to specialized fields such as marketing, accounting, and engineering outside of the mainstream of identical organization theory. That is changing rapidly today, but the truncated terminology of vertical integration and disintegration still dominates the discourse on organization in economics and geography (e.g., Williamson, 1985; Scott, 1988a,b).

### Static integration

Integration means more than simply pulling together disparate pieces of production systems; it requires sets of actions which are usually loosely

[7] In the manner of certain recent neo-Ricardians entranced by Sraffa's insights into complex production matrices (e.g., Steedman, 1977; Roemer, 1982).

grouped under the heading of "coordination". It is easiest to begin by posing the problem in static terms.

The simplest element of productive integration is physical linkage among work units. Materials, machinery, labor-power, and information – the life-blood of production – must flow to the various limbs of every production system, and commodities must reach final consumers. One cannot assemble a car without the requisite dies from the machining department, engine blocks from the casting factory, or windshields from the glass company. One cannot run a modern office without telephone calls, photocopies, and paper supplies. The movements of materials and labor must be accompanied by information about their quantities, handling, installation, and attributes. Information theories try to capture this facet of production (e.g., Hepworth, 1990), but information flows are not the sum total of integration (Hodgson, 1988).

Labor processes also require cooperation among workers so that the various parts of a larger system operate in a unified way. A component must be cut to specification to fit properly in a later assembly process. In extended production systems, designers, production engineers, and marketing departments need to mesh their efforts in order to come up with products that can be made in the company's factories and fit into its sales effort. Adequate information flow is essential to such coordination, but so is physical movement and intelligent effort by technicians, sales people, and other workers.

A third aspect of integration is scaling. Prevailing opinion has long held that unitary scale economies were central to the operation of modern heavy industry; that is, large equipment must be utilized at near-capacity levels to avoid wasting capital. Recent thinking holds that economies of scale are overrated because most machine systems and labor processes are divisible into constituent parts, each with its own optimal scales of operation (Teece, 1980). It may pay to run these different processes independently, bringing in or putting out work in line with capacity constraints and demand, unless there are concurrent scale economies or what are now called "economies of scope" (Panzar and Willig, 1981). Economies of scope also include shared knowledge, as in a patented invention or general know-how that extends from one process to another (Caves, 1982).[8]

A fourth element is continuity, embracing the flow of materials and investment through production, as well as the flow of working time at every work station or workplace. Time is the principal measure: if continuity is poor, the time in which materials, machines, and workers are idle increases, the production cycle lengthens, and the turnover of capital slows (Marx, 1886). In addition, product quality often depends on continuity, as with perishable foodstuffs or rust-prone metal parts. Economies of speed are therefore as important as economies of scale and scope (Chandler, 1990).

[8] If knowledge is considered as overhead, however, then, arguably, shared knowledge is a true economy of scale not of scope.

Integration also requires that labor, materials, and machines be effectively regulated according to calculations of cost, revenue, and profitability. That is, one must produce in line with the law of value: socially necessary labor time, average turnover time, average rate of profit, and comparable product quality. This means machines must be monitored, materials tracked, tasks charted, and the results evaluated in a constant effort to meet social standards. Quality must be checked, products tested, and the relative profitability of different products calculated and kept in line. Labor and capital must be properly allocated to parts of complex labor systems. Materials need to arrive as and when needed. These matters are often referred to as production balancing.

Finally, there is a need for balance between production and final consumption, or adjustment to demand. Whatever the origins of wants and needs, the problem of coordinating and regulating the link between supply and demand still looms: one cannot produce more than will be absorbed at the price, fail to meet competitive standards, or turn a blind eye to consumer wants, needs and practices. Neoclassical theory's blissful assumption that supply and demand meet through smooth adjustments in the market is no more a guide to this aspect of integration than Marxist theory's habitual dismissal of consumer demand.[9]

There are, of course, no perfect methods for integrating production (it is safe to say that no one in business schools has for the last thirty years believed in the Taylorist utopia of the "one best method") and explorations of better management systems continue. Certainly, awareness of the virtues of alternative systems of management and industrial organization increased sharply in the Anglo-American business world in the 1980s. But the dominant theory of industrial organization, transactions costs, is not an able guide to the multi-dimensional problem of integration because it collapses all the preceding elements into only two terms: costs of exchange and economies of scope.[10] The result is that production ceases to be a process and becomes a flat choice between making something or buying it. This assumes away all the difficulties of how things get made at all. In the transactions costs model, all the really interesting questions about production – such as proprietary know-how, labor relations, and product quality – must therefore be spirited in by the back door (Englander, 1988).

### Dynamic integration

Transactions theory also begins and ends with static efficiency, or cost minimization (Perrow, 1986). It offers no insight into the dynamic prob-

---

[9] Business historians and institutionalists have been the most attentive to these issues (e.g., Veblen, 1899; Porter and Livesay, 1971; Tedlow, 1990).

[10] Indeed, the transactions school seems to underestimate economies of scale in coordinate systems by calling coordination itself a form of *scope* economy. This renders the scope of the concept so capacious as to annul its distinctive value.

lems of integrating production in an environment of changing demand, advancing technologies, and fierce competition driven by the desire to accumulate capital, national rivalries, or warfare.

Dynamic interaction refers, first, to the crucial goal of technical innovation. Despite the importance of individual creativity, innovation is a fundamentally social process built on collective knowledge and cooperative effort. It flourishes where scientific, technical, and market information is readily exchanged and practical interaction is frequent; where users readily benefit from advances made by suppliers and suppliers gain from the feedback from users; and where pluralistic patterns of collaboration are the rule (Håkånson, 1989). Learning by interacting (Lundvall, 1988) is thus complementary to learning by doing (Arrow, 1962) and learning by using (Rosenberg, 1982). Without adequate integration, it is hard for people working in different specialized areas to come up with new or improved products or, more importantly, to implement good ideas through workable production methods (Morgan and Sayer, 1988). Research divisions cut off from marketing departments may invent things no one wants, eager salespeople are prone to promise products no one has yet invented, and both may think up schemes that manufacturing people cannot produce (see e.g., Freiberger and Swaine, 1984). In short, innovation requires many actors, open communications, and social networking (Freeman, 1991).

Regulation, too, changes in a dynamic setting. It requires more than monitoring, feedback, and evaluation of performance; it demands some ability to anticipate and plan for future conditions. Again, the problem is not static efficiency, but the development of the forces of production. Successful planning is not confined to perfecting existing techniques but includes exploring the possibilities for improving on known methods (Hayek, 1949; Auerbach, 1988). This requires investment – or the dynamic allocation of resources – releasing enough people to work on nascent projects so that future inequalities between segments of production systems do not jeopardize the implementation of new technologies (Walker, 1988). Capital investment alone will not suffice, however; managers and workers must learn and think creatively, which means people must be able to put their heads together to share knowledge and to act on good ideas without useless interference (Aoki, 1986; Florida and Kenney, 1990).

None of this means a thing if the final results are not up to competitive standards in product markets, capital markets, or military encounters; the global rat-race cannot be evaded.[11] Bad performance will be shown up

---

[11] One must attempt, however, to separate those difficulties attending all modern industrial production from those attached firmly to a particular social system. For example, the most general principle adduced for the choice of organizational form in the conventional literature is usually the need to protect property rights in technologies or brand-name goods; but private property is not a universal of industrial economies, and cannot explain as much of organization behavior as property theorists seem to believe.

even across modes of production, as witness the poor innovative capacity of Soviet industry in the wake of the contemporary industrial revolution (Amman and Cooper, 1987; Davies, 1989). Small but positive innovation and growth rates may not be good enough, if one's capitalist rivals are doing better.

Uncertainty is the watchword of capitalist innovation, planning, and competition. Yet the term uncertainty cannot, in the fashion of transaction theorists, be employed promiscuously to cover all effects of shifting demand, investment cycles, technical change, and the rest. The sources of production and consumption dynamics should be explored in their own right, as problems which organizational innovation must address.

### The social fabric of integration

Conventional economics has long been accused of working with concepts violently abstracted from the real context of society. Indeed, it is fair to say that the whole field of sociology is grounded on the attempt to recover social relations (and social oppression) from the dustbin of theory after the radical metamorphosis of classical political economy into neo-classical economics in the late 19th century.[12] Critics have repeatedly pointed out that the conventional axioms of atomistic individualism, rational action, and economic optimization cannot be sustained in the face of the evidence (Veblen, 1908, 1909; Myrdal, 1954).

Individuals are imbricated in a social fabric of practices and institutions, outlooks and morals, which it is impossible to dissolve (Bourdieu, 1977). Rational action is limited by habit, incomplete information, and the constraints of human calculation (Simon, 1947) – though to speak of limits on reason puts matters backwards. Human reasoning powers operate rather well in a hopelessly complex environment, just not in the logical positivist manner idealized by mainstream social science; rather, they depend on generative and prototypical categories, useful routines and explorations, and a wealth of tacit knowledge (Lakoff, 1987). Economic optimization is unachievable as well, on account of such things as inflexible fixed capital, uncertainty about the future, and the continual shifting of parameters in a dynamic system (Harcourt, 1972); indeed, in an exploitative and expanding economy efficiency is neither the principal objective nor necessary for growth (Storper and Walker, 1989).

Nonetheless, the violence of economic abstraction is not simply an error

---

[12] Despite their differences, Marx, Durkheim, Simmel, and Weber all had this in common. Institutionalists such as Veblen and Polanyi tried to keep the flame alive within the discipline of economics, without too much success, although the recent work of Hirschman, Nelson and Winter, Hodgson, and Auerbach has breathed new life into the institutionalist corpus. Ironically, Williamson calls himself an institutionalist.

of the economists; it captures, in an ideological inversion, a real process of transformation to market integration, commodity circulation, and bourgeois behavior (Sayer, 1987). Without question, the rise of capitalism ushered in a triumvirate of possessive individualism, ruthless calculation, and furious profit-making that fundamentally altered the way people thought, behaved, and treated one another and nature (MacPherson, 1962; Habermas, 1971). Neoclassical theory is thus no mere description of the world, but an active brief for capitalist dominance, a heroically one-sided idealization of bourgeois society. As Marx, Durkheim, and Polanyi argued passionately, from different angles, pure self-interested behavior, naked commodity relations, and unbridled exploitation could not be tolerated even where the bourgeoisie were triumphant; the social fabric had to be reknit to contain the worst monstrosities of the system.

The current debate in the realm of industrial organization is a pale reflection of the mighty battle of ideas between capitalist revolutionaries, liberal reformers, and socialist combatants over the last two centuries. But it contains the germ of the same idea that economic affairs depend on the surrounding tissue of sociality, morality, politics, and institutions. On the one hand, markets turn out to be utterly dependent on relations of authority and trust, and cannot subsist on free and equal encounters by self-interested strangers (Arrow, 1974; Granovetter, 1985). Williamson (1975) began by assuming that deception and opportunism between parties create transactions costs in the market; but these costs are no more than a residual after the greatest dangers of exchange have been dampened by the heavy hand of the state, as embodied in various laws of contract, tort, liability, and regulation, and by customary sanctions regulating tolerable behavior and mutual responsibility (Hodgson, 1988). At the same time, hierarchies within the firm do not depend on sheer command and formal structures, either: information contacts and custom remain essential to the smooth functioning of every large organization, oftentimes in spite of formal lines of responsibility; these systems of personal interaction can be remarkably persistent, even across corporate boundaries (Barnard, 1938; Chisholm, 1989). Similarly, studies of technical innovation repeatedly demonstrate the vital role of informal relations and exchange of know-how among key participants (Von Hippel, 1988; Freeman, 1991).

We are therefore justified in speaking of the social relations of integration which knit together divisions of labor. Production integration may be achieved with the help of such devices as authority, coercion, persuasion, moral stricture, reciprocity, planning, religious conviction, common language, national solidarity, and representative democracy (cf. Lindblom, 1977). Burawoy (1979) has employed this kind of sociological corollary to good effect in an essentially Marxian treatment of industrial work, by what he calls social relations *in* production, or systems of motivation, rivalry, and class conflict on the shop floor. The efficacy of such systems of integration depends on how well they control individuals and groups of workers, and this rests heavily on qualitative aspects of social relations. Feelings of participation, respect and moral rectitude may matter as much as or more than quantitative payment and penalty schemes. A measure of trust and cooperation between workers and management has been found to be essential to successful production. Indeed, "without . . . a community of interests and compatible objectives, problems cannot be resolved by coordination" (Seidman, 1980, p. 205). Lest we wax too romantic, however, relations of competition and adversity also work quite effectively to knit people into contending groups (Saxenian, 1988).

A dialectic of motivation also exists among capitalists, who not only compete with one another, but cooperate selectively to further their joint interests. A balance of competition and cooperation needs to be reached between firms that must work together (Sabel, 1991). Internally, no firm operates on authority and bureaucratic rules alone; all are beset with some degree of internal jostling for position and power (Downs, 1967). Conversely, competition with no sense of shared problems and prospects degenerates into open warfare. When this occurs – among fast-rising managers, between management and hostile would-be buyers, between rival firms, or even between countries – the results can be disastrous for everyone.

The varieties of social linkage and coordination, direction and regulation, stabilization and mobilization are difficult to formalize, though some interesting attempts have been made (e.g., Hirschman, 1970; Lindblom, 1977). This remains a stern challenge in the face of chauvinist reaction to the declining hegemony of Anglo-American business practices, managerial methods, and resolute individualism. When insight into the specific business methods of foreign competitors is lacking, the fallback explanation is the "national culture" of the mysteriously successful strangers. Yet every culture surrounding economic behavior and social order rests on material origins, whether long-standing traditions such as Confucian morality or recent innovations such as corporate paternalism (Morishima, 1982; Dore, 1987.)[13]

The rich social fabric of all functioning economic systems does not, however, obviate the tangible structures of industrial organization such as corporations, factories, and subcontracting networks. We still need to investigate the formal modes of organization available to capitalists who seek effective integration of production systems. Nonetheless, this glance toward social relations of integration demonstrates the acute need for a more open view of industrial organization than now exists, and a need to question the naturalness of the basic categories, such as market and firm, which come down to us as givens.

---

[13] The same may be said of relatively unsuccessful cultures of integration, such as the inefficiencies, torpor, and ultimate ossification which afflicted Soviet industry after a long period of reasonable performance – presently explained too neatly by reference to the evils of a transcendent communist oppression.

## The Basic Triad: Workplace, Firm, and Market

Industrial integration is achieved with the help of specific institutions which we call *modes of organization*.[14] The basic modes are the workplace, firm, and market, which we shall consider in this section; most industrial organization theory has been cast entirely in terms of these three common-sense categories. A mode of organization is the institutional framework within which capitalists, managers, and ordinary workers carry out the work of integration. These institutional envelopes limit and guide social action; they are structured social systems made up of formal rules and informal relations, habitual routines and creative initiatives. Every mode of organization has its characteristic principal of integration – legal ownership in the case of the firm, physical enclosure for the workplace, equal exchange for the market – but there can be a great variety of forms and continual innovation around each principal. Indeed, every extant mode of organization was at one time an invention (or radical adaptation) and their elaboration and improvement continues to this day.

### The workplace: the site of production

The workplace is the most palpable mode of organization: the site at which labor, materials, and machinery are brought together, where production literally "takes place". The dominant form is the *closed* venue, usually surrounded by fences and containing several buildings. The factory is the term most commonly used for large workplaces in manufacturing, and it has direct parallels in the mine, the warehouse, the department store, or the large office. The terms "establishment" and "plant" are commonly used to cover all these types, but we prefer the spatially explicit term, workplace.

Normally, discussions of modern industrial organization take the workplace to be a given, the elemental unit of the firm, and move on quickly to the way large firms are constituted on multi-plant and multi-locational entities (Pred, 1977). But the constitution of any single workplace is also an open process, involving choices about including or excluding even smaller work units organized around basic manufacturing processes, administrative departments, or support services. The scale and scope of activities within a single workplace can vary enormously. At the upper end are such leviathans as Henry Kaiser's Richmond, California, shipyards, which employed 125,000 during World War II, while the lower end is populated by

---

[14] We use this term in preference to Williamson's "governance structures", which he defines as contractual frameworks for transactions (1981, p. 1544), because modes of organization are much more than merely exchange systems (Williamson occasionally uses "mode of organization", too). We reserve the term governance to denote the exercise and restraint of power in organizations and organizational systems.

millions of small workplaces – workshops, boutiques, storefront offices, garages, etc. – embedded in the industrial fabric.

The basic organizing principle of the workplace is containment within a limited area. Direct connection and immediate access are its chief integrative effects: the workplace not only juxtaposes machine to machine and worker to object, but brings people into close proximity and facilitates social cooperation (Jones, 1982). The workplace is also a place of confinement, a piece of turf where the boss rules, a symbolic and a social world in which the capitalist's hegemony is normally reinforced. In short, the workplace is a system of labor control (Marglin, 1974; Burawoy, 1979).

The factory was once synonymous with modern industry, before the large corporation became the preeminent symbol of capitalism. The industrial revolution of the 18th century ushered in the factory proper, an event of incalculable consequence for working people, labor productivity, and the landscape. Even so, the factory did not sweep all of industry before it: Britain long remained a nation of workshops (Samuel, 1977), as did the United States up to the Civil War (Montgomery, 1967); factory production only became widespread in the 20th century (Nelson, 1975). Moreover, average size of factories has been shrinking since about 1930 (Gordon, 1988).

Why create large factories? One reason is the scale of the product: airplanes or ships are most easily constructed in one spot. Another reason is the need for continuous flow of liquids and gases, as in integrated chemical complexes. Integrated assembly lines are yet another. Before electric motors, belt-driven machinery had to be close to the prime mover. As important as such mechanical reasons are, however, communication and interaction between work groups, and easy oversight of workers at their various tasks, may be even more significant. Indeed, factories have made possible such production changes as an extended detail division of labor and large-scale mechanization (Marx, 1863).

Large factories have drawbacks, however. They require, in the first place, that industrialists have the power and organizational capability to build, staff, and operate such an entity. What Wedgwood and Arkwright were able to do in 18th century England required the King's minister Colbert in France (Armstrong, 1973). Some employers may not be up to the task, and so keep small shops. The factory also brings together hundreds of workers where they can develop a strong sense of collective power; hence, capitalists commonly split up factories, second-source, and subcontract partly to evade organized labor (Bluestone and Harrison, 1982). Technologies of product and process may also dictate a smaller workplace, where scale and scope economies are lacking. Small plants can allow more autonomy for groups of workers and managers, and hence more flexibility in the face of rapid technological and/or market change.

Not all workplaces are factories or workshops. The home is a workplace for domestic labor outside the circuits of capital, and has also served as the site of artisanal shops and family farms, and for taking in piecework, a practice that almost disappeared in advanced capitalist countries, but has recently been revived in garments, electronics, and insurance (e.g., Hadji-michalis and Vaiou, 1990). This system does not afford the benefits of worker coordination and control from bringing everyone under one roof, but rather takes advantage of systems of domestic and patriarchal control already in place (Beneria and Roldan, 1987).

A large class of *open* workplaces are not affixed or limited to one spot. Mobile workplaces, such as airplanes and ships, carry the workers on their backs. Nomadic workplaces occupy a site for a time and then move on, as in building construction, lumbering, and on-location filming. Extensive workplaces have rather loose boundaries, as on farms or airport tarmacs. Nodal workplaces such as railway repair yards or electric powerhouses, are found along infrastructural systems. Of course, some workers have no workplace at all, but rather places of call, as in sales or safety inspection; these raise special problems of surveillance and coordination.

The workplace is thus a fascinating element of industrial organization in its own right, to which far too little attention has been paid; mostly, it has been left to industrial sociologists to investigate (e.g., Woodward, 1965). In keeping with this neglect, the transaction cost and flexible specialization schools are exceedingly cavalier about whether they are referring to firms or workplaces in their models, an unwarranted identification that extends back to Marshall. Firms are a quite distinct mode of industrial organization.

## The firm: ownership and command

The central principle of the firm is impersonal ownership: the firm evolved as a legal shell under which assets could be assembled, contracts drawn up, workers employed, and capital accumulated. Capitalism arose without the firm, however: merchants were personally responsible for business which was carried out in homes, coffee houses, and public exchanges; agrarian capitalists were landowners and tenant-farmers; manufacturers were pro-prietors of their workshops. Over time, the legal umbrella of commerce and property developed (Tigar and Levy, 1977), including the earliest joint-stock trading companies, chartered merchant companies, and char-tered banks of the 16th and 17th centuries (Chalmin, 1985). Industrial firms only appear in the 19th century as partnerships, chartered corpora-tions, and finally, around 1850, as the modern limited liability and joint stock corporation (Williamson, 1951). Simpler forms remain active today, however.[15]

[15] It should be noted that the firms which concern most industrial organization theory are industrial companies. Chandler's histories are specifically accounts of the "modern industrial enterprise" though he suggests that the same principles of management extend to commercial and financial firms. We shall hold to this pattern for the time being, and consider commercial and financial capitalists' roles in industrial organization later.

The firm is, first, an entity with which others can do business: it has an ongoing existence, legal rights and obligations, an address. The firm is, second, a container for capital in its various forms: diverse assets – machinery, inventories, buildings, patents – can be held, money can be accumulated in company accounts. The firm is also an employer, which has the right to hire and fire, to set the terms of work, and to exploit labor-power. Fourth, the firm is an administrative structure for managing production, exchange, assets, and labor. Finally, the firm is a central actor in the competitive battle.

The focus of industrial organization research has long been on large firms. These arose partly due to economies of scale in mass production and are on average more capital-intensive than small ones (Marris, 1979). Larger firms are better able to administer larger operations and to integrate them with materials supplies and market outlets, to maintain volume and continuity of throughput. Size is also due to economies of scope. Diversified corporations combine both large and small workplaces, which they manage through multi-functional and multi-divisional systems of internal organization (Chandler, 1990). Large firms have also brought distribution and sales under their umbrella, improving access to customers and quickening the flow of commodities to market, and ruling effectively over hugely dispersed geographic empires (Dicken, 1986).

Big companies succeed in the competitive race by virtue of their ability to generate new technologies in R&D labs, to develop and retain experience (know-how) over long periods, to protect their flanks from challengers by guarding materials supplies and market share, to take advantage of a competitive edge by moving into far-flung areas of the world, and so forth (Marris and Mueller, 1980; Teece, 1986). In other words, they gain proprietary control over the means of competitive advantage. In a dynamic context, it should be added, growth through diversification is rarely random, because performance is normally better where activities are coherently linked to a "core competence" (Teece, 1988; Dosi and Teece, 1991).[16]

But size has less to do with productive capability than financial clout. Large firms can amass and allocate capital in unprecedented ways. They are better able to raise capital, borrow at lower rates, ride out bad times, stabilize profits by diversifying, and move capital rapidly to new areas (Marris, 1979; Taylor and Thrift, 1982). Big firms also have the financial muscle to move in on new areas of high profit by gobbling up successful firms.[17] Finally, those who once get ahead in the race to accumulate tend to stay ahead, buying out those who have fallen behind; inequalities in firm size thus cumulate in an irreversible way, and the big get bigger even if competitive advantages are distributed randomly (Nelson and Winter,

---

[16] Nonetheless, there are many examples of holding companies, conglomerates, *chaebol*, etc. operating without such technological coherence.

[17] Average profit rates of big firms are not noticeably higher than those of small firms, however, according to Marris (1979).

1982). Large firm expansion is strong, unidirectional, and ratcheted upward, in a way that the comparative statics of transaction cost analysis cannot capture (Hodgson, 1988).

Despite its achievements, however, the large firm is continually beset by the difficulties of integrating its internal division of labor. Size and diversification may open new markets, lessen risk, achieve economies of scope, and so on, but they come at a cost in organizational coherence and manufacturing competence; the business press is thus rife with stories of overextended companies busily shedding divisions and subsidiaries in order to "get back to basics" (*Business Week*, 1986). Centralized control, top-down coercion, and hierarchical ordering – long the hallmarks of the internal power structure of these enterprises – have distinct limits as means of enhancing communication, stimulating creativity, and evoking individual or collective effort; as Lindblom (1977) puts it, such authority systems are "strong thumbs, no fingers." Indeed, one can turn the transactions costs argument against the large firm, where opportunism, uncertainty, asset specificity, bounded rationality, and bargaining prevail internally and limit the effectiveness of unification and capitalist power even under dispersed management (Perrow, 1986). Coupling activities within a single firm can therefore prove inefficient and, worse, inflexible in the face of changing circumstance (Sabel, 1989).[18]

In recent years, large firms have been moving away from top-down decision-making toward more decentralized, cross-cutting, and informal forms of interaction among lower managers and work-units. As Martinez and Jarillo (1989, p. 489), reviewing the literature on multinationals, describe it:

> These new coordination mechanisms are rather informal and more subtle than the existing ones. Among them are: first, microstructural arrangements (lateral relations) that cut across the formal lines of the macro structure, such as teams, task forces, committees, individual integrators and integrative departments. Second, informal communication channels, such as direct and informal relations among all managers, without distinction of subsidiaries and headquarters, that supplement the formal information system and improve the communication process. Third, the development of a strong organizational culture that includes both a deep knowledge of the company's policies and objectives and a strong share of organizational values and beliefs.

The hardware available to managers has improved markedly, too, with microcomputers allowing more rapid handling and process of data (Hepworth, 1990). Yet what is most important in the shift toward a

---

[18] The limits of industrial concentration have perhaps been best illustrated by the Soviet Union, whose theorists drew the wrong lessons from the early 20th century capitalist trend toward bigger factories and firms (Auerbach *et al.*, 1988).

"network firm" is (contrary to Antonelli (1988)) not mechanical bridging but better attention to the quality of information available, the way people communicate, and how well managers work together, settle differences, and learn from the past, as we shall see in chapter 4.

One consequence of the limits of the modern industrial enterprise is that small firms have survived *en masse* – particularly in Europe and Japan, in a manner quite unexpected from either the Marxian theory of centralization or the business history school of managerialism. A principal reason is scope (dis)economies (Scott, 1988b).[19] The scope of activities encompassed by individual firms is exceedingly diverse. Some comprise small single plants and fill specialized product or task niches within an industry; others span several industries and have facilities all around the world (Eatwell, 1971). An erroneous presumption of traditional models of industrial concentration is that all firms in an industry are doing the same thing, and that small firms are just less successful versions of large ones (e.g., Averitt, 1968). In fact, most small firms survive by doing things large companies do not do well: serving local markets, making highly specialized goods, exploiting marginal labor forces, etc. They often content themselves with operating in lower profit areas that do not attract the sharks, but they can also be more flexible and adventurous in exploiting unexpected opportunities, and sometimes grow at extraordinary rates (e.g., Freiberger and Swain, 1984).

### Markets: institutions of exchange

Markets are a mode of transferring property which are open to the participation of many parties.[20] The principle of market integration is exchange of equivalent values, which come to be measured in a universal equivalent, money (Marx, 1863). Commodities therefore have a price and are ordinarily paid for with money. Under capitalism markets exist for an enormous range of things: goods, to be sure, but also labor-power, corporate assets, land, and paper claims to income. There are almost as many kinds of markets as there are commodities: a local farmer's market is one sort of arrangement, the Chicago Corn Exchange quite another. While we may speak loosely of "the world market", the practical handling of wheat exchanges is independent of movements of oil.

The reason for this specificity of markets is that the practical business of exchanging commodities has to take place within an institutional framework. This requires a measure of equality among parties, consensus, rou-

---

[19] Ironically, Scott and others have found Williamson's framework more fruitful in accounting for the viability of small firms than the large firms for which it was intended.

[20] A transfer exclusively between two parties, with no other potential buyers and sellers, so prices cannot form through multiple comparisons of value, is not market exchange, as Hodgson points out (also Marx, 1863). Nor is there such a thing as an "internal" market within a firm, despite common usage; such a situation would be a contradiction in terms.

tinized behavior, information-sharing, policing, and protection. But behind those conditions lie formal and informal institutions, built, like any other form of organization, of people, laws, and practices (Arrow, 1974). The elementary question "what is a market?" is rarely answered by economists, who take markets to be a sort of primordial ether in which firms swim about. "In the beginning," says Williamson (1975, p. 20), "there were markets", thereby recapitulating the bourgeois idyll of Adam Smith. Yet the existence of functioning markets is not to be assumed, contrary to Williamson, "the onus is as much to explain the existence of the market as it is to explain the existence of the firm" (Hodgson, 1988, p. 210).

As institutions, markets have authority structures that put constraints on free – in the sense of libertine, unscrupulous, or deadly – action. They cannot, therefore, be held up as fundamentally egalitarian in contrast to all other modes of organization (Lindblom, 1977). Markets as institutions are first of all legal systems, embracing contract, property, and tort law; market exchange is thus regulated not only by prices, but also by contract stipulations, government overseers, and personal relations of trust. Markets also need an infrastructure of intermediaries such as merchants, brokers, jobbers, shippers, and auctioneers to knit together disparate sets of buyers and sellers. Furthermore, ordinary retail markets require enormous numbers of sellers to provide access to goods and services for large numbers of consumers (see chapter 2). Markets have a geographic infrastructure, as well: historically, a market place, such as an agora or merchant hall, was essential in bringing parties together to make comparisons, set prices, and clear stocks. Stock exchanges still operate on this principle, and most natural resource commodity flows are orchestrated by traders and brokers concentrated in London, Chicago, and New York – possibly from a single building such as the Chicago Mercantile Exchange.

Commodity exchange was the historical starting point for capitalism, and long-distance trade was the first significant form of economic integration beyond the compass of very small regions (Braudel, 1979). Markets thus existed before firms or factories, in the modern sense, and are so much a part of capitalist life that it is little wonder they appear as a given, a natural substrate. Nonetheless, as socially constructed practices and institutions, serving specific purposes and places, markets have been built up over time, and have been modified or even abandoned as needs change (Agnew, 1986). The histories of wholesaling, brokerage, exchanges, and retailing are rich with innovation and expansion (Porter and Livesay, 1971; Tedlow, 1990). In the ancient field of natural resource trading, for example, a truly global market has only emerged since World War II with the growth of international trading companies, the breakdown of the old empires, the appearance of state marketing boards, and the formation of commodity futures markets (Chalmin, 1985).

The neoclassical concept of the market, due to Leon Walras, is nothing more than an idealization of a village square where all parties enter, make their offers, and leave satisfied. Prices form, markets clear, the deed is done. It is not hard to think of counter-examples: bidding for government contracts, photocopier lease-buy agreements, or labor hired through a gang boss all involve exchange of a commodity for a certain amount of money with some contractual stipulations on performance. Beyond this, just about everything else is up for grabs and subject to the most ingenious variations (Williamson, 1975). The type of market exchange depends on such variables as the number of sellers and buyers, number of commodities, time of sale, time of delivery, form and terms of payment, time of consumption, and quality standards. A particularly important choice for capitalists has been between direct buyer-to-seller interaction and exchange mediated by merchants. Almost no one operates in a market setting as a perfectly free, wholly detached individual. Given the imperatives of integrating production systems and institutionalizing and routinizing practices, industrial firms and traders invariably become caught up in webs of coordinate exchanges via markets, as in loose subcontracting networks or chains of traders.

Markets are distinguished by relatively individualized, voluntaristic, formally equal, and temporally limited transactions among many parties. They allow for remarkable flexibility through a changing mix of participants and commodities, and for sharp mobilization of personal energies via competition and the pursuit of self-interest. Nonetheless, the market mode is beset by certain difficulties. The formation of stable, workable market exchanges can be limited by uncertainty (incomplete current information, an uncertain future), small numbers (few parties, irregular transactions), bounded rationality (inability to handle all information and contingencies), and opportunism (misrepresentation, reneging) (Williamson, 1975). Equally important reasons for market failure are the lack of a suitably developed superstructure of law or infrastructure of commercial dealers; indeed, appropriate behaviors and institutions develop in tandem historically (Hodgson, 1988). Markets are also notoriously poor at mobilizing large collective efforts, securing stability, employing all available resources, and handling common property. Finally, the cash nexus has a moral and anti-social impact that is far from desirable in many areas of life; in other words, markets often fail even as they succeed (Polanyi, 1944). Because of what Lindblom (1977) aptly calls "the limited competence of markets", commodities may not be available, monetary payments may not appear, critical information may be withheld, labor processes poorly coordinated, capital flows blocked, and social life undermined. In short, integration among the parts of the division of labor may not be sound enough for production to proceed in a stable and effective manner.

The problems of market integration, then, follow from limits to the exercise of power, lack of permanence, poverty of knowledge, and indirectness in the relations among the parties involved. Competitive individualism has drawbacks as a way of organizing social action. It may be necessary, therefore, to avoid markets by internalization within a single workplace and firm or to work around them in ways we have not yet considered. Markets can also be manipulated by key actors, such as merchants, or by small groups of oligopolists tacitly coordinating price and production policies. Such behavior is usually regarded as sheer profit-taking, but it is sometimes a response to ruinous conditions of glut and instability, and is partly aimed at improving the market's performance. The same may be true of government boards, used to steer market prices in accord with national policy, as well as to extract surplus for the state. Where the control of monopolists or the state is complete, the market mechanism may, of course, be subverted entirely.

### Beyond markets and firms

The classical dichotomy in economics between market and firm goes back at least to Marx, who contrasts the anarchy of the market with the despotism that obtains inside the capitalist's gates, and the irrationality of the former with the rationalization of the labor process in the latter.[21] The neoclassical economist puts a brighter face on it: interior to the firm is a rationally operated technical production function; exterior are efficient price-fixing markets. Chandler, Barnard, and other students of the modern corporation have depicted the firm chiefly as an administrative system for rationalizing production, distribution, and investment. All these approaches have taken the line between market and firm to be clear and solid.

The great achievement of the transaction cost school is to have broken down that line. Coase (1937) was the first to see firms and markets as *alternative* means of coordinating production, each with certain advantages and disadvantages. But how does the economy arrive at a particular balance of internal administration and external market exchange? This ought to depend on the relative efficacy of bureaucratic command and open-market transactions (Arrow, 1969), or, as it is now formulated, the degree of internalization/externalization depends on (1) economies of scope among related processes and (2) transactions costs of either coordinating those processes among several dispersed firms (or workplaces) or of bringing them under the wing of a single firm. The optimal degree of integration/disintegration depends on such things as the relative scale of coordinate processes and the institutional conditions for enforcing contracts.

With all this, however, Williamson's analysis does not add all that much to prior accounts of organizational development, whether it is Chandler's interpretation of the modern corporation, Marx's discussion of the rise of the factory system, or Hymer's treatment of multinationals.[22] The trans-

---

[21] For a useful discussion of Marx's views on production and exchange – which nonetheless cannot solve the problems raised – see Levine (1980).

[22] On Chandler, see Williamson (1981); on Marx, see Williamson (1980); on Hymer, see Buckley and Casson (1976), Williamson (1981) and Teece (1985). Their criticisms of Hymer's reliance on

actions school is right to argue that industrial organization cannot be reduced either to technology, as Chandler is wont to do, or to control of labor, as Marglin implies, and the institutional matrix of workplaces, markets, and firms must be taken into account. Nonetheless, one simply cannot stretch the transaction cost framework far enough to encompass the full richness of industrial integration and organization, without the term ceasing to have any definitive meaning (Perrow, 1986).

As the dichotomy of firm and market breaks down, it becomes possible to see that the world outside the firm needs to be managed and the world inside the firm needs to be regulated in light of external conditions. Paradoxically, Chandler's paean to modern management called attention to the way the large corporation sought to imitate the market by dividing divisions and playing them off against each other like competing firms, and this has only intensified as management has introduced such devices as profit centers and *ad hoc* work teams to try to emulate market conditions of profit equalization, and competition, within the large corporation (Eccles, 1985).[23] Conversely, there is irony in the fact that the explorations into market failure – by Arrow, Alchian and Demsetz, and Williamson – have helped uncover the need to form stable, reliable, and multilateral relations among firms.

One outcome of market failure will be efforts to upgrade market institutions, a point to which we shall return later in this chapter. Another outcome will be efforts by capitalists to stabilize relations, join strengths, and build defenses through interfirm alliances and networks. Transactions cost theory, however, gives us only two organizational choices: the market and the firm. It has recently come to recognize a middle ground of relational contracting (Aoki, 1984; Williamson, 1985). This turns out to be a vast area once we prise open the tight dualism of market and firm and cast a little light on the alternatives – so vast, indeed, that the void in the initial dualism cannot be patched over.

In a dynamic setting, the failure of transactional analysis is even sharper. The organizational problem is not simply one of choice between existing alternatives in a setting of markets and firms. What is needed to produce, innovate, and compete in a dynamic capitalist system often simply does not exist and must be actively created (Storper and Walker, 1989). One cannot assume that another firm can be found to supply a needed machine or that the necessary infrastructure is in place to move a finished good to market (Auerbach, 1988). Capitalists therefore build up an organizational

---

oligopoly theory are valid, but do not depend on transactions costs per se, only on paying attention to production considerations over market power. For a defense of Hymer, see Kindleberger (1984).
[23] This does not mean the firm is literally a system of internal markets as proposed by Alchian and Demsetz (1972). Williamson and his followers do not go to this extreme, but they still portray firms as systems of internal transactions, in which transactions costs are weighed against market procurement in a "make or buy" decision. As Hodgson (1988, p. 207) observes. "The function of the firm is ... not simply to minimize transaction costs, but to provide an institutional framework within which, to some extent, the very calculus of [transaction] costs is superceded.

framework at the same time as they produce commodities, and one way they do so is by probing outward, forming alliances and networks with other firms – a form of industrial integration that for a very long time lay entirely outside the compass of the organizational literature.

## The Middle Ground: Inter-Firm Alliances and Networks

As Williamson was developing his transactions cost framework, a few people began looking into the cracks *between* the market and firms.[24] In a provocative article that provides a touchstone for current thinking about networks and alliances, Richardson (1972, p. 883) explored "the dense network of cooperation and affiliation by which firms are interrelated" beginning with a pointed rejection of his own long-held faith in firms as "islands of planned coordination in a sea of market relations".[25] At the same time, Blois (1972) took note of looser alignments than full acquisition, which he called "quasi-integration". MacNeil (1978) recovered the idea that contractual relations did not fit the neoclassical model of exchange, but secured relatively complex and durable affiliations between parties. Meanwhile, there was growing interest in Japan and its interwoven business groups, as in Ouchi's (1980) characterization of "clan" relations among firms. Finally, Piore and Sabel's (1984) flamboyant exposition of the flexible specialization model brought the small firm networks of Italy's industrial districts into the discourse.

Inter-firm cooperation, alliances, and networks are now the rage in industrial organization theory (e.g., Powell, 1990). They represent a second tier of modes of integration for complex production systems. Yet the literature is still in a rather chaotic state, so some clarifications are in order. First, there is no standard terminology as yet, only a jostling of concepts such as coalitions, cooperative relations, and affiliations; we use "alliances" to denote two-party arrangements and "networks" to indicate multiparty systems. Second, one must classify inter-firm alliances and networks by their structure not their functions, such as technology-sharing, marketing, or risk-spreading (e.g., Harrigan, 1985; Tyebjee, 1988); these are the purposes to which all industrial organization is addressed. Third, inter-firm coalitions are not distinguished by their social embeddedness or reliance on trust, because all modes of organization entail cooperation as well as competition, and implicit relations as well as formal agreements. Finally, we want to rectify the excessive attention given to alliances and networks among industrial firms, to the neglect of the vital role played by commercial capitalists, financiers, and other third parties.

---

[24] Literature on inter-organizational relations goes back a long time, of course, but was almost wholly cut off from economic theory (see Whetten, 1981).

[25] One can forgive Richardson his shortcomings in other respects: his rather slim list of cooperative relations and his use of the term division of labor for the split between markets and firms.

*Relational contracting and subcontracting*

Relational contracting refers to ongoing relations of exchange, interaction, and mutual development between two or more firms. These relations violate the tidy boundaries between firms by giving one party access to the secrets and activities of another, and they intrude on market exchange by building bridges between firms that do not lead onto the streets of commerce. Relational contracting and subcontracting are therefore to be distinguished from arms-length, off-the-shelf, or open bidding types of market relations. All supplier and subcontracting systems do not fall under this heading: simply passing work along a production chain, the traditional way of securing auto parts in the United States, is a linked system of market exchange and does not constitute relational (sub)contracting.

The most direct, formal, and limited type of relational contract is the two-firm alliance for manufacturing, marketing, or research. These evade institutions of true markets by establishing restricted channels of distribution, and make the boundaries of firms permeable to the sharing of know-how. One type of alliance is the marketing agreement, where a company tries to expand its sales outlets quickly, usually in another country. Such agreements can be forged with a manufacturing firm selling similar goods and wanting to fill out or upgrade its product line, as in the case of Chrysler's sales of Mitsubishi cars, or they can be undertaken through merchant firms (agents) with established lines of access to retail outlets or industrial users (Glasmeier, 1989). Another important type of alliance is the technology-use agreement, which includes both one-way licensing of process or product technology to a foreign firm for use in a limited market (Mowery, 1988) and two-way cross-licensing or technology exchanges between firms with complementary processes, national markets, or product lines (Steinmuller, 1986; Cooke, 1988). Research alliances involve product and process development based on the complementary technical competences of firms (Freeman, 1982). Licensing and marketing agreements have a venerable history going back well into the 19th century (Chandler, 1990).

A different form of interactive relation is the customer-supplier contract or subcontract.[26] (Sub)contracting is relational when it includes an enduring interchange and some degree of involvement by one firm in the production process of another, such as design specifications, providing materials and machinery, technical assistance, financial assistance, and on-site quality control. These relations may not require a formal contract, yet they depend on tangible commitments of resources, frequent meetings to coordinate policy, and sharing of knowledge. Where they are focused on

---

[26] The term "subcontracting" is the source of some confusion. Literally, it means undertaking some portion of work that has already been contracted for, implying at least a three-tier hierarchy; but it is also commonly used with respect to contracts drawn up between relatively equal firms – what we call simply buyer-supplier contractual relations.

innovative products and processes, such relations necessarily involve mutual intrusion and bridging for understanding and solving new problems, and mutual adaptation of techniques and routines (Von Hippel, 1988).

Relational contracts usually link sets of firms into coherent networks, in which firms depend on others to perform specific tasks, and which confront the world as loosely entwined units (Håkånson and Östberg, 1974; Håkånson, 1982; Håkånson and Johanson, 1988). While the extent and depth of functioning networks may be partly unintended, they invariably demand concerted efforts by key actors in the participant firms and are only built up over many years of experimentation and learning (Teubal et al., 1991). Networks have been an important facet of production and innovation since the early industrial revolution (Sabel and Zeitlin, 1984), providing an alternative to both factories and large firms (Scranton, 1983), and have made a dramatic comeback against the large corporation in the last twenty years (De Bresson and Walker, 1991).

Supplier-buyer networks can be divided into two broad groups. Associational networks are relatively egalitarian or horizontal, as in the case of the Third Italy. The Emilian-Tuscan model of an industrial complex is justly famous for the achievements of quite small firms operating in sophisticated product niches in machining, ceramics, and foodstuffs (Becattini, 1978; Brusco, 1982; Scott, 1988b; Goodman et al., 1989). Firms work in close alliance to solve specific problems of design, manufacture, and marketing and have long-lasting working relationships.[27] For example, innovation in specialized tile equipment in the Sassuolo district has been stimulated by the close interactions between tile producers and equipment producers (Russo, 1986). Silicon Valley in California is another notable locus of associational networks. As Saxenian (1988, 1991) has shown, the continued success of the area in the highly innovative electronics industry depends heavily on close working relations among firms, generally in the absence of formal contracts. Sun Microsystems and Cypress Semiconductor, for example, evolved in tandem by solving linked problems of workstation and integrated circuit design; Hewlett-Packard lets a small firm, Weitek, use its sophisticated equipment to manufacture a specialty chip the larger firm cannot produce as well or as quickly.[28]

A second species of relational network is the hierarchical subcontracting system, in which a lead capitalist puts out segments of a complex production system to many suppliers, under specific product or labor-service

---

[27] The enormous Japanese coordinate groups, known as Kigyo Shudan or *gurupu*, are a rather different species of industrial organization, yet they share some features of associational networks. Member companies trade with each other extensively and favor long-term, stable relations over open-market buying, even when external prices are lower (Sako, 1989). Imai (1987/8) argues that the *gurupu* have evolved into looser networks comparable to the Italian model, but this seems far-fetched. See chapter 5.

[28] At the same time, valley firms have also tied the knot in all manner of long-distance contracts and joint ventures with Japanese and European firms, which have expertise and equipment not found locally (Gordon, 1991).

contracts (Holmes, 1986). Usually, the subcontractors are smaller than the lead firm and pass on work to yet smaller firms; the relationship is strongly vertical. At the same time, lead firms very often come to rely on subcontractors making contributions through their specialized competences, close attention to the lead firm's technology and business practices, and innovative capacity. Relational subcontracting is advantageous both to lead firms which lack sufficient capacity, expertise, or power to drive down wages, and to follower firms which lack adequate market outlets, financing, technical skills, or managerial competence. But there can be disadvantages compared with internalization or open-market buying. For lead firms these arrangements can mean too little control over production, material flows, and the labor force, leakages of profits, failure to develop in-house production capabilities, managerial headaches, and so forth. For subcontractors the possible drawbacks are notorious: lack of marketing alternatives, loss of control, financial dependence, and drain of profits to the lead firm. For both there is always the danger of overcommitment to long-term contracts.[29]

Hierarchical subcontracting networks can involve producer subcontracting, commercial subcontracting, and commercial franchising (which might be called forward subcontracting). Producer subcontracting is the dominant form today; it is common in component-assembly systems of production, as in automobiles, aircraft, and computers, but only some of these systems, such as that of German auto-parts maker Robert Bosch, are organized as relational networks rather than loose webs of arms-length market transactions (Morgan and Sayer, 1988). While older batch industries, such as garments and toys, have been notorious for using subcontracting to cut costs and exploit labor, stable inter-firm relations are also commonly used to dampen the chaos facing capitalists, as in the dense small-firm subcontracting networks of Taiwan (Shieh, 1990), and where refined craft is required for customized and quality products, as in construction and film-making (Eccles, 1981; Storper and Christopherson, 1987). The most famous instance of organized relational subcontracting is the Japanese *keiretsu*, made up of three to five layers of subcontractors, knit together tightly by the lead firm's directives and interventions, as well as by knowledge-sharing; we discuss them in chapter 5.

The commercial subcontracting network is a distinct form. Here the lead firm is not a producer, but a wholesaler or retailer orchestrating suppliers. The merchant-directed putting-out systems of the 17th and 18th centuries were, in fact, the first specifically capitalist mode of organizing production (Dobb, 1947; Kriedte, 1983), and merchant subcontracting continued to predominate in places such as Paris throughout the 19th century (Harvey, 1985a). Commercial subcontracting, once thought to have died out in

---

[29] Despite a raft of new research, however, we still do not understand the complex reasons managers have for putting out work, nor is there a clear general trend to greater subcontracting in the United States and Britain (Harrison and Kelley, 1990).

the face of the modern integrated industrial corporation, still thrives in several forms. One is the supplier networks for giant department stores such as Marks and Spencer or chains such as Next and Ikea, which provide subcontractors with designs and specifications, technical and management counsel, financial aid and performance oversight (Rainnie, 1984; Gardner and Sheppard, 1989). Another form of subcontracting network is that established by grain merchants such as Cargill and Louis Dreyfus, the giant Soga Shosa that dominate Japanese foreign trade, and Taiwanese import-export firms, all of which are involved in orchestrating a certain amount of production leading into their trade pipelines (Morgan, 1980; Chalmin, 1985). One last type of commercial network involves small wholesalers, such as the *impannetore* who channel so much of the output of the Third Italy into foreign markets because they can catch the winds of fashion and direct small firms into the best and newest lines (Belussi, 1989).

A final form of subcontracting is the franchise network, which is especially popular in retailing. The fast-food chains are the leading contemporary example (Luxenberg, 1985), but the practice goes back to farm equipment sales in the late 19th century and includes car salesrooms in the early 20th. The relation is closer than a licensing or marketing agreement because the franchise usually carries the name of the parent firm, as in Burger King stands or Coca-Cola bottling plants. Moreover, the franchise normally carries with it a whole system of production and management to assure product quality under the brand-name label. As the franchise empires illustrate, inter-firm networks do not have to be localized and often cross national boundaries; indeed, they provide a way of overcoming distance and territorial limits on interaction (Camagni, 1990).

### Ownership, investment and management ties

Another form of network organization centers not on commodity relations but on relations of ownership, investment, and control. These are usually treated quite separately from alliances and subcontracting systems, but their effects for industrial integration can be similar or complementary to those of contractual relations between firms. In a system of impersonalized possession and constrained management, the most important question may not be who controls business but how business is organized through financial, property, and managerial ties that extend beyond the boundaries of the single firm.

In a long debate over corporate control, unleashed by Berle and Means (1932), one side has tried to demonstrate the independence of management while the other has sought to prove that corporations remain under the capitalists thumb. The managerialists appear to have won on the evidence: a clear decline in the United States of family ownership and bank control of large companies accompanied by growing control of operations by professional managers. Nonetheless, the ⸀financiers and owners of the modern corporation – lenders and stockholders – have not departed the scene; they appear on boards of directors, as buyout artists, and at stockholders' meetings to harness and harass top management in a way that belies the idea of managerial autonomy. While owner-financiers have allowed managers considerable leeway, managers have internalized the logic of capital: profits and accumulation remain the "bottom line". The resulting framework combines constrained managerialism and restrained finance in a system of control which has been called "impersonalized possession" or "institutionalized capitalism" (Herman, 1981; J. Scott, 1986).[30]

The simplest form of inter-firm alliance via ownership and investment is the parent-subsidiary relation. Virtually every large company has such satellites, either wholly or partially owned, which are allowed a degree of autonomy from the parent. There is no attempt to absorb the subsidiary completely because it might reduce flexibility, scare away skilled managers and technicians, or conflict with laws of some nations. In fact, an increasingly popular strategy is to spin off new firms to handle new product lines, while retaining substantial ownership and contractual ties; Japanese electronic and Italian craft firms are notable for this (Aoki, 1987; Lazerson, 1988; Florida and Kenney, 1990).[31]

Another straightforward alliance, the joint venture, occurs when two or more firms enter directly into a shared ownership agreement in a third firm (Harrigan, 1985). Joint ventures are usually new entities, but can involve the revival of an ailing unit of an existing firm (Tyebjee, 1988).[32] Joint ventures go farther than licensing and technology-sharing agreements in combining the strengths of independent firms – such as complementary product lines, marketing networks, and production skills – without sacrificing the remaining autonomy of the partners. Joint ventures have long been used as a way to enter foreign markets, set up producing subsidiaries abroad, gain access to natural resources, and undertake major capital investments (Mowery, 1988). Joint research and development ventures have also been very popular recently at the international level, no doubt because of increasing global competition, leveling of research capabilities,

---

[30] Capital is re-personalized, however, by the upwelling of new entrepreneurs in new sectors and growth regions, such as Silicon Valley, as well as in the new faces which appear in the financial realm, such as T. Boone Pickens and the other takeover wizards of the 1980s.

[31] Holding companies and conglomerates are usually considered as networks of independent subsidiaries held together by a single financial investor, and sharing only a modicum of central direction. They have appeared in three bursts in US history, circa 1900, 1920, and 1970, but have not survived well (Herman, 1981). The holding company remains surprisingly strong in Britain, however, where even supposedly unified corporations all too often turn out on closer inspection to be mere conglomerations of unconnected and warring units, with little coherence (Morgan and Sayer, 1988).

[32] Some buy-ins to (rather than buy-outs of) budding entrepreneurial firms and minority-held spinoffs might also be considered a form of joint venture, though the equity arrangements can be quite intricate (Powell, 1987).

and need to be on the technological edge (Gomes-Casseres, 1988; Ohmae, 1989; Freeman, 1991).[33]

Many looser forms of owner-investor networks exist as well, and are defined by overlapping lines of financial and managerial participation. These are generally effected by stock ownership, interlocking directorates, and bank ties. In the United States and Britain, such networks are much more diffuse than in continental Europe and Japan, even though all industrial groups' ties are considerably more relaxed than earlier in the century. In the latter countries, bank lending has been more important than securities markets in raising capital.

The pre-war *zaibatsu* of Japan, for example, held together as many as 294 companies (the Mitsui group) through family and bank majority stockholdings, and through direct appointment (and circulation) of managers (Imai, 1987/8). The post-war *gurupu*, in contrast, rely much more on inter-corporate minority stock holdings, shared directorships, a council of presidents, and favored lending relations with a single city bank (e.g., Mitsubishi, Sumitomo). In addition to the big six groups, controlling 118 of the top 250 Japanese corporations, there is a set of groupings with US-type corporations at the center of a constellation of subsidiaries (e.g., Matsushita, Yamaha) and a set of hybrids (the little six) with some reciprocal holding of the parent by the subordinate firms (e.g., Toyota, Nippon Steel), involving another 25 of the biggest corporations (J. Scott, 1986).[34]

European companies have also commonly been grouped around powerful multi-purpose banks, such as the Grossbanken of Germany, which provide crucial financing (particularly in capital-intensive industries), secure substantial equity ownership, use their leverage to secure positions on the boards of creditor firms, and give advice to their clients (Kocka, 1980b; Zysman, 1983; Chandler, 1990). Spanish and French banks have played a similar coordinating role in many industries.

In Britain, the United States and Canada, securities dominate over bank finance and stock holdings are diffuse,[35] but substantial networks of companies are tied by a number of key institutional investors, especially the top banks, insurance companies, and investment groups which dominate these countries' financial systems (J. Scott, 1986). Directoral interlocks are also common, and are again dominated by major banks and third party investors (Useem, 1983), and commercial (clearing) banks have considerable say in relation to their lending activity, to which many contractual

---

[33] Joint ventures have not been universally adopted in all industries, being far more common in aircraft than in pharmaceuticals, for example (Porter, 1986; Mowery, 1988).

[34] Scott further distinguishes two groups within the big six: those linked strongly by intercorporate holdings and those relying more on the group bank. Financial relations are by no means confined within the groups, which are not closed entities.

[35] So diffuse that it requires rather broad alliances to exercise effective control over management. Scott believes that "controlling constellations" of up to twenty top stockholders can be identified in the majority of corporations, but the subject is contentious. Such constellations can also be found, outside the big groups, in Japan.

covenants are attached.[36] All these connections provide means to gather information, monitor performance, and give advice, activities which have an integrative and constraining effect on industrial and commercial firms under a wide umbrella of financial hegemony (Mintz and Schwartz, 1985; Auerbach, 1988).

A new type of financial coordination that has emerged in recent years is the network forged between a new breed of venture capitalists and electronics companies in Silicon Valley (Florida and Kenney, 1988a). This model is being widely emulated around the world. It involves a kind of brokering and personal networking comparable to the classic role of the merchant capitalist, who was (is) also frequently a financier (cf. Porter and Livesay, 1971; Glasmeier, 1989).

### Independent associations

Networks of firms can also be coordinated by third-party associations, established independently of any one company, with powers to aid, abet, guide, and cajole participating businesses; these represent yet another species of industrial organization. Trade associations, such as the Electric Power Research Institute and the Chemical Manufacturers Association, are the prototype; in the United States they grew up alongside the trusts and modern corporations, becoming tacit price stabilization boards during the 1920s and early 1930s (Galambos, 1966). Today, trade associations represent virtually every industry and act to lobby, educate members about government policies, and generally encourage collective behavior. A similar instrument in the United States is the employers' association, invented about 1900, to create a collective voice in union bargaining. A highly specialized form of multi-firm industry association is the research consortium, such as the Esprit project of the EEC. The first of these arose in the 1930s, with the oil companies' massive effort to develop catalytic cracking, and their numbers have grown rapidly in the last fifteen years (Freeman, 1991).

Behind these industry groups stand a host of higher-level capitalist associations, such as the Business Roundtable and the Conference Board, where leading corporate executives meet, listen to experts, and formulate policies on major social issues. At a further remove are the major think-tanks, such as the Brookings Institution and American Enterprise Institute, which pump out research and policy guidelines for US business (Domhoff, 1979). The US government also sponsors hundreds of advisory councils

---

[36] US investment banks played a guiding role in the trusts, holding companies, and corporations that burst on the scene from the 1880s to 1904, and it is possible to identify four big constellations of firms around key investment bank-commercial bank allies, such as Morgan and First National City Bank of New York, up to the 1930s (Herman, 1981). British banks have almost never become involved in company management, except when industrial firms have suffered calamitous financial setbacks.

bringing together top corporate executives in areas such as military procurement or pesticide regulation (Herman, 1981). And behind all the
formal bodies lies still another layer of social connection through private
clubs, schools, and family ties (Domhoff, 1974).

Germany is well-known for close collaboration among firms within
industries, with the help of the state. Various arrangements were tried
there, as elsewhere, around the turn of the century, including sales syndi-
cates, trusts, conventions, and consortia which involved contractual agreements, profit-pooling, interlocking directorates, and even cross-purchase of
company shares (Chandler, 1990); eventually, most industries stabilized
around trade associations or communities of interest (Leaman, 1988). A
direct consequence of this greater association *within* industries in Germany
was that there were fewer diversified corporations on the US model. In the
Third Italy, a National Confederation of Artisans (CNA) actively supports
small firm networks in place like Modena, providing accounting services,
policing labor agreements, and supporting research (Lazerson, 1988). In
Barcelona, the Catalan government has helped set up a subcontractors'
exchange that acts as a combination trade association and mercantile cooperative for small firms. The EEC is now supporting such local government initiatives to encourage networking throughout Europe (Bianchi and
Bellini, 1991).

### Families and workers' associations

The family stands on its own as a fundamental system of organizing social
relations and production (Bott, 1971), and still represents an important
mode of industrial organization in capitalist economies. After playing a
central role in the rise of capitalism, the family is supposed to have been
displaced by professional management and institutional ownership; the
evidence certainly supports this if one looks at corporations in the US,
Germany and Japan.[37] Nonetheless, family stockholdings, directorships,
and management still play a role in the large corporate realm of advanced
capitalist countries (J. Scott, 1986). One thinks immediately of the staying
power of the Rothschilds, Rockefellers, and Baring Brothers. In Britain
family-based enterprise has been so prevalent that Chandler calls it "personal capitalism". The family also plays a dynamic role in transitional
economies such as Taiwan and India, where family networks stitch together
webs of firms (Shieh, 1990). Even the giant Korean *chaebol* and Japanese
*gurupu* are sometimes still orchestrated around family ties, as in the case
of Samsung (Bello and Rosenfeld, 1990), and the familial mode remains

---

[37] There is nothing anymore quite like the Mitsui family empire, which stood astride the largest
*zaibatsu* and even had its own political party. While the shift to impersonal corporation may indeed
be a natural evolution of industrialism, one ought to observe that the ten key families controlling
the *zaibatsu* were forcibly ejected by the US occupation in a thoroughgoing revolution-from-above
in Japanese capitalism (Aoki, 1988).

important in new firm formation at the growing edges of the advanced countries, particularly among immigrants to the United States (Sassen, 1988). One must also remember those pseudo-families of close personal allegiances that knit inter-firm networks and are the subject of intense myth-making, as in the case of the legendary descendants of the pioneers in Silicon Valley (Rogers and Larsen, 1984).

Workers' associations are often ignored, but they forge inter-firm communications, keep unruly competitors in line, and force employers to form a united front for setting wage and other policies. Engineering associations, in particular, have been an essential conduit for technical knowledge between firms in and out of government and between the university and private research laboratories (Clark, 1987). Unions have pressed for standardized wages and working conditions across trades and industrial sectors, and encouraged intra-industry unity and rationalization in the process (as in the US steel industry) (G. Clark, 1989). Political parties, too, may become a part of the fabric of industrial organization, as in the close alignment of employers' associations, unions, and local governments in Italy with the Communist Party in Emilia-Romagna or Christian Democrats in the Veneto (Sabel, 1982).

### The web of inter-firm relations

Inter-firm alliances and networks are ubiquitous in capitalist economies, forming a broad terrain on which the conventional triad of firms, markets, and workplaces operates. These modes of organization are by no means new and they have taken a wide variety of forms since the industrial revolution. Contractual alliances, consortia, and joint ventures, in particular, are on the rise, especially in leading sectors such as electronics, telecommunications, and aerospace (Hergert and Morris, 1988), while families, cartels, and bank holding networks are declining. But the fabric of inter-firm relations has always been strong, and remains so; in fact, networking appears to be characteristic of some of the most successful instances of contemporary national industrialization, such as Japan, Germany, and the Third Italy, and is much enlarged in the international domain.[38]

These arrangements have many advantages over classic administrative internalization or externalization on the market. Collaborative strategies let companies disperse financial responsibility and risk, react more quickly to market and competitive conditions, and launch new products or enter new markets without jeopardizing established business. Perhaps more im-

---

[38] Among US multinationals, cooperative arrangements involving licensing, partner or local shareholders outnumber wholly-owned subsidiaries abroad by the surprising ratio of 4 to 1 (Contractor and Lorange, 1988). Nonetheless, as Gomes-Casseres (1988) shows, there have been several upswings of international joint venturing over the course of the 20th century (around a secular upward trend), and the recent cycle does not necessarily indicate a permanent shift.

portant, they decrease costly layers of management by taking advantage of horizontal relations (both above and below top management), encourage autonomy and creativity among a large domain of managers and workers, broaden access to technologies and skills, and help suffuse learning through the production system (Aoki, 1986; Freeman, 1987; Håkånson, 1989). Indeed, it seems that some sort of systematic user-supplier relation and networking in complex production systems is essential to innovation (Lundvall, 1988; De Bresson and Amesse, 1991).

On the down side, alliances may not pan out as planned, can raise thorny management issues of their own, and can lead to conflict over partition of gains and losses. At times networks allow leakage of valuable knowledge and resources, trap a firm into association with the wrong set of partners, or become overly rigid – cross-fertilization can become cross-sterilization (Powell, 1987; Håkånson and Johanson, 1988). Enthusiasts of interfirm cooperation have tended to evade the role of power within networks (Storper and Harrison, 1991). Networking does not automatically imply equality; hierarchies among firms are pervasive. The Emilian-Tuscan model of relative equality appears to be the exception not the rule, as it grows out of a specific political substrate of working class and small capitalist organization (Cooke, 1983).

Some sort of inter-firm collaboration exists in every domain of capitalist enterprise, a middle ground "between markets and hierarchies" in the language of the transactions school (Thorelli, 1986).[39] In this view, inter-firm collaboration arises where there are correlative firm-specific assets which make open market transactions uncompetitive and internalization impossible (Teece, 1986). But this is the neoclassicals' usual backwards way of coming at problems of externalities and collective goods; they begin by assuming a universe of completely private affairs and discover, with much fanfare, that the real world of shared resources, social production, and human interaction slops over the partitions of these arbitrary boxes (Marshak et al., 1987). One ought, instead, to start from the assumption that technology, information, the division of labor, and the other constituents of modern industry cannot be chopped into private property without doing violence to the necessary connections. A universe of firms is caught in just such a bind. Markets and firm giantism can overcome many of the problems created by private property and competition, but they cannot solve them all. Hence the need for inter-firm networks.

Indeed, to say that collaboration and networks lie "between" markets and hierarchies is to miss the point: they are over, under, and around markets and firms, as they must be to sustain the necessary capitalist

---

[39] Other inventive, but equally unsatisfactory, terms have proliferated, including "quasi-disintegration" (Aoki, 1988), the "quasi-firm" (Eccles, 1981), "interpenetration of organization and market" (Imai and Itami, 1984) and "hybrids" (Powell, 1987).

fiction of a divisible industrial universe. This web of formal and semi-formal institutions operates as a safety net for the isolated firm. It is as necessary and predictable a part of economic life as the general fabric of social life alluded to by the concept of embeddedness, discussed previously, or as the substructure created for markets and firms by the institutions, infrastructure, and coercive powers of territories and states, to which we now turn.

## The Larger Field: Territories and States

We began exploring modes of organization at the micro level, then moved to the middle ground, and now push into realms normally excluded from the discourse of industrial organization and left to disciplines such as geography and political science. We bring territorial and state systems to the forefront as the third and last level of modes of industrial organization.

Capitalism and industrialism are impossible without cities and regions, yet cities and regions are usually seen simply as places to be filled by the "real" economic actors, business firms. This will not do. Territories are deeply implicated in the ability of industries to function and the way industrialization unfolds (Storper and Walker, 1989). One role they play is as modes of industrial organization. Ordinary discourse provides few concepts for thinking of "Chicago" as a living institutional complex that regularizes social relations in a way that allows production (and life in general) to proceed, yet an institutional fabric of established social relations, rules of behavior, channels of interaction, and command structures operates in a particular place, just as it does within a firm or a factory. Territories operate in a loose and informal manner, akin to inter-firm networks, and students of industry often jump from those networks to territorial clusters without recognizing the two things are independent. Networks take a spatially concentrated form only when there are particular kinds of inter-firm (and intra-firm) relations and under the impress of territorially-based systems of governance.

A similar evasion marks the treatment of the state in conventional economics. Nation-states (and imperial states) are territorial, of course, but they are territories with a difference, thanks to their unique powers of coercion and consensus. Williamson and the neoclassicals depict the state as a sort of transcendent referee, floating somewhere behind the active institutions of the economy (Hodgson, 1988). This is, of course, a goal of bourgeois societies, to keep relations between civil society and the state at arms length so private enterprise is free to pursue its goals. Yet the state is an active entity, not only creating and recreating a legal framework, but selectively administering and coercing civil society through government, agencies, and armies (Jessop, 1986). The state does more than "make policy" and intervene from time to time in the economy to correct (or distort) market performance – indeed, any corporation, merchant, or union does no less. The state, among its many other purposes, uses its enormous powers and capacities to act as a mode of organization.

## Cities and regions

Territorial entities such as cities and regions are extensive modes of organization which allow many and varied production activities to be assembled within reach of each other. Buildings, machinery, workers, and other material things are bound to particular sites and linked by roads, wires, and other infrastructure, thereby achieving some spatial order and coherence. The territorial complex is, in one sense, an extensive work site which brings disparate production activities into advantageous relation with each other, at a different scale and scope than the workplace or the firm. Other modes of organization are also rooted in particular venues, where they become part of larger territorial configurations, even if they extend across boundaries. But territories are not just congeries of material objects, workers, and firms; they are systems of social relations embodying distinct cultures and practices.

Territorial formations thus offer a way of organizing production systems above and beyond workplaces, market contracts, or legal ownership. The principle of territorial integration essentially involves three aspects of spatial interaction: simple propinquity minimizes the costs and effort of movement, maximizes access, and pools resources; locational fixity of infrastructure and daily life provides a built-environment of resources, lowers uncertainty and information costs of access, channels movement, and reduces social distance; and geographic boundaries limit movement, turn social interaction inward, and solidify (and differentiate) social relations. In very general terms, territorial complexes not only lower tangible costs such as transport and communication, but also allow information sharing, permit pooling of labor and fixed capital, stabilize physical and social relations, help people identify with each other (and against "foreign" competitors), and generate distinct cultural practices over time.

To be sure, territories are extremely loose, open, and "disorganized" compared to the factory or the firm. Like markets, their institutional fabric is quite thin, they rely heavily on voluntary interaction, have few formal rules, little administrative apparatus, and allow extreme flexibility among many parties. Nonetheless, territories serve as huge production and circulation complexes, allowing a degree of integration with a minimum of central control and a maximum of flexibility (Scott, 1988a). They are, in that sense, a highly suitable form of social organization for the anarchic side of capitalism. Yet the coherence of industrial territories is such that they slip readily into everyday language with monikers such as "Route 128," "Wall Street," or "The City" (of London).

It is truly astonishing that the territorial dimension is so widely considered an epiphenomenon of industrialization. It is impossible to render a cogent history of the industrial revolution without situating it in the particular context of Britain and following its geographical spread across Europe and North America (Pollard, 1981). Within nations, industrialization has always been profoundly regionalized, from the German Ruhr to the Basque country of Spain. Cities have been synonomous with capitalist development from the Middle Ages (Merrington, 1975). Industrialization in 19th century France was virtually identified with Paris; in England, with Manchester and Birmingham; in the United States, with Lowell, Philadelphia, and New York (Vance, 1977). Today, greater Los Angeles boasts an economic output larger than India's (Soja, 1989).[40]

Cities are clusters of production activities, people, and infrastructure in a densely built environment. Spatial proximity increases accessibility, thus broadening the number of buyers and sellers; this, in turn, provides sufficient demand for economies of mass production as well as minimum thresholds needed for specialized production. The city makes comparison easier, it pools diverse buyers, sellers, and information; it is where one goes to find the merchants and brokers who know market conditions, and to develop personal relations of trust. Spatial concentration facilitates worker interaction and managerial oversight, evaluation and feedback across related activities in different factories and firms. It heightens competitive emulation in business. Urban centers aid technological change not only by making it easier to stay abreast of the latest information, but by providing a pool of knowledgeable workers and diverse and creative suppliers. Cities are also great labor markets, whirlpools of humanity drawing immigrants to replenish the stock of cheap, pliable, and diligent workers; they are vast built environments of collective infrastructure, the principal nexus of the money economy, centers of credit (Pred, 1966; Harvey, 1985a; Sassen, 1988).

In addition, cities have well-defined institutions of order and control, such as real estate markets, municipal governments, and special authorities. Urban industries are often regulated by geographically-defined councils set up by local firms (Piore and Sabel, 1984). Every town in the United States has its local Chamber of Commerce representing small business; in larger cities, multi-industry groups form under the tutelage of leading merchants and banks, such as the San Francisco Bay Area Council. Political parties may also provide loose governance of powerful local interests, as in Italy (Lazerson, 1988), and educational and research institutions may serve as important reference points and information exchanges, as in Silicon Valley (Saxenian, 1988). In the United States local

---

[40] Even in geography the prevailling approach is to regard industrial location patterns as the result of prior organizational forms and decisions. Weberian theory works out of the calculus of single plant siting decisions; enterprise theory looks for the intra-firm logic of giant corporations; oligopoly theory argues that the behavior of a few large firms determines location. Against these views we take the position that territorial complexes of industry stand on their own as modes of organization.

governments regulate land use, install infrastructure, and lead "local booster" efforts to attract industry (Logan and Molotch, 1986). In Italy, one finds municipally-sponsored industrial parks and associations of small firms. Occasionally, local governments try to enhance inter-firm trading networks, as in San Jose, California, or the Japanese *ugoka* ("fusion") program in over five hundred local production districts (Sako, 1989). State governments in the United States have additional responsibilities for industrial coordination, including the right to pre-empt local powers that fail to serve business adequately (Heiman, 1988) and to create quasi-private special districts for such critical functions as ports and water supplies (Walsh, 1978). On the other hand, the city's implanted infrastructure, government structures, and patterns of social life impose a rigidity that must be periodically overcome through planned destruction and rebuilding to keep up with changing economic circumstances, as illustrated by the repeated redevelopment efforts from Hausmann to Robert Moses (Berman, 1982; Harvey, 1985a).

Regions are systems of cities and towns in a rural matrix, networks down whose channels flow deep and swift currents of goods, labor, information, and money (Pred, 1973). Major transportation and communication arteries cement these linkages, but so do the filaments of personal knowledge, institutional ties and cultural practices. We can distinguish five varieties of these territorial complexes: the local district, based strongly on labor submarkets (Scott, 1988a); the metropolitan region, such as greater San Francisco, made up of a system of specialized districts (Walker *et al.*, 1990); the city-satellite system, where a number of subordinate towns are tied closely to a dominant city, as with the textile and shoe towns around 19th century Boston (Pred, 1980); the cluster in which no city dominates, as occurred in the metalworking towns of the Connecticut Valley; and the large manufacturing belt as developed in the American midwest (Page and Walker, 1991).

The spatial organization of capitalism consists of an uneven but highly interconnected web of places, rather than a set of discrete regions. That is, there are networks of regional economies, which themselves contain networks of firms (which contain networks of plants within firms, and so on). Some territorial complexes will be tightly knit, as in Modena; other regions consist of loose coalitions of activities more tied to other places than to the locality. In short, everything that has been said about inter-firm networks applies to spatial patterns of integration, and the combination of these two levels of organization makes for hugely complex production systems (Storper and Harrison, 1991).

## Nation-states

Nation-states have enormous powers to effect industrial integration within their boundaries, and should therefore be considered as modes of economic organization. This begins with the state's basic framework of law and commercial policies. Commerce is fundamentally constrained by customs, duties and import-export limitations; despite the dramatic increases in global investment and commerce, the overwhelming majority of trade in the world remains internal not international. Labor markets are principally bounded and regulated at the national level, despite internal variations and international migration. Currencies are almost entirely national and are regulated by central banks; capital creation and movement always face national restrictions. These are the terms of analysis in international trade theory, which acknowledges that factors of production are not highly mobile across national boundaries (Ohlin, 1939). Although capitalism has spread industrialization over large portions of the globe, a fundamental tension remains between this and national sovereignty (Lipietz, 1987). Japan, for example, owes its success in forging a coherent industrial system to tight barriers on commodity imports and close restrictions on capital exports, which have allowed it to keep interest rates low and loans abundant, propelling a high rate of industrial reinvestment and giving backward industries room to develop (Okimoto et al., 1984; Yamamura and Yasuba, 1987). This is no more than Germany or the United States tried to do in their own time (Gerschenkron, 1962).

States have been a means of forging national cultures and social cohesion internally. This is an active process and includes standardizing laws, inventing shared traditions, elevating national heroes, and ferociously suppressing non-conforming peoples (Anderson, 1983; Hobsbawm, 1990). National cultures provide shared values and morals, habits and predilections that go a long way to normalizing economic relations between buyers and sellers, or between managers and engineers. National cultural practices in business and consumption are still strongly idiosyncratic, and create subtle barriers to entry by foreign firms. Finally, one should not underestimate vigorous nationalism as a motivating force for hard work, selective purchasing, and competitive spirit. At the extreme, states whip up fervor for war, the cauldron in which millions are steeled for sacrifice to a unified purpose and from which have come the most dramatic economic mobilizations ever known.

States also engage in direct forms of industrial organizing, administration, and planning to advance the fortunes of key sectors. The simplest form is state ownership, in which a government agency supplants private capitalists as proprietor, director, and investor of an enterprise. Common candidates are the postal service, telephones, airlines, and oil, though practices vary widely across capitalist nations. Prevailing theories of natural monopoly and public goods are virtually useless in explaining which spheres of production will be brought under the public thumb. In some cases, such as airlines, national prestige plays a part; in others the working class has a hand, as in the formation of the national health system in Britain. At times, there is a political score to be settled, as in the French nationalization of Renault to punish Nazi collaborators. But practices differ widely among countries in a way that liberal-conservative, statist-privatist labels do not capture well (Lindblom, 1977). Japan, for example, has little state ownership despite very active state planning, while Brazil, long embracing free-market policies, has massive state ownership.[41]

Broader economic planning and coordination is commonly undertaken by the nation-state as well. The state can encourage concentrated arrangements as the Meiji government did for the *zaibatsu* in late 19th century Japan (Moore, 1966) or more dispersed and localized forms of integration, as is common in Europe today (Bianchi and Bellini, 1991). Supplanting the market by government direction has proven feasible on a wide scale, even if global planning of the Soviet kind never worked without an enormous amount of political and personal coordination among managers and party members (Kornai, 1986). One much underrated tool of planning and coordination is government procurement; military weaponry has been the object of the most active efforts to push industrial suppliers onto a higher level of performance. Even in the laissez-faire context of the United States, the Department of Defense has been instrumental in setting standards, subsidizing suppliers, restricting competition, training managers, and generally forging a well-defined and well-organized military-industrial complex (Markusen *et al.*, 1991). The British Navy and French Army had this kind of effect much earlier, while wartime governments have implemented the most draconian forms of economic planning and administration.

Two much-discussed cases of national planning for successful capitalist development in recent years have been the Korean and the Japanese. The Japanese government, under the auspices of the Ministry of Industry and Trade (MITI), has skillfully used a mix of research, policy formulation, financial aid, and persuasion to cajole firms in key sectors to work together on common, agreed-upon projects (Johnson, 1982; Okimoto, 1989). Korea, after the coup of 1961, embarked on a successful effort to follow the Japanese model. Its Economic Planning Board and the ministries developed five-year plans to steer the economy away from its traditional core of textiles and into new areas such as steel, shipbuilding, and electronics. The state has carefully tended the growth and diversification of the giant firms, or *chaebol*, supporting favored projects such as Hyundai's shipyards with investment funds, tariff protection, and labor repression, while allowing unsuccessful companies to fail in carefully controlled circumstances. It also created new companies, such as Pohang Iron and Steel, entered into

---

[41] State ownership offers an alternative legal framework for production, but does not sunder the enterprise from the capitalist economy. Nationalized industries usually produce for the open market, and often compete fiercely across national boundaries. Finance capital circulates freely through government operations, such as public works or mortgage insurance, in a way that elides the boundaries of public and private investment quite thoroughly (Harvey, 1982). State operation brings politics into the picture, of course, but there are many ingenious ways to insulate against popular input and keep the capitalist class in the key positions, as in the Irrigation Districts of California.

joint ventures with foreign corporations, as in chemicals, and linked corporate suppliers and buyers in crucial cases such as the chain from basic steel to mini-mills to auto plants (Amsden, 1989).

As a consequence of national commercial policies, national cultures, and national industrial planning, there is a distinctiveness and coherence to national firms, industries, and economies in the face of the world market. It remains cogent to speak of Japanese, American or Swedish capitalism as structured industrial systems to this day, despite their internal variations and the increasing level of interpenetration by multinational capital.[42] Most industries still consist of several national centers which do battle in the international arena, as in automobiles and aircraft.

International cooperation among nation-states has also played an important integrating role in global economic development. Pairs of countries have consummated any number of specific trade, lending, and aid agreements but of greatest significance are multi-national conventions that diminish boundaries and actively manage international economic affairs. Coordinated monetary and trading spheres, of which the British Commonwealth and Sterling Area were long the leading examples, widen the fields of operation for investors, producers, and merchants. The Bretton Woods agreement stabilized the greater part of post-war world finance and trade around the US dollar, under the aegis of the International Monetary Fund, the principal central banks, and the General Agreements on Tariffs and Trade. The United States presided over the whole, using its hegemonic power to keep things in line (Block, 1977). With the breakdown of that system and leveling of the capitalist powers since 1970, annual meetings of the Group of Seven have been held to arrive at some degree of coordination. The most impressive example of international integration today is of course, the European Community's trade and monetary union of 1992.[43]

### Politics and markets in the larger field

We have drawn from an expanded palate of organizational modes to show that the process of industrial integration goes on at every level of economic activity. Williamson's neoclassical faith in microeconomic analysis as the gateway to industrial organization leaves out a huge swath of the economic institutions of capitalism he purports to account for. A truly

---

[42] Anderson (1987) calls Japan the last example of a truly national capitalism, but this is perhaps a hasty conclusion and very much reflective of his concern with the peculiar capitalism of Britain, which has been the most internationally open economy in the world from the outset of the modern period.

[43] The Soviet planning system and Comecon trading group stood aside from the general drift of capitalist internationalism for many years, representing an important alternative to the market and private enterprise model of industrialization (Nove, 1983).

multidisciplinary view of industrial organization ought to include geography and government; how odd of Williamson to speak of governance structures and leave out arenas in which the term is most commonly used. State and territory have always been fundamental to international studies of economic systems, but most of this literature is sealed off from industrial organization theory. Those who have crossed the barrier, such as Schumpeter, Gerschenkron, and Lindblom, are more honored in the breach than taken to heart. Fortunately, a revival of interest of these connections has taken place recently in political science and geography.

A view of industrial organization widened to include geography and government makes it clear once more that markets and firms are not the end-all of industrial organization. Cities and states are not embedded in the almighty market so much as the other way around. Again, we have to conclude that the greater part of the organizational fabric of modern capitalist production is woven outside and around markets and firms. Surprising as it may seem, capitalist economies are not predominantly market systems (Lindblom, 1977). Unfortunately, the economic institutions of capitalism do not come in tidy packages, and we shall have to continue to stretch our imaginations to comprehend Los Angeles in the same sweep as firms, factories, markets, and inter-firm networks.

The farther we move from the restricted plane of markets and hierarchies, the less the tidy analytics of transaction costs and scope economies serve as useful tools for understanding industrial organization. Our concept of industrial integration emphasizes the unification, coordination, regulation, and development of social labor. Integration, in this view, is not about lubricating transactions, choices, and efficiency, so much as it is about planning, directing, and orchestrating immense divisions of human labor enmeshed in a thicket of social relations. At the level of the region or the nation, it becomes patently obvious that politicking, culture, force, and conflict are what mostly propel economic coordination and such united action as is possible. Moreover, if we are after visions of a better world, such as the kind of democratic economy hinted at by the "craft districts" of the Third Italy or "market socialist" Yugoslavia, it requires a good deal more than clever management or a lot of small firms. Non-hierarchical social relations are very difficult to build without egalitarian community structures and democratic state power (Horvat, 1982). To put it bluntly, economic organization is not a matter of choice so much as a matter of politics.

Economic organization at the macro-level also makes it clear that static efficiency is not the core problematic of industrialization. Cities, regions, and nations have grown up around a furiously dynamic process of industrial growth. Korean state planners and *chaebol* did not reach their goals by close attention to costs, but by actively investing, learning, building up production, and developing organizational capabilities (Amsden, 1989). It is to this last topic, organizational dynamics, that we now turn.

## Organizational Dynamics and Economic Development

One cannot be satisfied with presenting a set of organizational choices as if economic interactions were a static matter of finding the best mode or modes of organization. It is equally important to see how the organizational capabilities of capitalism have improved virtually across the board. In this section we consider the way these advances have been made, where they have unfolded, and for whom they have been put to work.

The development of organizational capabilities has been essential to the forward advance of capitalism, and accounts for a substantial part of the overall growth of the productive forces. These capabilities have been promoted for several reasons, but a fundamental one has been to cope with a steadily expanding division of labor. As the problem of integration has grown more complex and challenging, the competitive advantages that accrue to capitalists (and capitalist nations) who can innovate organizationally have become sharper. An ever-expanding division of labor forever challenges capitalists to come up with innovative organizational solutions.

We look first at the advances made in the technologies of management – administrative capabilities attendant upon the tasks of integration within firms and across markets, inter-firm networks, territories, and states – and the institution-building that has continued since the dawn of the capitalist era, bringing us the present form and extent of all modes of industrial organization. Organizational development is not a one-track history of steady improvement and expansion, however, as one might imagine from some accounts. It is an uneven process which has seen many surprises along the way, not least the way the once-unassailable dominance of leading US corporations has been slipping in the face of challenges from abroad, as British industry was formerly bested by its rivals. To comprehend this situation, we need to grasp the sheer variety of organizational solutions extant in the industrial world at any time, even within the same industries or countries, and hence the possibility that organizational innovation can come from unexpected quarters and take unanticipated shapes. Finally, we take note of the fact that industrial progress remains two-edged, thanks to the dominance of capital over production and over the purposes to which the economy is directed, i.e., exploitation and accumulation.[44]

### Progress in organizational technologies and institutional capacities

Managerial methods and organizational capacities have been improved over time chiefly because effective integration of production systems is a fundamental way of establishing a competitive edge and hastening capital accumulation (Morgan and Sayer, 1988; Storper and Walker, 1989). Ad-

---

[44] This double-nature of industrialism has also been true of Soviet economies, of course, but with substantial differences in forms and effects, a topic we cannot take up here.

vances in management techniques and organizational capabilities have been documented most intensively in the case of the large corporations, which have been leaders in improving the sales effort, stabilizing resource supplies, generating new technologies, insulating their workers, raising capital more effectively, influencing government policy, and so forth. The heart of the rising corporation has been, in Chandler's terms, a managerial revolution, consisting of greater competence and technical facility among managers, better systems of bureaucratic organization, and more reliable methods of controlling diverse activities and flows of information (Beniger, 1986). This story is so well known that it does not bear recounting.

One example of better management will suffice: progress in the field of accounting (Parker, 1969). Not until the early 20th century did company accountants learn to calculate such basic measures as value of assets, depreciation, marginal cost, and rate of profit on fixed capital (Chandler, 1962). Only with post-war business school training have managers learned the worth of calculating net present value and risk-benefit ratios, of strategic budgeting and planning, or of disaggregating performance by projects and profit-centers (Auerbach, 1988). And only very recently have steps been taken to monitor closely the rate of turnover of materials or cost of set-up times.

The wrong lesson is normally drawn from the saga of modern management, however. If the large firm has become so prominent it is assumed that other modes of integration have lagged behind – virtually every mode of organization, including subcontracting, cities, and merchant trading, has been consigned to the dustbin of history at one time or another. This triumphalist view of the large corporation is blind to the broader development of the organizational forces of production. In fact, many of the powers and effects attributed to the giant firm were visible long before the modern corporation ever appeared. In Britain and the United States modern management began in the late 18th century (Pollard, 1968), both as a process of work rationalization in the factory, and as systematic product design and marketing of factory-made goods (Forty, 1986). By the middle of the 19th century the telegraph, steam packets, newspapers, canals, and merchant networks had already knit together fully national economies (Pred, 1973, 1980; Pollard, 1981).

Markets have certainly improved over time. They have not only benefited from transport and communications advances, but they have developed institutionally via commodity and securities exchanges, correspondence and branch banking, computerized tracking, and international materials traders, among other things (Chalmin, 1985; Thrift, 1987). Market transactions now reach farther around the globe, move more goods and money faster, and penetrate more deeply into everyday life than ever before – and one can now buy, through the market, the kind of managerial knowledge, marketing strategies, and technical advice once available only to the state or the large firm (Auerbach, 1988; Moulaert et al., 1989).

Similarly, the performance of individual workplaces has continued to rise thanks to a more carefully orchestrated labor process, especially in large factories. Detail division of labor, improved machinery, and fierce supervision are not the whole secret to greater labor productivity, however; it has benefited equally from more rational ordering of work stations and work flow, better balancing of labor time and machine utilization, closer monitoring of effort and materials, and finer evaluation of results (Walker, 1989b). Closer attention has been paid in recent years to problems of scale and scope, material flows, product quality and intelligent effort, and learning by workers, unleashing dramatic gains in both mass and batch production economies, as we shall see for the Japanese case in the next chapter.

Inter-firm alliances and networking have become more commonplace with increased understanding of how to benefit from them. Not only are Japanese subcontracting practices widely imitated, big Japanese firms are themselves steadily using more subcontracting and controlled decentralization (Aoki, 1988). Silicon Valley venture capitalists and high-tech entrepreneurs have learned how to stimulate innovation and enrich networks by systematically packaging new firms (Saxenian, 1991). In recent years, joint ventures and collaborative agreements have proliferated and become more sophisticated; even IBM is no longer going it alone and was, by 1985, involved in over 50 collaborative agreements in Europe alone, many with small firms (Freeman, 1991).

Cities and other territorial production complexes have been able to continue growing in size by virtue of improvements in transport, communication, water supplies, and other infrastructure (Teaford, 1984). Advances in the key elements of capitalist property development, such as real estate markets, mortgage finance, and large-scale building have played an equally important role. Private innovation has been accompanied by better public governance, including administration, finance, and land-use planning (Hall, 1984). The minimal conditions of urbanism generally work on a scale far beyond that possible a century ago and this is why all predictions of "natural" size limits to cities have proved wrong. In addition, advanced capitalist regions like California have developed articulated education, transport and water storage systems that would be the envy of many countries.

Since the 18th century, national governments have played a progressively more competent role at managing state industries and national economies (Armstrong, 1973). They have had to do so both to keep up with the achievements of private business (Hays, 1959) and to press forward economic development where adequate markets and capitalist organization have been lacking (Gerschenkron, 1962). This has required better training of state managers, more advanced systems of bureaucracy, a deeper understanding of economic affairs, and better mechanisms to monitor, regulate, and plan for private enterprise.

Many of the advances first employed by managers of large companies have diffused widely into small and medium-size enterprises (Auerbach, 1988). Indeed, the large corporations themselves have been hit hard by the global restructuring of industries over the last twenty years. The contemporary managerial revolution is not as yet well documented, but appears to rely on more decentralized decision-making, more autonomy for competent managers and skilled workers, more complex and flexible integration among various work units, and closer response to demand and control of work flow from suppliers through assembly to distributional outlets (Drucker, 1986, 1988; Sabel, 1989; Carmagni, 1990). *Ad hoc* organizational forms, such as special project groups and internal ventures are now very popular for moving quickly into promising new product and technological arenas (Friar and Horwitch, 1985). Computerized information networks also allow better communication and more dispersed decision-making (Hepworth, 1990). All these make a firm better able to manage a larger variety of smaller-scale production processes, respond to changing demand and competitive challenges, and to catch fresh technological winds.

As a consequence of across-the-board improvements in integrative capability, capitalist production has steadily expanded around the globe. Too often globalization is attributed to improvements in transport and communication alone, or to the sheer speed of capital movements. Rather, industrialists have increased their power to integrate and organize increasingly far-flung and complex labor systems. Moreover, innovation in organization and management has itself become more systematic, in the same way as innovation in the realms of materials and machine technologies. All this has profound implications for organization change, since it brings more varieties of industrial organization into the world market, intensifies international competition, and occasionally allows the ascendance of previously unheralded forms of integration. It is to this situation that we now turn.

### Industrial organization and uneven development

Organizational advance takes place in a context of widely disparate organizational forms across industries and places, and the combination of divergent organizational practices and differential rates of innovation is central to the overall process of uneven geographic development under capitalism.

Every industry necessarily consists of a matrix of organizational forms, or an *organizational ensemble*. The Marshallian model of industries made up of a plurality of small, single-plant, single-product firms is clearly inadequate, but conversely, the corporatist model of industries dominated by oligopolies of huge, multi-factory, fully-integrated corporations has also been found wanting. Both views are impoverished in light of the reality of organization modes and forms: some large firms, some small; some big factories, some tiny workshops; some plants clustered in cities, some scattered in rural areas; some core financial and mercantile players, some intermittent lenders and traders; some relational contracting, some sub-contracting, some licensing; some state intervention, some laissez faire reliance on the market. It is not hard to construct possible worlds from the rich mine of existing integrative schemes (Storper and Harrison, 1991). Modes of organization are not exclusive sets from which one must choose on the basis of comparative costs, as in transactions models, but are complementary ways of solving a range of problems arising from production, distribution, competition, and investment. Finding the right combination, or organizational architecture, is critical to competitive success (Foray, 1991).

The US semiconductor industry, for example, comprises a wide variety of technologies and product niches, from bubble memories to customized integrated circuits to standardized random access memories. As a consequence, a complex patchwork of firm sizes and strategies has emerged. Texas Instruments (TI) and Motorola are among the country's largest corporations, and the in-house chip makers, IBM and ATT, are gargantuan, yet the industry is famous for its many vigorous small firms. Some chip fabrication plants involve several thousand workers, many specialty shops are tiny. The industry relies heavily on venture capital for financing and managerial expertise. In Silicon Valley it forms one of the most famous urban industrial agglomerations in the world, drawing engineers, entrepreneurs, and existing firms; even IBM does some of its most advanced R&D at its huge plant there to take advantage of the rich environment. Yet there is also considerable dispersal from this center. TI and Motorola have built separate complexes in Dallas and Phoenix, for example, and there are outlying growth peripheries all over the western United States and southeast Asia. Silicon Valley firms have formed a rich web of relations within the region, but they have also reached out through innumerable joint ventures and special arrangements to foreign companies.

Our view of organizational ensembles differs considerably from the transaction cost theorists' effort to come to grips with industries and territories, such as Scott's (1988a) argument for integration-disintegration along parallel axes of large and small workplace, large and small firm, and industrial districts versus dispersed production. The urban territory is not a direct outcome of systems of small workplaces (or firms); a city like Los Angeles will contain an assortment of large factories and firms, as well as local manufacturer's councils, specific governmental interventions across local industries, and banking-industry networks. At the same time, Los Angeles has had proportionately smaller financial or mercantile sectors than San Francisco, which is why venture capital has not set up shop as munificently there as in Northern California. Scott's own analysis suggests that even large plants and firms are likely to depend on the dense fabric of external linkages found in disintegrated production complexes, and the evidence he adduces for the claim of dispersal of large plants is simply not convincing, especially as all the industrial clusters he refers to – semiconductors in Silicon Valley, garments in Los Angeles, aerospace electronics in Orange County, auto plants in Tokyo – show the persistent presence of very large factories.

Different industries are put together in configurations appropriate to their particular production problems. Shipyards are a far cry from garment factories, and shipbuilding companies have been much larger, on the whole, then clothing producers, which rely on merchants and big retailers for much of their coordination. Putting-out is common in garment work but precious few parts of a ship can be made at home; shipyards tend to be located on the outskirts of cities, garment work in their centers. Shipbuilders are financed mostly by banks or governments, garment makers largely by merchants and commercial credit.

Yet organization cannot be reduced to a simple outcome of other forces, such as technology or labor control. The same industry may assume different organizational garb at different times and places. While Chandler is right to note the link between mass production and modern management, his own evidence shows a substantial variation between countries and across industries in the prominence of the large enterprise, depending on national circumstance and serendipity; for example, large firms never dominated the food sectors in Germany as they have in the United States and Britain, while heavy machinery makers in Germany have generally been much larger and more diversified than in the United States. Dependence on market and state also varies: US electronics has gone commercial much more than its British counterpart, which is almost entirely locked up in the armaments industry-military contracting nexus (Morgan and Sayer, 1988). The use of territory may differ, as well: in the early electrical industry the Germans utilized spatial concentration, in the Berlin suburb of Seimensstadt, in a way unmatched in any other country. Today, despite the worldwide enthusiasm for alliances and networks, German companies have not joined the rush to cooperate externally (Hergert and Morris, 1988).

That industries and their organizational ensembles, including their geography, evolve together is clear from such histories as military electronics in Orange County, California (Scott, 1988b), just-in-time auto assembly in Toyota City, Japan (Cusumano, 1985), and flexible machining in Tuscany, Italy (Becattini, 1978). Industrial development is not a rational choice among a menu of alternatives, but an open process of search along unknown paths (Storper, 1989); hence, the history of every industry is replete with organizational discoveries and discards as capitalists seek better ways to compete. Successful firms, networks, or industrial districts must be actively created and sustained, in the same way as product and process innovation. The struggle over oil in pre-war Europe provides a marvelous tale of strategic moves and countermoves by such participants as the Rockefellers, Nobels, German banks, Dutch and English interests, and

Austrian merchants, growing and shedding institutional skins, the better to survive in the snakepit (Chandler, 1990). And organizational development can become self-reinforcing and increasingly specific to the firm or the network, like all technologies (Foray, 1991). One can even discover significant differences in organizational culture between two such closely identified areas as Boston's Route 128 and Silicon Valley (Saxenian, 1989; Weiss and Delbeqc, 1988).

Industries must occasionally undergo radical reorganization to meet shifts in technology, demand or labor relations. For example, US meatpacking has gone through a revolution, led by Iowa Beef Packers, to buy beef from feedlots, slaughter beef, and box it for supermarkets with non-union labor (Page, 1992). Los Angeles's garment industry has become less integrated to take advantage of an influx of low-wage immigrants (Scott, 1988a). Change may be forced by external competition from new centers of industry using new methods of production and new organizational ensembles (Storper and Walker, 1989). Or an economic crisis may compel state intervention, as in the financial disclosure laws enacted in the United States during the New Deal that made companies keep their books more honestly and improve the quality of information reaching the market. Even more dramatically, change may follow conquest or social revolution, as in Germany after World War II or when much of China's industrial bourgeoisie moved to Taiwan after 1949.

National systems of industrialization are perhaps the most striking instance of alternative organizational ensembles. National capitalisms present significantly different competitive faces to the world, and new challengers have arisen over time based on organizational models substantially at odds with dominant ones. In Britain, for example, the characteristic approach to industrial organization has consisted of small firms, family control, and little reliance on professional management. The British have favored the classically educated amateur on the old aristocratic model (Armstrong, 1973) and Britain's rich networks of commercial middlemen allowed its firms to survive on a smaller scale and scope for much longer than those in other countries (Hannah, 1980). In Germany, by contrast, the most advanced technical university system in the world before World War II provided industry with ample cadres of engineers, researchers, and managers, so that Germany developed a system of professional management independently of the US version (Chandler, 1990). Finance has come chiefly from commercial banks, and intra-industry cooperation and cartelization has been legally sanctioned, in strong contrast to US or British practice. German industry never went over as fully to the large factory and mass production methods as the United States, however, leaving it with a greater reservoir of alternative practices to build on in the present era (Katzenstein, 1989). Of the 1,000 largest firms in the world in 1990, only 30 were German, compared to 345 from Japan and 353 from the United States (*Business Week*, 1990).

International groupings as well as nation-states compete for economic supremacy. The fierce inter-imperial rivalries of the late 19th and early 20th centuries took place between empires based on quite different systems of colonial administration and economic relations. The imprint of Japanese methods of management can still be seen in Korea and Taiwan (Cumings, 1984). After the dissolution of their empire, the British tried to compete via their Commonwealth ties, when industrial might alone no longer sufficed. Today, the European Community is moving to levels of continental unity never before achieved under capitalism, as the Soviet bloc breaks up, the Japanese expand their ties throughout southeast Asia, and the United States draws its wagons more closely around the Americas.

## Capital and industrial organization

Market exchange does not define capitalism, nor set the ultimate terms of production. Capital in circulation is the supreme arbiter of industrial activity, the ultimate linkage mechanism, the key regulator of social labor, and the driving force behind industrialization. Capital flows in and out of all the pores of production, whatever the mode of organization: between industries, among firms, and across workplaces, across national boundaries, and from city to suburb. As it circulates, capital weaves its way through the warp of an immensely complex system of production to knit a tapestry of extraordinary richness. This flux of value in search of surplus value animates the far-flung labor systems of the capitalist world, throws them into competition with one another, compels them to exploit their labor forces, and drives them to accumulate – and lays waste those that do not perform well enough.

Capital is not bound, therefore, to any one form of organization, certainly not to the firm. The firm is usually seen as the unit of competition and accumulation (e.g., Auerbach, 1988), but competition equally takes places between capitals, or invested funds, each quantity of which needs to extract a satisfactory rate of profit. Competition may therefore occur within firms, between divisions, profit centers, or individual projects. At the same time, investments sunk into specific places mean that competition will equally be manifest in local boosterism between cities or in the global clash of nations. Similarly, capital accumulates not only within firms, but in banks, merchant fortunes, the infrastructure of cities, and the savings of nations, and exploitation is not confined to the shop floor or the envelope of the firm; surplus value is also siphoned off through wide channels of rent, interest, and taxation.

Capitalist competition has a developmental history, as well. Competition has not been reduced by the centralization of capital in large firms, but increased by the advances in all forms of organization. In fact, early industrial capitalism was distinguished chiefly by the poor development of its economic institutions and competition was cramped by limited transport, weakly developed markets, and haphazard measures of profit, among other things. The emergence of nationally unified markets, futures markets, multinational firms, global banking, and the rest have brought capitals into more direct confrontation everywhere (Auerbach and Skott, 1989). Advances in industrial organization have smoothed and accelerated the circulation of capital, refined the calculation of gain and loss, heightened the perception of market opportunities, and hastened the compulsion to keep up with the pack.[45] As Weeks (1981) has said, it is the quality of competition not the quantity of competitors that is the most important variable and there is no reason to think the total number of global competitors has shrunk. Many attempts have been made to restrict competition through horizontal mergers, international cartels, and national autarky, with substantial short-term effects, but they have all failed in the end to contain the power of competition in a dynamic economic system, backed by the political power of states committed to the capitalist ethic (Auerbach, 1988).[46]

We have said nothing about class power in light of the evolving economic institutions of capitalism. This is a common failing of industrial organization theory, as in Williamson's massive evasion of the question of power (Perrow, 1981). The integrative function of industrial organization should not obscure the power of a capitalist class that still prevails by virtue of its ownership of property, including corporate assets and monetary instruments. Yet the growing complexity of industrial organization has had an effect on the ruling class and how it rules, even if the wizard behind the organizational curtain is still the capitalist (see chapter 1). This change was once thought to be simply one of substituting the corporation for the capitalist, yet even within corporations there are contending sites of capitalist control, just as there are between industrial companies and financiers, or between private enterprise and state managers and politicians. In short, there is no single site of capitalist power, and the elevation of one faction over another depends on the opportunities, on the weapons mustered by various parties, and on the historical emplacement of rules and institutions – banking laws, securities regulations, reorganizations by occupying armies, etc. Yet even as the power brokers squabble, it is necessary to take into account the integrative effects of organizational connection for capitalist class formation as a whole (Useem, 1983; Mintz and Schwartz, 1985). Moreover, we should expect that as the organizational fabric of capitalism evolves and changes over time, the locus of capital and capitalist power may be dramatically altered.

---

[45] Behavioral changes may be as important as the strictly technical and institutional in this process, as Auerbach (1988) has argued; that is, the revolution in human thought and practice ushered in by capitalism continues on its way.

[46] As competition has increased over time, so has the regulation of worldwide production by the circulation and competition of capital, and for the same reasons. That is, industry is compelled to operate according to the law of value, even outside the compass of capitalist production, as the Soviet bloc has discovered.

The recent success of industrial districts on the Third Italy model has triggered some false hopes for a more egalitarian, petit bourgeois future for capitalism (e.g., Piore and Sabel, 1984). While it is wholly salutary to break with the old Fordist-Stalinist fetish of giantism in industrial organization (Auerbach *et al.*, 1988), small firms and flexible networks are still capitalist enterprises: they neither eliminate the imperatives of accumulation nor solve the problem of democratic rule versus class prerogatives in the workplace, the firm, the city, or the nation. The utter futility of union organizing in Silicon Valley speaks to the secure class power of the entrepreneurial business class in a classic disintegrated production complex (Walker *et al.*, 1990). Furthermore, the degree of egalitarian or innovative social relations in places like Silicon Valley and Emilia depends very much on prior class relations in the region, rather than being a simple result of small firms, networks, and industrial districts. Indeed, innovative forms of industrial organization depend on the state of class relations, as with Japanese post-war innovations following on the palace revolution engineered by the US occupation (Cole, 1971).

We still need to think out the implications of the relation between expanding forces of production, including the division of labor and organizational capability, and continuing capitalist relations of production. One implication of this conjunction has been the increasing power of capital to orchestrate labor systems over ever-larger geographical areas, which occurred in spite of more dispersed production sites and more disintegrated organizational forms. If large factories are less useful today, it may be that the division of labor has expanded so that the factory is insufficiently large to encompass entire production systems, so another way must be found. If capitalists can effectively master larger production systems through subcontracting networks or politics, then they may not require those fortresses of private property known as factories, and more dispersed forms of capitalist production can ensue. The same principle applies to firms. Today, the buying and selling of entire firms, and the assembling and dismantling of giant conglomerates, is an everyday occurrence; capitalist empire builders are treating megacompanies in the same terms the latter have treated their subsidiaries. Simultaneously, companies large and small are eagerly forming alliances and networks across the globe in order to compete with other constellations of firms, and these alliances are multiple and overlapping. Firms are becoming nodes along a neural network of production and circulation rather than bounded entities ricocheting off each other like billiard balls. Perhaps, then, we have come to a time when capital is outgrowing the 20th century corporation.

### Divisions over the new industrial divide

Not only has the whole field of industrial organization theory been unceremoniously thrown off its foundations by the discovery of vibrant alternatives to the long-dominant US multinational corporation, global capitalism appears to be perched on the crux of what Piore and Sabel have aptly called an "industrial divide", in which the old verities of production and management, from the Fordist assembly line to the Sloanist corporation, are in retreat before innovative new practices. What direction will global capitalism take? In Europe, the Italian model has caused the greatest uproar, and comparable patterns have been documented throughout the continent, including southern Germany, greater Paris, western Denmark and Portugal.[47] The general term of art for these arrangements is flexible specialization, which has had the virtue of putting industrial networks, districts, and regions high on the research agenda. In the Americas, greater attention is being paid to the challenge emanating from Japan and East Asia where the principal challenge is the hierarchical network, the industrial group, and national strategies of development.[48] A fierce debate has broken out over whether the Asian or Italian model will define the next epoch of capitalist development (e.g., Piore and Sabel, 1984; Scott, 1988b; Florida and Kenney, 1990). And there is a third group which thinks the big US and European companies, after learning some new tricks from Japan and the small firm networks, will come roaring back on top of the heap (Harrison, 1990).[49]

There are good reasons to think that the Japanese system of production and organization is the most formidable challenger for global hegemony, as we shall indicate in the following chapters. At the same time, smaller firms and industrial districts have proven to be more robust than critics imagined, and global networking appears to be here to stay. And surely most older corporate giants will survive and prosper, whatever modifications they have to make in this new environment, because of the continuing value of financial might in a harsh world of capitalist competition. The exact mix of surviving organizational forms into the next century is difficult to predict, and we leave that debate to others.

We prefer to draw three different conclusions about the new industrial divide, on the basis of the arguments made in this section. First, the organizational innovations of our time represent a revolution in production integration that is quite general, and this revolution is a rising tide that will lift many boats. Second, organizational innovation and restructuring are deeply implicated in the shape of uneven development, but the

---

[47] Sabel has been the great popularizer of these developments and his enthusiasm has swept up many collaborators (see Brusco and Sabel, 1983; Piore and Sabel, 1984; Sabel and Zeitlin, 1985; Murray, 1988; Hirst and Zeitlin, 1989). But note that some of the most compelling case studies for the new industrial districts have come out of California, one of the places one would have least expected them.

[48] Important writers on Japan are Dore, Imai, Aoki, Ouchi, Okimoto, Morishima, Johnson, and Kenney and Florida.

[49] We cannot swallow Sabel's (1989) theory that big firms are simply learning from the industrial districts, however. The shift in large firm management toward more flexible and network-type arrangements dates from the mid-1970s, about the same time as the industrial districts were coming into their own (and before they were widely touted) (Martinez and Jarillo, 1989).

constellations of production systems, capital flows, and places are sufficiently complex that the outcome will not be as simple as a general triumph of just-in-time systems or small firm industrial districts. Third, organizational advance and restructuring, and the geographic shifts that accompany them, are profoundly unsettling to workers, organized labor, and working class politics, which must reorient themselves to new ways of working, new forms of interacting, and, most of all, the new shapes and guises in which the capitalist class appears.

Finally, we should add that the organizational revolution is not the direct reflection of the micro-electronics or "information technology" revolutions, as Freeman (1991) maintains. While expanded capabilities for communication, programming, and regulation granted by computing are undeniable (Perez, 1985; Forester, 1987), organizational innovation has been pushed by, above all, the ever-expanding division of labor – a development in the forces of production that is as striking as the command of microcircuitry. But the organizational revolution is reducible to neither information technology nor the division of labor, for the creation of better solutions to the growing puzzle of integrating production systems is a field of technological advance in its own right.

## Conclusion: The Political Economy of Industrial Organization

We have tried in this chapter to lay out a comprehensive framework for understanding industrial organization as one of the key elements in the new social economy – one which need not take a back seat to technology or class exploitation in accounts of modern economic history. Our view of industrial organization goes well beyond prevailing theories of managerialism, centralization, networks, or transactions costs. Of course, such an expansive perspective makes it exceedingly difficult to reduce industrial organization to a tidy analytic model, nor have we attempted to settle the many disputes over the precise weight of various organizational modes, their changing force over time, or their uneven appearance across space. Instead, we have tried to set out as ample a conceptual palate as possible, from which future researchers might cover the industrial canvas more satisfactorily than in the past, vanquishing at long last the drab portrait drawn by conventional industrial economics and sociology.

Three themes have been sketched in this review. The first is that the integration of the division of labor is a pivotal problem in economics, central to (though not exhaustive of) the problem of production and its development. Modern production poses a fundamental problem of social labor and social governance, from top to bottom. Ours is thus a brief for cooperation as well as competition, equality as well as command, coordination as well as specialization – over the whole living tissue of social relations in production. Most of all, we recognize that the answers to economic questions are not miraculously provided by either markets or managers; they remain to be worked out in the process of creating more humane societies.

Our second theme is that there exists a very wide range of modes of integration. Every branch of industry consists of nested and interpolated layers of institutions orchestrated through diverse points of control. Just as workplaces are systems of specialized work units, and firms are systems of workplaces, so there are systems of firms embedded in a delicate fabric of collaboration, contracts, ownerships, families, and the like. In addition, regional and national systems of firms and networks, even systems of states, must be woven into the full web of industrial production.

Third, we have argued that growing organizational capabilities are absolutely crucial to economic development. These cover a wider range, and have been evolving for a longer time than is conventionally recognized. Organizational advance is central to industrial revolution, past and present. With capitalist expansion, moreover, innovation has not been confined to those already ahead or safe in core areas. This implies that less prosperous countries, including the ex-socialist ones, can learn from those now leading without slavishly imitating what may become tomorrow's obsolete organizational techniques. Local strengths may well be compatible with sensible strategies of development.

If industrialization in the broadest sense is still the fundamental economic problem in the world today, then economic development requires a policy for industrialization that includes organizational strategies of institution-building and guidance of economic actors. The market produces nothing by itself; it is no more than a tool – a powerful one, to be sure – for guiding producers. Not only are internal direction and planning needed within the capitalist enterprise, but there is need for economic policy and planning all through the industrial system. The network theorists and the flexible specialization school have shown that a little cooperation and social governance outside the bounds of firms and markets can go a long way. At the same time, Japan and Korea have shown that national planning is still essential for rapid industrial catch-up, as List, Lenin and their followers have long argued.

Finally, neither capitalist industrial organization theory nor communist planning theory is very satisfactory on the question of power and politics in the social economy, that is, in treating governance in its original sense of citizen and sovereign, democracy and government, civil society and the state. The managerialists fell under the 20th century spell of experts, the transactionists have been unduly obsessed with cost efficiency, and the flexible specialization school has romanticized the virtues of small firms. None has come to grips with the reality of capitalist power and the inequalities that follow from class control of the means of production and organization in the industrial system. Every one misses the persistent fact that capital operates behind all the forms of industrial organization, weaving in and out of the various modes, setting the conditions for regulation of the division of labor, driving fierce competition among all blocs large and small, and hurtling down the path of accumulation for accumulation's sake.

In other words, the socialist idea of bringing the industrial system under more conscious control, lessening inequality and redistributing surplus, and pushing democracy into the hidden abode of production remains on the agenda. Yet the traditional left portrayal of the capitalist economy as anarchy is no more satisfactory than the right's blithe paeans to the free market. Similarly, critiques of monopoly and corporate power all miss the point about the need for organization, often of a rather large scale. Yet the kind of worship of concentration and top-down command one finds among top managers and capitalists or, on its flip side, among Soviet cadres and party leaders of old is equally indefensible. What then is left? We shall explore some of the new developments in capitalism, and the questions these pose for socialist alternatives, in the remaining chapters.

# 4

# New Developments in Manufacturing: The Just-in-Time System

Having looked at the problem of organization of the division of labor in general, we now want to home in on the labor process within manufacturing plants and their tied subcontractors. This is the most highly organized area of the new social economy and one which has seen dramatic technical and social innovations in the last two decades. Most of the academic interest in these changes has concerned restructuring associated with major new technologies such as "information technology" and "systemic automation", including "flexible manufacturing systems" and "computer integrated manufacturing", which some argue underpin a new form of labor process organization termed "neo-Fordism" (Palloix, 1976; Aglietta, 1979). Strangely, less attention has been paid to some major changes in the organization of manufacturing which are not necessarily tied to new machine technology, in particular the just-in-time system of work process organization developed in Japan. This involves new types of relationships among workers, between workers and management, and between firms and their buyers and suppliers.

As has happened repeatedly in the history of capitalism, such managerial innovations have shown deficiencies in practices which formerly appeared to be the acme of capitalist rationality. For example, some observers have claimed that Japanese forms of production organization are as significant as Henry Ford's innovations 60 years ago (Monden, 1981). And, given the effect of the law of value in enforcing the adoption of the most productive techniques among competitors, they are already diffusing outside Japan. Once again in the history of capitalism, innovations in the organization of production are changing the processes and patterns of uneven development.

Many features of Japanese manufacturing have been cited to explain its competitive success – relations between industry and the state or between industry and banks; low wages, tame unions, workaholism, and the catch-all category of "Japanese culture" (Johnson, 1982; Morishima, 1982; Itoh, 1990). While many of these are important, we focus on the circumstances most directly responsible for the superior productivity of many Japanese industries, the organization of the labor process and its supports.

In this chapter we shall explain these organizational innovations, their origins, preconditions, and effects, and assess how much they can be exported and with what implications. We shall also discuss some critical implications which have eluded much past research on the labor process, with its restricted focus on the relationship of individual workers to a given technical division of labor. We shall consider the ways in which technical divisions of labor are constructed and related to the social division of labor, and the ways in which workers, managers, machines, materials, and products are divided and combined at work. Just-in-time systems are a powerful testimony to the crucial nature of this broader vision for capitalist development.

We begin with some theoretical points regarding the labor process and the diffusion and evolution of new forms of organization, then show how the Japanese just-in-time system differs from the so-called "just-in-case" system; we then go on to discuss their origins and preconditions and finally assess the implications of both systems for uneven development and for labor.

## Approaching the Labor Process

One of the main conclusions of literature on the labor process since Braverman has been the recognition that increased managerial control over labor and deskilling are just two among several means to the end of profit, rather than goals in themselves as Braverman (1974) implied (Kelly, 1982; Littler and Salaman, 1984; Manwaring and Wood, 1984). Empirical research has made it clear that product technology, product market conditions, employment relations, and state employment policy all affect the type of labor process (Littler, 1982; Burawoy, 1985; Kelly, 1985). This suggests that labor process research needs to cover a wider front than just the organization of the technical division of labor within direct production and the relationship of workers to management control and machines. Japanese methods are particularly difficult to comprehend within such a restricted view, for they rely heavily on the way in which work is divided and integrated, on the flow of working capital, on the relationship with suppliers and buyers, and on the individual worker as a creative force.

Recent literature on the labor process has also been more receptive than earlier efforts toward the idea of bureaucratic control and so-called "cultural" features, such as status systems and authority relations. Marxist researchers have come to acknowledge – instead of dismissing – the importance of worker consent and motivation and it is now widely appreciated that naked coercion is an ineffective way of getting high productivity (Burawoy, 1979).

While Japanese innovations represent a radical departure from Anglo-American managerial wisdom, they have important elements in common with previous managerial and process innovations: the search for time economies in the use of circulating capital and machinery and in the application of labor, and the search for dynamic economies as product and process evolve. Insofar as the pursuit of these goals is part of the nature of capitalism we can apply the abstract theory of capital to any particular capitalist firm, American, Japanese or whatever.

In concrete instances, it is important to recognize that the particular organizational forms of capital bear the imprint of the social formations in which they develop: capital is never born by immaculate conception; hence there is no such thing as normal capital. Actual capitalist firms are profoundly shaped by the characteristics of labor and product markets, labor organization, ethnicity and gender, employment legislation. All of these circumstances and many others affect labor-management relations, management techniques, the organization of the labor process – even the way technology is used. Particular organizational forms such as Taylorism are not universal stages through which all capitals must pass, not simple unmediated expressions of the developmental logic of capital, but local responses to local and transient contexts. In the United States, the exceptionally high rates of labor turnover at the turn of the century (100–300% per year), helped to make Taylorism – deskilling and extreme task specialization – attractive to capital (Foster, 1988). Similarly, the scope for Fordist organization and pursuit of economies of scale in mid-century was much greater in the enormous American market than in the much smaller Japanese market. Thus while new forms of labor process organization are shaped by the general class character of capitalism and driven by capital accumulation, they bear the imprint of historically- and spatially-specific conjunctures, both in their genesis and diffusion.

The point becomes clear when we reflect upon American and European attitudes to Japanese capitalism. Characteristically, westerners bracket it as a special case, distorted by the peculiarities of Japanese culture. But this is thoroughly ethnocentric, for if Japanese capital is shaped by its cultural contexts, then so too is American, British, etc. Once we recognize this, we can ask whether certain characteristics we had previously assumed to be normal, or intrinsic to capital as such, are really effects of parochial, national, or regional contexts, and indeed we shall later suggest that an appreciation of the nature of Japanese industry helps us look at our own practices with new eyes.

Although new forms of production organization are invariably modifications of pre-existing ones adapted to suit local contexts, the adaptations can sometimes accumulate into a major new framework of far-reaching potential, as in the case of just-in-time (Sahel, 1981; Walker, 1985). If the new forms permit super-profits, they are likely to diffuse outwards to other contexts, as the innovating firms invest in production overseas, license to competitors, or outcompete them, and then are imitated (cf. Littler, 1982; Storper and Walker, 1989). In particular, multinational firms act not only as bearers of market forces (Murray, 1972) or new technologies, but as bearers of new social relations. It is this diffusion process which has aroused popular interest in cases of Japanese direct investment over-seas, such as that of the new Nissan plant in the Northeast of England (Garrahan, 1986) or Honda in Ohio (Mair *et al.*, 1988). These cases show that the diffusion process is limited by the degree of compatibility bëween the new methods of production and existing practices in the areas where they are implanted (see also Morgan and Sayer, 1988). Inevitably there is a process of mutual adjustment between the two sides. In a rather Lamarkian manner, the innovations are either adopted and adapted to the local environment or rejected; occasionally, the resulting mutants turn out to be more vigorous than their parents. Indeed this was true of the previously dominant US methods earlier in the century, as Littler (1982) has shown.

It is perhaps easiest to grasp the nature and significance of the Japanese practices if we contrast them with the forms of organization familiar in other established industrialized countries. We shall do this by contrasting two extremes – the "just-in-case" (or JIC) system and the "just-in-time" (or JIT) system (Abernathy *et al.*, 1983), the former common in the United States and Europe, the latter emerging in Japan and beginning to diffuse outwards. To a certain extent, these descriptions are idealized, even cari-cature, but they are easily recognizable from the business literature. As with many of the most striking innovations in production organization, both JIC and JIT are most common in mass production, but both have some features which can extend to small batch production and perhaps even project production. The two forms of social organization imply different material arrangements of production equipment, though the differences in hardware between the two systems are not dramatic.

## The Just-in-Case System

Many of the characteristics of the "just-in-case" (JIC) system are familiar, but we shall describe them critically as particular ways of organizing the labor process, rather than as the only possible face of "normal" industrial capital at a particular stage of development.

Just-in-case is shorthand for a bundle of production characteristics in certain leading industries outside Japan, such as automobiles and appli-ances. These include particular approaches towards volume and specializa-tion, flexibility and demarcation, skills, quality control, bureaucratization of procedures, and relationships between groups, management techniques, innovation, and the labor market. This is a diverse range of aspects, not all necessarily present in any one case, but they have a certain unity. In some respects JIC resembles "Fordism", but for reasons explained later this term will not be used.

The starting point of JIC production is the establishment of a rationalized sequence of processing or assembly steps in which workers and machines are dedicated, as far as possible, to only one simplified task. The tendency is to give priority to maximizing the utilization of fixed capital and labor by increasing the speed of each operation as much as possible, and by pursuing economies of scale. So, unless limited by restricted markets, high volume and standardized output are seen as the route to lower costs per unit. Similarly, at the pre-assembly stage, long production runs of components are sought in order to minimize the amount of time machinery is down. Even in small batch production, runs of a particular product may be long in relation to orders at a given time, in anticipation of future orders. In other words, if it takes six hours to set up a machine for producing a particular component, it seems rational to maximize the length of runs so as to minimize the amount of time spent re-setting the machine, rather than re-setting the machine repeatedly for short runs. Both the kind of machinery and the organization adopted therefore tend to be rigid since they are dedicated to executing a single operation repeatedly. Under this system, nothing must break the continuity of production. The imperatives are well summed up in the phrase used in US manufacturing, "getting metal out the door" (Abernathy *et al.*, 1983; *Business Week*, May 14, 1984).

Long runs in turn entail large inventories, and, in addition, ample "in-process inventories" or "buffer stocks" are maintained so that the rest of the system can continue to function should any stage of production be disrupted or any component prove defective. In any case, the fact that each task or each machine operation is pushed as fast as possible creates inevitable gluts and shortages – a situation which tends to both produce and require buffers. Similarly, stocks of parts from suppliers are kept at a level sufficient to insure against supply interruptions, and a reserve stock of labor may also be hired to prevent absences causing interruptions to production. All these practices explain the "just-in-case" tag.

The rigidity of the hardware of the system matches the rigidity of the social organization of production which has come to be associated with it; indeed, there has almost certainly been a process of mutual adjustment between the technical and social sides through the responses of management and labor. Several features of this system tend to encourage demarcation: particular parts of the factory are specialized for particular operations; long production runs; moving parts on a conveyor between specialist workers instead of moving workers between several activities; separating tasks and workers by large buffer stocks – all insulate different groups and tend to generate coordination problems, though the buffers themselves give some flexibility. In turn, coordination problems reinforce the need for a deep vertical hierarchy of supervisory and managerial labor to link and control the specialized activities.

Typically, suppliers are chosen primarily on the basis of price competitiveness and key inputs are obtained from more than one source to protect against possible disruption.[1] Quality of goods supplied from other firms is checked by testing samples to ensure that reject levels do not rise beyond a certain percentage of deliveries. The suppliers themselves use JIC methods, and deliver in large but infrequent batches to their customers' warehouses. Both functionally and geographically the relationship between firms and their suppliers tends to be distant; the purchaser is not concerned with how the suppliers operate provided the price is right, and this lack of inter-firm contact together with the infrequent nature of deliveries allows sup-pliers to locate at a considerable distance from the purchaser, if by so doing they can minimize labor costs and other expenses. Consequently, horizontal interaction between different groups of workers is weak not only within firms and plants but between them, though interplant relation-ships are at least flexible. Since management does not extend between firms, production chains within the social division of labor are therefore unorganized and wholly exposed to market regulation.

So familiar are many of these characteristics that it might seem strange to comment on apparently obvious things like the fact that suppliers tend to make bulk deliveries to warehouses. But as we shall see, this is by no means the only way industrial capital need operate, nor a feature of nor-mal capital. We are also accustomed to seeing the JIC type of organization presented as an ideal realization of the capitalist logic of time economy, in combining extreme specialization of tasks, long runs, and maximum util-ization of machines, and with its moving lines delivering materials to workers at speeds determined by management. In practice, however, the JIC system has many problems, often mutually reinforcing:

- The system is geared towards uniformity and standardization, and is inflexible and unresponsive to changes in the market. This problem increases in slumps where the fact that production is geared to maxi-mum speed and output rather than to demand levels becomes more un-satisfactory. Exacerbating this lack of responsiveness is the preference for full pre-planning of the production process rather than learning-by-doing.
- For complex products like cars or televisions it is difficult to balance the various flows of parts and sub-assemblies into the main assembly process without gluts and shortages arising. Myopic concern with speeding up *individual* machines or lines and *individual* workers' actions allows the creation of serious imbalances. Any attempt to introduce product diversification heightens these problems, or requires expensive information systems to monitor processes.
- Large inventories and buffer stocks are expensive in terms of interest charges, storage and monitoring costs, and wastage when specifica-

---

[1] Although buyer-supplier relations for the sourcing of very specialized inputs are often very durable even in JIC systems.

tions change (Estall, 1985; Hay, 1988; Williams *et al.*, 1989). One of the most striking things about factories organized in this way is the large volumes of capital tied up at any one time in stocks which are needed to support a single flow line. On average, it has been estimated that materials or parts are only being worked upon for 5% of the time they are in the factory (Ballance and Sinclair, 1983, p. 148), and that 30% of production costs in industry go on warehousing, carrying, and monitoring inventory (*Business Week*, May 14, 1985). Similarly, overhead costs – including a large element of information transaction costs – make up an estimated 35% of total manufacturing costs in the US, though only 26% in Japan (Miller and Vollman, 1985).

- Rejects and the effects of other production problems tend to be concealed by the presence of buffer stocks. The possibility of using a part from the buffer stock instead of the defective one, and the imperative of keeping the line moving lowers the priority given to dealing with the source of the problem, as does the use of rectification lines. Also the time lapse and the physical distance between the discovery of the reject and its source conceals the origin of the fault, often a supplier.

- "Testing quality in" is far more expensive than building it in from the first. With testing, significant quantities of labor and materials are squandered on producing rejects and then on identifying and rectifying problems. This means quality control departments are non-productive; their very existence is an acknowledgement of waste, and an excuse for not ensuring that work is done right the first time.

- Distant and non-interventionist relations with suppliers have disadvantages: large, infrequent deliveries imply heavy warehousing and related costs; permissive attitudes to suppliers' quality control increase costs; and arms-length relations fail to encourage harmonization between supplier and customer, for example, in the design of components.

- As noted, the JIC system requires a deep vertical hierarchy of control to co-ordinate different tasks, given that each worker generally only knows a single, specialized, and often deskilled, job. This in turn is both cause and effect of numerous but rigid demarcation lines, highly complex pay structures, and an overweight bureaucracy. Attempts to push specialization to the extreme, eliminating all overlap in activities or competition between internal divisions – sometimes termed "hyper-rationality" – has frequently been found to be false economy as it encourages empire-building, poor coordination between activities, inflexibility, and a lack of competitive stimulus (Peters and Waterman, 1985; Aoki, 1985).

- Restricting workers to single tasks underuses their abilities, reduces motivation, increases boredom and hence fatigue, absenteeism, soldiering, and resistance. (This is not to imply that flexibility necessarily solves all these problems.) JIC suffers from many of the deficiencies of Taylorism: rapid turnover of labor encourages deskilling; management's withdrawal of trust and responsibility from labor tends to generate behavior which justifies that withdrawal. Separating workers by buffer stocks conceals real interdependencies between them and this reduces feedback and cooperation between workers. (Again, production workers need not worry about quality if that is the job of a separate department.) Demarcation also inhibits flexibility in restructuring because it freezes a particular technical division of labor and makes it difficult for capital to reduce the number of workers on particular tasks or to re-design and re-allocate jobs.

- Several features combine to inhibit innovation. "Getting metal out the door", coupled with the priority of quantity over quality, inhibits process development, thereby missing possible economies; dedicated machinery and low-skilled workers limit adaptability; excessive division and separation between related groups creates scope for coordination problems and increases innovation lead times.[2] The division between engineering and manufacturing is especially significant in this respect, not least because it presupposes an ability to pre-plan and engineer a perfect system before work begins, an assumption which implies a corresponding disregard of the value of learning-by-doing.

- Poor coordination, poor quality, low skills, and restricted shopfloor discretion mean that it is common for the line to go down for long periods. Thus, notwithstanding the apparent rationality of the system, considerable wastage is almost inevitable. For example, in 1979 at British Leyland's plant at Longbridge, the line only worked for 65% of the time due to hold-ups for materials and breakdowns (*Guardian*, May 16, 1984; see also Shaiken, 1984). Consequently, despite the "flow-line" tag frequently used to describe the JIC system, it has proved far from fluid, but thoroughly rigid and mechanical.

In short, the JIC system cannot be seen as the end-point of capitalist rationalization of the labor process. While it achieved a higher labor productivity than earlier forms of production, it nonetheless came up against a series of limits: hyper-rationalization of one part of the system may come at the expense of overall efficiency; maximizing current output is not sufficient in a changing environment; efficiency of the factory may come at the expense of efficient linkage to suppliers; degradation of common labor can cause withdrawal of worker interest and intelligence; cost minimization can be counter-productive; and so forth. These limits could be overcome, as it turned out, but that revolution in mass production occurred largely in Japan.

Appreciation of JIC's problems increased in the late 1970s, when, after a decade of falling productivity growth in Europe and North America, the

---

[2] Such problems are partly due to a failure to overcome longstanding divisions between craft and other workers, for which capital was not wholly responsible.

export performance and extraordinarily rapid growth of Japanese manufacturers, particularly in cars and consumer electronics, began to arouse
interest in the west. Initial accusations of dumping held little water, and it
was found that Japanese car firms had a $1,900 cost advantage per car
over American firms in the early 1980s. The Japanese firms had markedly
higher productivity, were more flexible in response to market opportunities, and were more innovative in product design. Most extraordinary of all
was the realization that the higher productivity was, in general, not due to
greater and more sophisticated equipment; indeed, Japanese workers were
often using smaller amounts of machinery and less sophisticated technology
than their American counterparts. In addition to well-publicized aspects of
Japanese manufacturing, such as quality circles and lifetime employment,
some startling features emerged, such as the practice of assembling completely different models on the same line. Gradually, it became clear that
what was involved was not simply a few cultural peculiarities but an
innovative and highly integrated system of production organization.

## The Just-in-Time System

"Just-in-time", like "JIC", is shorthand for a group of related practices.
This system emerged in the post-war period as Japanese car manufacturers
– particularly Toyota – attempted to adapt US practices to Japanese
conditions. Strictly speaking, JIT refers narrowly to a way of organizing
the immediate manufacturing labor process and buyer-supplier relationships between firms, but it is normally surrounded and supported by a
wider set of practices regarding skills, labor-management relations, and
labor market conditions (we discuss these later).[3]

Just-in-time is first and foremost a novel form of integrating the parts of
a manufacturing system, involving an approach to time economy different
than that of JIC. It is literally a system in which tasks are done just when
needed, in just the amount required to meet desired output levels. If tasks
are executed before required, waste in the form of idle capital – particularly circulating capital – is created. Ideally, all machines, even the most
expensive, should be run no faster than the speed necessary to produce the
required output. Although this means that utilization of an individual
machine may not be maximized, this does not matter for eventually a
machine that is running too fast must be switched off, and both the excess
inputs it has consumed ahead of time and the idle output have to be paid
for. Fixed costs are not absorbed any more quickly by running machines
faster than demand requires, for costs are recovered only by what is sold.
In other words, the question is not how many people are needed to make a

---

[3] The following account is based primarily on Hay (1988), Monden (1983), Schonberger (1982),
and Suzaki (1985).

line run as fast as possible, but rather how fast the line has to run to meet demand, and how many people are needed to achieve this (Hay, 1988, pp. 41–42).

The most startling feature of the whole system is the one which gives rise to the just-in-time label. Instead of pushing production through at maximum speed in long runs, each operation is done just-in-time to meet projected orders. Buffer stocks are very small and are only replenished to replace parts removed downstream. Workers at the end of the line are given output instructions on the basis of short-term order forecasts, and they instruct the workers immediately upstream to produce the parts they will need just-in-time, and those workers in turn instruct workers upstream to produce just-in-time, and so on. Sometimes the instructions are communicated by means of tags or boards called *kanban* (hence the "*kanban* system") which are passed to the worker upstream when required. In practice, this means that buffers between workers are extremely small. In short, it is a pull rather than a push system.

By this simple method, the *kanban* system increases responsiveness to market changes while markedly reducing planning, information-handling, and supervision costs and increasing the utilization of capital (Ohno, 1982). According to Suzaki (1985), Japanese car firms only require one-fifth of the indirect workers needed by US firms. *Kanban* works more cheaply and effectively than expensive computerized production control systems, known as Materials Requirement Planning (MRP) systems, in which workers must update information continually and, since they operate on a "push" basis, are liable to cause pile-ups when something goes wrong. However, MRP can cope with larger output variations than *kanban* and it is the former which Nissan uses most.

Notwithstanding the advantages of this pull system, JIT is not primarily a demand-response system in which actual demand drives production without the need for pre-production planning. Short- to medium-term planning is still vital and, for reasons which will be given below, fluctuations in actual demand (an unwelcome kind of flexibility) need to be smoothed out as much as possible. Rather, JIT is primarily a way of achieving a smoother, better-integrated production process.

JIT systems undermine many of the principles of just-in-case time economy. As noted above, an important element of JIC is that long setup times of machinery justify output in long runs. But this can produce mismatches between the output of parts and the demand for them downstream, so reduction of machine setup times is a key target in JIT systems. Achieving this is greatly assisted by the close relationship common in Japan between management, process engineers, and shopfloor workers. Sometimes firms will use several small, simple machines rather than one large, highly sophisticated one in order to give more flexibility. While CNC machines and the like can reduce setup times, quite low-tech, non-computerized modifications will often do the job (Schonberger, 1982).

Buffers are regarded as evidence of waste – wasted labor (in producing more than is needed at a given time), waste through imbalance between workers and between processes, production problems, "idle time, surplus workers, excessive equipment capacity and insufficient preventive maintenance" (Sugimori *et al.*, 1977). Therefore, key targets of management are to reduce buffer stocks towards zero and to eliminate errors and rejects in each task. In fact these targets are related. The smaller the buffer, the more sensitive the system is to error, and hence the more visibile the source of the error and the greater the incentive to remedy it and ensure that it does not happen again. Conversely, the more quality is "built in", the less the need for buffers and the more responsive the system. To this end, machines are designed to stop automatically as soon as a defective part is produced. By contrast, in JIC systems the time lag between the occurrence of errors and their detection is so large that their sources are very difficult to trace.

To be most effective, this process of error elimination requires both management and process engineers to be highly knowledgeable about the details of the work process, to keep work under close surveillance, and to be able to elicit the workers' own knowledge of the process in order to improve it. Managers and supervisors also need to give individual workers rapid and clear feedback on their performance and to train them to get quality right before going for quantity. It is not for nothing that JIT's full title is sometimes given as the Just-in-Time/Total Quality System: quality is essential.

Accordingly, establishing a new production facility is often not seen as a matter of setting up a standardized line as quickly as possible and then "chasing volume". Rather it is a matter of building it up slowly, making piecemeal improvements in the process. The skills of the workers, particularly the behavioral skills of cooperativeness and self-discipline, have also to be developed as part and parcel of this process, and this has the effect of making the firm less mobile in its investments than JIC firms where workers are seen as standard, substitutable, low-skilled inputs into fairly standard production setups.

Obviously, JIT needs a more cooperative, involved, and adaptable workforce than is expected in the JIC system. Workers are expected to do their own regular preventive maintenance and to take some responsibility for remedial action should a problem arise. Where appropriate, workers are trained to switch between jobs as and where needed, on their own discretion, helping fellow workers who are overloaded. This also exposes those jobs which do not need one worker's whole attention, thereby highlighting further room for intensification. Often workers carry out a fixed cycle of different operations: Suzaki (1985) reports the case of a Toyota worker who handled 35 different production processes in a cycle lasting eight minutes and 26 seconds (plus or minus two seconds) and who walked six miles a day in the process! To facilitate each worker's ability to execute a range of different tasks and reduce idle time, U-shaped lines are often preferred to linear layouts.

The reduction of buffers therefore not only keeps capital more active, but stimulates a continual learning process. Ideally, the production process never becomes entirely standardized and learning curve economies continue long after those of orthodox firms have leveled off. For the latter, the only way to further lower costs, aside from more extensive automation, may be to shift production to a location with cheaper labor. It is for these reasons that JIT is not simply a low-inventory system of production as some commentators have implied (e.g., Estall, 1985). It is a particular and sophisticated method of learning-by-doing. And this is perhaps the primary reason why Japanese firms have had so much success in competing with established western firms which had treated their industries as mature and saw relocation to cheap labor countries as the only way of improving competitiveness. Hence, when the learning curves of western firms in consumer electronics, air conditioners, cars, and office machinery were thought to have reached a plateau, those of the Japanese continued to improve (Rosenbloom and Abernathy, 1982; Abernathy *et al.*, 1983).

The kinds of machinery used need not be very different from those used in JIC systems. Often quite simple modifications, rather than microprocessors and robots, can dramatically reduce set-up times and increase flexibility of machinery. But the layout of the factory must be changed radically from clustering machines with the same function (e.g., lathes) to grouping equipment in line with material flow (Suzaki, 1985; Schoenberger, 1988). Consequently, parts and materials do not need to be stored in batches and moved by expensive fork-lift trucks or the like, but can be moved by simple chutes or trolleys. Thus, as we shall see, the different time economies of JIC and JIT imply different ways of economizing in the use of space.

Advanced automation, such as flexible manufacturing systems, has the greatest effects on productivity where it is applied to a production system that is already rationally organized; otherwise it is likely to perpetuate the inefficiencies of the old technology and working practices, as has been found with many major new technologies, including computer-integrated manufacturing and office automation (Riley, 1985; Sayer, 1985; Schonberger, 1987). The JIT system, and more generally the involvement of managers, process engineers, and workers in continual study of the work, suggest that the ground will be particularly well prepared in leading Japanese firms. In this light, it is interesting to note that Japan is the only OECD country to have sustained an increase in (so-called) capital productivity over the past decade (Lipietz, 1987).

Capital can get most benefit from the JIT system when suppliers also adopt it, so the superior integration of the division of labor can extend beyond individual workplaces to inter-firm relations and indeed to entire production systems. In fact, just-in-time production is greatly assisted by just-in-time deliveries. Orders to suppliers and subcontractors are small and frequent; indeed, deliveries may be made several times a day. Ideally, supplies are delivered not to warehouses but to the precise location within the workplace where they are needed, thus reducing the need for expensive materials handling equipment and storage space. Also, it drastically reduces the labor needed to receive supplies, to store and keep track of them, to authorize their release from the warehouse and to move them to their place of use; at the extreme, the warehouse should not be needed.

For a single firm or workplace, the just-in-time principle applies not only with respect to deliveries of inputs but to outputs. If a firm has to deliver to a customer infrequently and in large batches (even though the customer may use the supplies on a day-to-day basis), then changes in the pattern of demand are liable to be large and lumpy rather than smoothly incremental. JIT sacrifices the protective flexibility of large stocks, which enable firms to ride out demand shocks, but it benefits from a constantly, or only marginally, changing daily or weekly output.

This highlights the need for smoothing the work pattern as much as possible, for a taut, pull system such as this is obviously poorly protected against bottlenecks. And unlike push systems, sharp drops in demand produce not gluts but idle capital and labor – a different kind of waste. If workers upstream of final assembly are to be fully and continuously occupied, then it is important that the level of hourly and daily output be as constant as possible. So a smoothed output is both a condition and an effect of deploying workers on fixed cycles of tasks, otherwise the rule that workers must not act unless so instructed from downstream would result in enforced periods of zero productivity. Therefore, although it is possible to deal with minor fluctuations in volume (plus or minus 10%) at short notice, output has to be smooth enough, and production planning accurate enough, to minimize both the number of workers and the size of work-in-progress inventories without risking holdups (Sugimori *et al.*, 1977; Monden, 1981, 1983; Hall, 1982; Aggarwal, 1985).

So JIT, while more fluid than JIC, has its own kinds of rigidity in terms of the need to repeat standardized cycles of tasks. Thus, contrary to the impression given in the literature on flexibility, one of the first tasks facing a firm wishing to institute JIT is to ensure that deliveries of inputs are not only just-in-time but are as constant in content as possible (Hay, 1988).[4] This is quite different from the situation of retailers, who can demand flexible, rather than constant, inputs on a just-in-time basis, for their lack of a production process means they escape the constraints of JIT production. For this reason it is quite misleading to generalize from retailing to manufacturing (e.g., Murray, 1988).

One of the biggest problems of integrating any complex labor process derives from the inevitable inequality of cycle times of different operations, i.e., the time it takes to complete each task. If one task which takes 60 seconds to complete requires an operation upstream which takes 80

---

[4] Hay cites the example of Xerox, where the first step toward JIT was to persuade suppliers to deliver 20 days of repeated orders, not day-to-day flexibility.

seconds, and another which takes 90 seconds, then their coordination is liable to be poor unless the proportions of workers assigned to each operation are very precisely calculated, especially where protection from buffers is minimal. Such problems bring home the necessity of flexible working practices, and of arranging the detailed layout of the factory so workers can move quickly between jobs. In other words, smooth coordination of a JIT system requires increased attention to the micro-geography of the factory: indeed, there is no area of modern life where time geography and time-space coupling constraints are subject to such scrutiny. It also brings home the need for learning-by-doing for perfecting these arrangements, for it is inconceivable that the practical constraints and opportunities specific to every task could be anticipated before production begins. These coordination problems occur in JIC systems, too, but there the need to smooth production and to optimize layout or space economy is reduced by buffers, for these allow – at a cost – the temporal and spatial disarticulation of interdependent processes. Closer articulation creates big problems in production engineering but brings considerable benefits in terms of faster turnover of capital.

Earlier we mentioned the practice of processing different models on the same production line and in a mixed sequence rather than in homogeneous batches. Whereas this would be a nuisance under the JIC system, it is a necessity under JIT. To illustrate, assume that a firm makes just three different models, A, B, and C. If a firm using JIT were to make As in the morning, Bs in the afternoon, and Cs in the evening, it would have disruptive consequences for suppliers who must make just-in-time deliveries since, for example, a supplier of parts for model A would get an irregular work pattern which would make for inefficiency. If, however, the main firm mixes its sequencing of models as much as possible, within the constraints of demand patterns for different models and the time required for different operations, there will be less tendency for output irregularities to get amplified upstream. Suppliers will be faced with a smoother demand schedule, and output of different models will always match the pattern of final demand rather than being dictated by a need to produce in large, homogeneous batches.

Japanese manufacturers tend to make greater (and increasing) use of subcontractors than their western counterparts, and major companies usually draw upon several layers of subcontractors (cf. Ikeda, 1979; Sheard, 1983; Aoki, 1985). (We shall have more to say about this in the next chapter.) Subcontractor firms may also use JIT and total quality control so that it extends across a system of linked plants, and this is obviously facilitated by geographical proximity between firms and their suppliers and hence by industrial agglomeration. The most extreme example of such a localized, complex, multi-layered production system is Toyota City on the edge of Nagoya, where most of Toyota's three million vehicles per year are made. In such cases the same JIT principles operate between firms as operate within them. Relationships between buyer and supplier firms are accordingly much closer than is usual in the west, with the main buyer firms exercising considerable influence not only over pricing, but over product development and the day-to-day internal functioning of the supplier, in order to ensure security of quality and supply. Sometimes one firm may "lend" some of its workers and managers to another firm with which it works.

In other words, where the JIC system relies upon large stocks to provide security of supply, and upon competitive and sometimes multiple sourcing, the JIT system relies upon close management surveillance, cooperation, and overlapping ownership. This system is said to enjoy the organizational advantages of vertical integration without the financial obligations (Altshuler et al., 1984).

Foreign visitors to Japan rarely see the smaller firms which do subcontract work for the major companies, and until recently many accounts of the Japanese industrial phenomenon ignored them (but see Dore, 1987; Yamamura and Yasuba, 1987). Attention has rightly been drawn to the differences in pay and conditions between the two sectors: lifetime employment is absent in the subcontract sector; the lower the "layer" of subcontract work, the greater the proportion of female workers and the lower the pay – indeed up to 40% lower than in the major firms (Guardian, May 11, 1982). This is obviously to the advantage of the large firms; for example, components purchased by Nissan have been estimated to be 30% cheaper than those bought by British Leyland and Ford (Sunday Times, May 23, 1982). Some observers have attributed Japanese industry's success to this hidden pool of cheap labor, but while it obviously contributes to competitive advantage, it does not account for the markedly higher output per worker that Japanese firms frequently have over their overseas rivals.

To summarize then, whereas JIC is a system of mass production based on a collection of large-lot production processes, separated by large buffers and feeding into a final assembly line, JIT is a system of mass production consisting of a highly integrated series of small-lot production processes. Further, JIT is a learning system which generates economies by making fabrication and assembly more closely approximate a continuous flow line, by reducing the amounts of machinery, materials, or labor power which are at any time inactive or not contributing to the production of saleable (i.e., adequate quality) output. So with JIT, economies do not follow simply from major technological developments, nor from simple speed-up of individual tasks, but from a different way of organizing the labor process, coupled with piecemeal changes to machinery.

Adoption of JIT produces dramatic economies. Aggarwal (1985) claims that Japanese firms which have used JIT for five years report a 30% increase in productivity, a 60% reduction in inventories, a 90% improvement in quality acceptance rates, and a 15% reduction in floorspace use. Williams et al. (1989) argue that western accountants and management have tended to be obsessed with the utilization of fixed capital and labor, while paying little attention to the turnover of working capital, stocks being regarded as an asset rather than a liability, whereas their Japanese equivalents are more concerned with the turnover of working capital.[5] And whereas manufacturing inventory levels in other industrializied countries have remained steady or fallen only slightly over the last thirty years, Japanese levels have been halved.

We have seen that JIT has a remarkable unity, its various facets providing mutual reinforcement. Yet in its full form, as described above, JIT lends itself only to mass production, because small batch or project production, with their more irregular activities and demand, offer too little scope for the necessary fine tuning of repetitive tasks (although as Hay (1988) points out, some of the basic principles are applicable in these other kinds of production). Full JIT production, therefore, can only be used by a minority of manufacturers, for – contrary to the impression given in much of the Marxist literature – most manufacturing employment is in small batch and project production.[6]

However, JIT depends on a set of conditions which are much more pervasive and can be applied to a variety of types of labor process. These include multiskilling, flexible working, and job rotation (facilitated by simple payment systems), low turnover of managers and key workers, close involvement of managers and process engineers on the shopfloor, zero defect or total quality control, quality circles (not only for reducing waste due to defects but for securing active involvement and for rationalizing the labor process), and strong horizontal links between interdependent groups within and between firms (Aoki, 1988). Obviously some of these conditions already exist outside Japan. While JIT is unlikely to work without these conditions the converse does not apply. Firms may make significant economies by achieving these conditions and, given their complementary character, the gains to capital will be greatest where they are introduced together; for example, attempts to get active consent and flexible working will be impeded if there is a lack of job security and complicated payment systems based on a rate for each job.

While we have focused on the material organization of production, we do not wish to deny the effects of culture, such as attention to interpersonal relations, dislike of error that discommodes another, nationalistic solidarity, etc. Nevertheless, there are some common dangers in explanations which stress these factors: "culture" can easily become a "dustbin cate-

---

[5] Williams *et al.* (1989) cite a British survey which found that 50% of respondents did not know their warehousing costs, and most drastically underestimated the costs of holding stock.
[6] Kaplinsky (1984, p. 131) estimates that about 75% of manufacturing output is produced in batches of fewer than 100; Ballance and Sinclair (1983, p. 150) estimate that 80% of all manufactures are made in lots of fewer than 15 units; Littler (1985) estimates that only 1.4 million of Britain's 20.4 million employed work in factories concerned with mass production and only about 0.7 million on direct mass production lines.

gory" for anything we can't explain; further, culture is often misrepresented as something ethereal and eternal, divorced from historical material practice, or misconstrued as a self-perpetuating tradition that determines contemporary actions. We are not qualified to make specific claims on just how cultural practices affect production, but we would insist that those who make such claims demonstrate exactly how cultural characteristics translate into high productivity, for productivity depends not just on individual attitudes or even on types of authority relations, but on specific forms of material social organization which make these qualities yield economic results.[7]

## Just-in-Case, Just-in-Time, and Fordism

We can now explain why we prefer the term "just-in-case" to "Fordism" and make that effort because, as we show in the next chapter, the many different uses of Fordism provide fertile ground for confusion. Moreover, Fordism is neither wholly in conformity with JIC nor wholly opposite to JIT. This can be illustrated by relating JIT to the innovations of Henry Ford himself.

As Hounshell's (1984) authoritative study shows, Ford's innovations were not limited to the marriage of Taylorism with a moving line for mass production. Ford's system was primarily about rationalizing work flow, though this applied only to assembly. This required the elimination of "fitting", and its makeshift use of poorly-made parts, and the introduction of truly standardized parts for the first time (compare JIT's emphasis on total quality control). It ended the isolation of workstations operated by craft workers working at their own pace (compare JIT's replacement of maximizing speed of individual operations with regulation according to output requirements). Ford tried to bring the workers into a more systemic relation, rationalize the steps, improve work flow, and produce a standard quality for a good price. The assembly line followed from this, and the moving line thereafter. Arguably, then, there are some analogies with JIT, and it is significant that Toyota's production handbook has a major section devoted to Ford, for Ford himself aimed for the kind of flowline principle that JIT achieves.

However, Ford made only limited progress in countering the problems of coordination created by Taylorist specialization and deskilling. And upstream of assembly, different jobs were still poorly integrated; in the 1920s, for example, right-hand door panels would be stamped in runs of half a million and stored until needed (Cusumano, 1985, p. 270). Contrary

---

[7] At the same time, it is important to guard against a certain tendency to dismiss cultural explanations by "rationalists" on the spurious grounds that culture is not susceptible to science, whereas "rational choice" is. Whether something is a cause of certain social phenomena has nothing to do with whether it fits our preferences as to the character of science.

to the literature on Fordism, the resulting problems of extreme rigidity and massive inventories surfaced not in the 1960s and 70s, but in the late 1920s when Ford was overtaken by General Motors. GM used smaller lot production in body stamping and final assembly, designed machinery to be adaptable for new models, and relied more on product innovation and differentiation in their market strategy ("keeping the customer dissatisfied"). GM could therefore restructure far more quickly than Ford; while Ford changed from extreme Fordism, it never caught up again (Hounshell, 1984).[8] Yet by modern standards GM's methods are still rigid and lacking in integration and until recently GM has been widely regarded as a classic example of the JIC system.

The relationships of JIT and JIC to Fordism therefore involve neither simple contrast or equivalence. Failure to recognize this has caused considerable confusion.

## Origins of the just-in-time system

An important factor in the development of Japanese manufacturing was the extent to which it escaped the influence of Taylorism. In practice, Taylorism can be broken down into separable elements, and it was chiefly its emphasis on the study of work practices that made a lasting impact on Japanese manufacturing. Littler (1982) argues that deskilling and specialization were limited because the work groups (oyakata) used in early forms of labor process organization based on contracting, proved, unlike those in Europe and the United States, resistant to fragmentation and were generally incorporated rather than broken up by employers. This, together with the prevalence of family ties within and between firms made group working and inter-company collaboration more congenial to capitalists than in the west.

Nevertheless, after the Second World War, some aspects of western methods were widely imitated, under the encouragement of the Japanese Productivity Centre (Cole, 1971, p. 55). An early example of this – in retrospect ironic – was Nissan's licensing from Austin of Britain to learn advanced production techniques in the 1950s (Haruo, 1981). Quality circles were actually another import, introduced by US business advisors during the occupation; in this case, the Japanese took them more seriously than their American counterparts, and in the long run the adapted version proved more successful than the original. Early forms of JIT, or the "supermarket principle", had been tried in the US aircraft industry before the war, though with little success. Moreover, the methods used by the Japanese for making machinery more flexible were not unknown to US manufacturers, but the latter ignored them.

---

[8] Unfortunately, a fixation with Ford's methods and a pervasive blindness to the role of product innovation in capitalist competition has allowed this development to escape the notice of most analyses of capitalist manufacturing (see chapter 2).

In Japan, JIT and its supporting conditions emerged as a way of making a virtue out of necessity. The tiny size of the Japanese domestic market after the war ruled out the possibility of pursuing economies of scale on JIC lines – for example, an entire year's output of the Japanese automobile industry equalled only one and a half days of production in the United States. In addition, expensive resources and limited space made a low-waste, low-stock system imperative. According to Cusumano, there were also important limitations on the amount of capital available for inventories and storage facilities; the form that time economies took was shaped by the need for space economy. Taiichi Ohno, the Toyota production engineer most responsible for the development of JIT, began limited experiments in 1948 and gradually extended them through the company; by the mid-1950s the first suppliers had begun to adopt JIT (Cusumano, 1985). By the early 1970s, productivity had overtaken levels elsewhere and the assault on export markets had begun.

So, unlike neo-Fordism, JIT arose neither from the problems facing Fordism in the recession of the late 1970s and early 1980s, nor from the possibilities offered by microelectronics, but as an attempt to adapt western manufacturing practices to the context of post-war Japan. Foreign commentators – often reluctant to concede that the Japanese have any significant innovatory capability – have sometimes described the development of JIT simply as an application of the US-originated supermarket principle. However, this does no justice to the twenty years of development of JIT, and ignores the failure of US manufacturers to develop it.

Many of these developments took place in a turbulent context as regards labor organization; this was certainly not simply a story of management learning imposed on a compliant workforce. Wartime defeat and post-war US reforms brought into being a whole new cadre of managers and capitalists. In the late 1940s and early 1950s, Japanese trade unions were under strong left-wing control and posed a serious threat to the reconstruction of Japanese capitalism. There was a series of strikes, culminating in the decisive Nissan strike and lockout in 1953 (Cusumano, 1985; Gordon, 1985; Kenney and Florida, 1988).

Among the issues in these strikes were resistance to the new production methods introduced at Toyota, and demands for greater employment security and a reduction in pay differentials. Resistance at Toyota was quelled rather more easily than elsewhere, largely because it was located in a rural area where workers had no previous experience of industrialization and no alternative sources of employment. Nissan, by contrast, had factories dispersed throughout the Tokyo-Yokohama conurbation and a more militant workforce. What most alarmed the capitalist class was the prospect of a different kind of organizational innovation – the attempt by Nissan trade unionists to introduce western-style shopfloor committees. While the story is complex, it is worth noting that lifetime employment was among the concessions made by Nissan at this time in order to buy off key workers: it was not some eternal cultural characteristic.[9] And it was the breaking of this 1953 strike in particular which paved the way for the displacement of the industrial unions by supine company unions which avidly collaborated in management initiatives. Meanwhile, having weathered the storm (and without needing to make so many concessions to workers), Toyota pressed on with the development of JIT and slowly other firms began to adopt the system (Cusumano, 1985).

Thus, the development of the just-in-time system is a good example of the importance of place-specific practices in industrial innovation (Storper and Walker, 1989). This rests on the necessity of practical activity and experience for developing workable methods, which takes time; working out solutions to specific problems facing people in one country or even one region; the collision of outside ideas and technology with local conditions, traditions, and problems; and a degree of distance from dominant centers which allows new approaches to be nurtured. The JIC system, developed most fully in the US, adapted its own internal economy of time and space within production to that of its great continent. In Japan, a different kind of economy of time and space in the organization of production was constructed, also well adapted to its context but proving superior to established methods.

## Diffusion of just-in-time

The uneven development of the social organization of production is reflected in differences in productivity and profitability between firms in different countries. As new production methods and forms of organization have developed, the leadership in productivity has passed between major capitalist countries.[10] In 1937 it took nine times as many workers to produce a car in Japan as in the United States (Ohno, 1982).

While it is not yet clear how far particular elements of the Japanese manufacturing system can be detached from the whole and still yield an advantage, the fact that they are diffusing and being imitated elsewhere suggests that they are not as firmly rooted in features unique to Japan as many commentators imply (Mair *et al.*, 1988). Direct diffusion of JIT by Japanese firms has been slow. One obvious reason why major Japanese firms have been reluctant multinationals is that competitive advantage in Japan derives from unique social relations in production and favorable relationships with the state and financial institutions. Another derives from the nature of the JIT system itself with its need for close working relationships with, and proximity to, suppliers, and a long period of incubation. While localized complexes of firms might be set up abroad, this is inevit-

---

[9] The *nenko* or lifetime employment systems also had important effects through making firms take more seriously the need for retraining and the need for long-term strategies.

[10] See Storper and Walker (1989) for a general analysis of this phenomenon.

ably more difficult than setting up a single overseas branch plant on JIC lines (Morris, 1989).

Particularly in mass-produced goods, Japanese firms have been able to undercut and outsell overseas producers through exports, even with high transport costs. Many of the usual gains of foreign direct investment are likely to be offset by losing the productivity advantages enjoyed in Japan. Consequently, the Japanese have made less use of the "runaway industry" strategy of setting up export-oriented manufacturing plants in Third World countries than have American firms, although the rising value of the yen is beginning to change this. Generally, they have preferred to forego the once-and-for-all advantages of cheap labor locations for the longer-term advantages of developing automated production at home (Sayer, 1986). (However, productivity levels are still likely to be higher than for local companies in the host countries.)

On those occasions where Japanese firms have invested in production in other advanced countries, the overriding reasons have been to pre-empt protectionism, escape problems created by the rising yen and labor short-ages at home, and find outlets for the explosion of surplus capital – a process facilitated by the growing financial integration of Japan with the US and British markets. So far only a few such firms, such as Honda in Marysville, Ohio, have been able to establish anything more than a watered-down version of Japanese practice (*Fortune*, June 15, 1987). This is not only because of unfamiliarity, and resistance from workers and managers, though these are important, but also because it takes time to build up a network of local suppliers to provide JIT deliveries. This is especially difficult when, as is often the case, host country suppliers cannot match Japanese ones for quality and cost. Without high quality, reliable local suppliers, a strict and comprehensive JIT system cannot be operated, although quality control and piecemeal process improvements can be sought through other means, without the discipline of JIT. In general, Japanese overseas mass production plants are obliged to compro-mise between JIT and JIC, in their relationships with both suppliers and workers. As might be expected, diffusion is also limited by cultural bar-riers, such as the US resistance to paternalism and close inter-company ties, by union opposition to new forms of contracts, and – often under-estimated – by management resistance (Holmes, 1987; but see Morgan and Sayer, 1985; Brown and Reich, 1989).

Imitation by western firms is also widely reported. Obviously, this can reflect recessionary pressures which both force them to restructure and ease the process by weakening labor resistance. Nevertheless, the specific form of reorganization often seems to owe much to Japanese influences – as one would expect, given their productivity lead. It is also clear that European and North American firms are combining this social reorganiza-tion with investment in new technology, often with convergent effects on flexibility and inventories.

We have argued that the supporting conditions of JIT are much more pervasive than JIT itself, and indeed features such as flexible working and quality circles have been developing in the west among leading firms, and some, like IBM and Hewlett-Packard, have developed similar measures independently of the Japanese. In Britain, while as yet there are probably very few mature cases of just-in-time (Turnbull, 1986), the supporting conditions are now widely sought by indigenous and foreign firms alike (Morgan and Sayer, 1988).

There has been considerable hyperbole in the business world regarding quality circles, JIT, etc., and frequently these are extremely limited or passing fads, because managers erroneously believe they can simply be implemented immediately. But quality circles require lengthy planning and training of workers and supervisors to produce results (significantly, few Japanese firms in Britain have felt ready to start them yet; Morgan and Sayer, 1988). Similarly, one suspects that many US and British managers have interpreted JIT simply as lower inventories, or ensuring that normal deliveries are just-on-time rather than late, ignoring its wider functions and preconditions – suspicions compounded by the widespread practice of confusing JIT deliveries with JIT production; in fact, firms retaining JIC production can make JIT deliveries.

As they have with factory and office automation, many US firms are trying to implement JIT without first rationalizing their production processes (Miller and Vollman, 1985). Yet both JIT and quality circles must be tailored to suit individual plants. Clearly, we must remember the ambiguity of the injunction "beware of imitations". Indigenous western firms are also more likely to be hampered by precedent, existing commitments, and a lack of greenfield sites than Japanese inward investors. For this reason, and others already given, imitations are likely to be diluted.

One of the best-known examples of direct imitation is Ford's "After Japan" (AJ) strategy. In 1981, productivity at Ford's Halewood (UK) plant was reported to be half that of Ford at Saarlouis (Germany), where, in turn, it was reportedly a third that of Toyota (*Financial Times*, February 28, 1984).[11] Under the AJ strategy, inventories were reduced from a three-week supply to one week. Redundancies of workers in repair and maintenance (once one in five of European manufacturing workers) were eliminated by training production workers to do simple maintenance, and a third of the workforce was organized into quality circles. The hierarchy was squeezed by eliminating two supervisory levels and the number of different grades of workers drastically reduced, for example, from 12 grades of fitter to one or two. However, labor resistance has held up the implementation of this strategy in Britain, especially at Halewood where workers have demanded West German pay levels for matching West German productivity (*Financial Times*, Nov. 13, 1985; *Fortune*, Oct. 18, 1982). This

[11] Nevertheless, as Cusumano shows, such figures exaggerate the Japanese productivity lead.

particular example illustrates how vulnerable JIT is to industrial action, for in the 1987 strike, the whole of Ford Europe was brought to a standstill far more quickly than would have been possible hitherto.

Again in the automobile industry, the joint venture of General Motors and Toyota at Fremont, California is also interesting. The former GM plant at the site, which had a history of militancy and rigid demarcation (10–15 maintenance grades alone), was closed in 1982. The new joint venture operates a just-in-time system and some of its suppliers have moved to the area. It has no demarcation within maintenance; there are just three grades of skilled workers, and a single grade for assembly workers. The latter are organized into five- to seven-person teams covering a range of jobs and taking a number of responsibilities formerly the preserve of middle managers. To a significant degree, the UAW has had to accept these changes and GM is using the plant as a benchmark against which to compare other plants (*Fortune*, July 9, 1984; Altshuler *et al.*, 1984; Brown and Reich, 1989).

At another GM complex – "Buick City" at Flint, Michigan – Flexible Manufacturing Systems and JIT are reportedly combined and many suppliers are moving onto the industrial park (*Business Week*, May 14, 1984). However, they are unlikely to go to Toyota City extremes in their localization of production and to a certain extent JIT has been diluted in order to accommodate suppliers within a one-day trucking radius. The nature of GM's relationships with suppliers is also changing, with more interventionism and attempts to involve suppliers in design decisions so as to improve coordination (Altshuler *et al.*, 1984; Aggarwal, 1985; Estall, 1985).[12]

### Industrial clustering under just-in-time and just-in-case

The just-in-time system suggests local agglomerations of interacting plants and relatively immobile investment, as we have seen. However, the high degree of localization in Japan also reflects the poor infrastructure for freight movement and the mountainous terrain; another example of making a virtue out of necessity and of adapting time-space economies within production to the wider space economy (cf. Cole, 1971; Ikeda, 1979). There are, however, the beginnings of some local complexes of firms developing JIT methods outside Japan. In Northeast England, Nissan is bringing over some of its suppliers to locate near its Washington assembly plant, and similar moves are under way to serve the Honda plant

---

[12] Examples of JIT have been reported, outside the automobile industry as well. The computer firm Hewlett Packard has 11 of its 40 manufacturing divisions using JIT, sometimes combined with MRP. Besides reducing waste, these have enabled HP to produce a more flexible range of products, including a minicomputer that can be made in 6 million variants (*Business Week*, May 14, 1984). Just-in-time purchasing of components is also being introduced – and HP is now selling a new minicomputer especially designed to control MRP and JIT systems (*Computing*, Sept. 20, 1984).

in Ohio (for example, Asahi glass).[13] Nissan has also persuaded some British suppliers to establish bases close to its plant, only these are often warehouses – an example of just-in-case practices. Schonberger (1982) also cites several other examples of agglomeration in the US, particularly near the Kawasaki plant in Lincoln, Nebraska, which is relatively advanced in its development of just-in-time methods.[14]

At the international scale, the adoption of JIT principles may prompt a limited amount of "de-internationalization", with firms abandoning the policy of using far-flung overseas plants and centralizing activities in a few locations. One semiconductor firm found that the 14-day in-transit loop of partly processed chips meant that errors introduced in early stages of fabrication were slow to appear. As the reject rate is perhaps the most critical cost factor in the production of integrated circuits, they decided to co-locate fabrication with assembly to get more rapid feedback on quality defects and lower costs. Similarly, Apple Computer assembles its Macintosh personal computer in a highly automated JIT plant in Fremont, California, instead of shipping components out to Singapore for assembly, which, despite lower labor costs, proved expensive on inventories, inflexible, and difficult to coordinate (*Business Week*, May 14, 1984). Other American firms following suit are Motorola and IBM, which has adopted a policy of localizing suppliers around its main manufacturing plants, though it needs to maintain a manufacturing base in its major overseas markets and so this will not entail de-internationalization (*Business Week*, Oct. 3, 1983).

Perhaps the most important force working against re-centralization outside Japan is the tendency of large agglomerations of related industries to provide labor with a strong base for organization. The huge concentrations of Fiat workers in Turin are a good example (Murray, 1983). Looking back at the Japanese Toyota City syndrome, it must be said that such agglomerations depend on labor being weakly organized. For these reasons, adoption of the new work practices is most likely in greenfield sites, not merely to escape traditional and militant labor, but to make a fresh start with new management (Mair *et al.*, 1988).

We can now look at the taken-for-granted characteristics of JIC and its locational patterns in a new light. The development of highly decentralized, global production systems has often been seen purely as a consequence of the expansionary dynamic of capital, rather than as partly a consequence of the historically and spatially specific JIC form of organizing in production. In particular, it is now clear that large, infrequent deliveries and extensive warehousing were special features of this system which were actually sought after as sources of economies of scale in transport and

---

[13] Dan Jones, personal communication.
[14] For further information on Japanese plants in the USA see Mair *et al.* (1988) and Florida and Kenney (1990).

handling; yet the JIT system shows that these scale effects can be reduced, not only within plants but between them.

However, as many of these examples remind us, such developments are always set within a mass of different, often contending, forces, and the tendency towards re-centralization should not be exaggerated. For example, Toyota has invested in a Canadian plant for making wheels in order to get cheap aluminum; the wheels are then shipped back to Japan for assembly in familiar fashion (Holmes, 1987). Capital accumulation has never led to a single, universal type of spatial outcome, and there is no reason to believe that future capital accumulation will be any different, though non-random changes will surely be discernible.

### Just-in-time and labor

Given the characteristics of the just-in-time system and its supports we can summarize the implications for labor in firms adopting them outside Japan.[15] (Remember that while JIT is unlikely to be feasible without some of these supports, they can exist and aid productivity without full JIT organization of the labor process.)

- As output per worker is likely to increase markedly, employment of both direct and indirect production workers is likely to fall, other things (such as demand) being equal.
- Reducing the porosity of the working day, by eliminating idle time and requiring workers to switch continually between jobs, plus the internalization of disciplinary pressure within work groups or production teams, will increase the intensity of work and associated stress.
- The emphasis on flexibility implies multiskilling, the elimination of demarcation, and reduction or abolition of job descriptions. To facilitate this, only one union is likely to be recognized if unions are allowed at all, and bureaucratic systems of control will give way to more informal, diffuse and paternalistic methods.
- Individualized, discretionary merit payments instead of a rate-for-the-job are often introduced in order to facilitate flexibility.
- Flexibility means that workers are given much longer and more diverse on-the-job training so they can, for example, maintain and repair as well as operate machinery. Yet training in "behavioral skills" is probably more important than technical skills; i.e., the skills of collaborating with others and taking personal responsibility for work are given priority, for the worker is expected to exercise much more discretion than under Taylorist methods. This need for a more knowledgeable workforce (including management) is especially strong as JIT is par-

---

[15] This section collates findings from Morgan and Sayer (1985; 1988), Turnbull (1986), Brown and Reich (1989), and others.

ticularly vulnerable to disruption. The early labor process literature held, fallaciously, that control and knowledge of the production process are a zero-sum game between management and workers; in reality management also has to become more involved on the shopfloor.

- The need for long periods of on-the-job training, the greater number of acquired, company-specific skills, and the lack of employment reserves make it important for firms to minimize labor turnover and make much more use of overtime and the "internal labor market" than traditional JIC firms, which have high labor turnover and make heavy use of the external labor market. Fluctuations in demand are met by heavy use of overtime, among both primary and secondary workers, and of subcontract firms, as shock absorbers. While this may mean employment in the large firms is more stable, the instability of subcontractors is liable to increase. "Core" workers therefore have more job security than most traditional manufacturing workers, though they must be willing to be redeployed where management deems it necessary. However, lifetime employment applies only to a minority of workers (male only), and can be like a prison sentence because other firms are extremely reluctant to employ anyone who has left such a job (Kumazawa and Yamada, 1989).

- There is a greater realization that profitability can depend quite heavily on the performance of workers who are technically unskilled or semi-skilled, but behaviorally highly skilled. Unlike traditional firms, JIT organization defines core workers to include these as well as the usual professionals and technically skilled workers, and provides them with better employment security (cf. Morgan and Sayer, 1985).

- The need for low absenteeism and behaviorally skilled workers leads to very careful screening in order to ensure that recruits will have minimum distractions from the domestic sphere; in a patriarchal context, this means an even stronger preference for male workers than under a JIC system. JIT employers also avoid recruiting workers or managers who have acquired traditional habits which they would have to unlearn. This, and the emphasis on adaptability, biases them towards younger workers, often school leavers.

- As JIT is a "high trust" system, ways must be found of motivating workers to use their discretion in an acceptable manner, i.e., not to use their collective strength against capital. Individualized systems of communication between management and workers – as opposed to communicating through foremen or union representatives – or company unionism are therefore preferred. Much greater efforts are made to encourage identification with the company in its competitive struggle against other companies, and to this end more information on profitability, performance and competition is given to workers (Kamata, 1982; Ichiyo, 1984; Dore, 1987). Single status employment conditions (i.e., uniform conditions and benefits for all grades) within firms also help retain and motivate workers. However, the line between consent and coercion is thinly drawn, and it is not surprising that interpretations of Japanese work practices are polarized according to the political sympathies of the author. On top of this is the problem of explaining how consent is won with employment conditions as different as those of the primary (large company) and secondary (small company and subcontract) sectors in Japan.

- Hierarchies within factories are flattened by the elimination of certain lower and middle management jobs, particularly those involved in supervision, quality control and production planning, and regulation. Greater horizontal communication between interdependent groups of workers is encouraged (Aoki, 1988).
- The formation of work teams or quality circles acts both to reduce costs and motivate workers, although again the line between consent and coercion is unclear. In one sense, quality circles represent a reversal of Taylorist orthodoxy; and yet, as Ichiyo (1984, p. 46) observes, they could be said to involve the internalization of Taylorism by workers themselves in terms of systematic analysis and rationalization of the labor process.[16]

Japanese labor-management relations can be appallingly oppressive, as one Japanese journalist found when he was a contract worker on an assembly line in Toyota (Kamata, 1982). There are also reports of compulsory participation in quality circles and suggestions schemes (Ichiyo, 1984), and of workers foregoing holidays and working overtime without pay. Workers tend to have to operate more machines, and are allowed little rest time. Employers strongly control union activity and expect identification with the firm. Again, these are not simply the effects of management strategy, but must be understood in the context of the history and balance of class forces, of divisions within the work force, peer group pressure, company unionism, and Japanese concepts of the individual and authority relations.[17]

As to the question of whether there is anything potentially progressive in the new practices, we can only give an ambiguous answer. What we have treated as a single organizational form can obviously work out in different ways, and workers' reactions differ accordingly. Not surprisingly, they often prefer flexible working, with accessible management and some involvement in decision-making, to demarcation, deskilling, and highly bureaucratic organization (Morgan and Sayer, 1988; Brown and Reich, 1989).[18] Certainly, such judgements must be qualified; in particular, grant-

---

[16] One can understand why Schonberger (1982) says that the Japanese "out-Taylorize us all".

[17] For example, company unions – so often seen as indicators of managerial autocracy – were prefigured, ironically, by the activities of militant unions in the immediate post-war period, which sought to make them into factory soviets (Kenney and Florida, 1988).

[18] This more positive view is also borne out by research on GM-Toyota conducted by Rebecca Morales (personal communication).

ing workers some measure of "responsible autonomy" has nothing to do with altruism but is as much a managerial strategy for making profits as Taylorism. Yet this qualification in no way falsifies the workers' attitudes. And it would be no surprise to anyone who has not read Braverman that under the right conditions and with the right methods, management can get more out of workers by giving them more tasks and discretion and by enlisting active consent than by deskilling and coercion.

However, we would suggest that while Japanese capital may gain from such a situation, the JIT system and its supports are not wholly dependent on them; suitably modified, they can work in contexts with a different balance of class forces, and with different implications for the experience of work. So whether there is anything progressive for labor in the new practices depends on the form they take, or the form which labor lets them take. For example, provided that they do not lead to company unions, single-union plants have at least the potential of overcoming the divisions in the workforce supported by multiple-union plants. Certainly the familiar organizational forms do not represent a golden age for labor, even if, at their height, labor was stronger than now, for it was a highly selective and ambivalent strength, one which actively reproduced divisions in the workforce, particularly between men and women (Cockburn, 1983; Bannon and Thompson, 1985).

In any case, trade unions are increasingly obliged to confront the new practices in fixing agreements with leading firms. In the electronics industry in Britain, unions are finding that to get a foothold in new foreign firms they have to accommodate a whole package of conditions which effectively make way for the development of the new management practices. Arguably, securing these conditions (flexibility, simplified payment systems sometimes with an individual merit element, and so on) presents organized labor with a more significant challenge than the much-publicized no-strike clauses (Morgan and Sayer, 1988). In the final analysis, those who want workers to have more control over their lives in industry can hardly spurn opportunities for them to gain some of the skills which might help them achieve that goal.

## Conclusion

In part, we have not avoided the trap of homogenizing Japanese (and European/American) industry, as Cole (1971) warns, and have relied too much on our ideal types of JIC and JIT. But we have also stressed the continuities between JIC and JIT, and the ways in which new types of production evolve and diversify as they adapt to the different contexts presented by existing patterns of uneven development. There is much more to be said about the wider Japanese context, particularly concerning cultural characteristics, subcontracting systems, industrial groups, relationships between industry and financial capital and government. We deal with some of these in chapter 5, but again ideally, and from an academic point of view, what is needed is nothing less than a major theorized history of Japanese industry, one which links the internal organization of capital (both big and small) to its wider context. However, JIT and its supports are now extending beyond Japanese shores, evolving in a way which severs their roots in the particularities of Japan, and our primary purpose has not been to give a representative picture of Japanese industry but to explain the rationale of the most advanced forms of production organization, and to demonstrate the centrality of innovation in the organization and integration of the division of labor.

# 5

# Beyond Fordism and Flexibility

There is currently much speculation that capitalist industry is undergoing major qualitative change. "Flexibility" has been the most common motif in writing on this subject, so much so that "flexible" has almost become a synonym for "new". But there are many aspects to what's new in the new social economy – many having little to do with flexibility. Indeed, we contend that what's new can be better grasped in other ways, most importantly in terms of new forms of integration and organization of the division of labor.

The best-known contribution to this debate is the *flexible specialization* thesis, deriving from the work of Piore and Sabel (1984).[1] There are now many variants of this, including hybrids that combine it with theories of Fordism and neo-Fordism deriving from the regulation school (Aglietta, 1979; Coriat, 1983; Boyer 1986; Lipietz 1987; Scott 1988b; Hirst and Zeitlin, 1989). From such sources there has emerged a consensus about what is new in some parts of the left (see, for example, *Marxism Today*, particularly Murray (1988)). According to this view, from the inter-war period to the mid-1970s the advanced capitalist countries were dominated by Fordism – by mass production of standardized goods in huge factories within large, vertically integrated global firms, churning out standardized products for mass consumption. This model was imitated across wide swathes of the economy. Towards the end of that period, the system had begun to exhaust its potential: mass markets had begun to break up as consumers tired of standardized products; labor resistance had built up; industries met inherent technical barriers, such as line balancing problems (i.e., shortages and gluts in the lines feeding into the main assembly line); and the system was too rigid to cope with the uncertainty of the recession. Meanwhile a new model of flexible specialization, or post-Fordism, has allegedly been emerging: small batch production in interlinked, specialized small firms, flexible in organization, work process and output, and tending to concentrate spatially into industrial districts. Microelectronics and information technology play a key role in allowing more flexible production and closer coordination between markets and producers targeting new

[1] We shall not dwell upon Piore and Sabel's influential but romanticized *The Second Industrial Divide*, since the debate has now moved on, but merely refer the reader to the critique by Williams *et al.* (1987).

market niches. Firms are supposedly becoming more responsive to changing tastes, and to the tendency for consumer products to support the construction of personal identities.

The literature on flexible specialization and post-Fordism has helped shake up the stale categories of radical theories of capitalist industrialization. It has not only shown that there are alternatives to continued industrial concentration and monopolization, but has brought the question of industrial organization to the forefront in theoretical discussions. However, the value of this opening has been obscured by an obsession with flexibility and an extraordinary attachment to dualistic conceptualizations of the recent history of capitalist industry. Faced with the bewildering variety of changes many have seized upon simple polemical contrasts, often in the form of binary histories: industrial v. post-industrial (Bell, 1973), modernist v. post-modernist (Harvey, 1989), mass production v. flexible specialization (Piore and Sabel, 1984), or organized v. disorganized capitalism (Lash and Urry, 1987). Inevitably, we risk ending up with overburdened dualisms and overly elastic concepts; worse, we invite a diminution in the richness – and therefore the power – of our conceptual equipment. The trouble with concepts like Fordism, post-Fordism, and flexible specialization is that they are overly flexible and insufficiently specialized.

In this chapter we shall demonstrate that the literature on post-Fordism is long on speculation and short on coherence, based on selected examples whose limited sectoral, spatial, and temporal range is rarely acknowledged. Strangely, this literature has paid little attention to the country which most threatens established capitalist industry in Europe and North America. We shall argue that, whatever the condition of mass production elsewhere, it is alive and well in Japan. The academic attention given to the Third Italy in the post-Fordist literature seems bizarre when one compares the number of firms which feel threatened by the Third Italy with those under threat from Japan, or the number of delegations from western firms sent to Japan with those sent to the Third Italy, or the numbers of books for managers on "learning from Japan" with those on "learning from Italy".[2]

Although we argue that mass production is thriving, we do not want simply to reverse the terms of the post-Fordist debate and argue on the same terrain. Nor do we wish to deny the reality of industrial complexes such as those of the Third Italy or Los Angeles.[3] Rather, we suggest that while Japanese industry has some features not unlike those described as flexible specialization,[4] it also has characteristics which cannot be usefully

[2] Some might reply that industrialists are mistaken in largely ignoring the Italian phenomenon, but given the trends we think the onus is on such critics to prove this.

[3] We would contend that the Third Italy and Los Angeles do not fit the boxes of flexible specialization or neo-Fordism, and need to be seen as different varieties of industrial organization, as argued in chapter 3.

[4] Dore (1986a, p. 190) notes certain parallels between the Japanese textiles industry and the woolen weaving district of Prato in Italy. See also Nishiguchi (1987).

subsumed under Fordism or post-Fordism. Thus, while Japanese industry has been responsible for some remarkable innovations in labor processes, we think it equally important to recognize first, forms of organization which cut across different kinds of labor process and second, broader contextual features which give the country a lead across a wide range of sectors.

The structure of the argument is as follows. In the first section we draw attention to the problems of explaining change in terms of dualistic frameworks. Next, we question the concepts of Fordism, and the crisis of Fordism, on which the debate about post-Fordism pivots. Following that, we examine two issues central to arguments about post-Fordism – flexibility and industrial organization – particularly with regard to the explanation of vertical disintegration. Then we discuss how Japan's apparently rigid industrial organization gives it significant advantages over less organized western capital, particularly in long-term flexibility in restructuring and innovation. Finally, we speculate about what these findings suggest for priorities in future research on uneven development.

## Dualistic Rhetoric and "Binary Histories"

The rhetoric of flexible specialization and post-Fordism has an underlying dualistic framework:

| Old Times | New Times |
| --- | --- |
| Fordism | Flexible specialization/post-Fordism |
| rigidity | flexibility, responsiveness, resilience |
| mass production | small batch production |
| dedicated machinery | flexible machinery |
| standardized products | differentiated products |
| JIC/large stocks | JIT/minimal stocks |
| Taylorism/deskilling | post-Taylorism/enskilling |
| vertical integration | vertical disintegration |
| global firms | industrial districts |

The main attraction of this kind of thinking – encouraged quite explicitly by *Marxism Today* and authors such as Moulaert and Swyngedouw (1989) – is its simplicity and symmetry. Yet, although dualisms are powerful ways of registering difference, the credibility of this rhetoric becomes strained when a whole series of these lists are claimed to run in parallel and reinforce one another, straddling a single divide, so that the future is presented as the opposite of the past – a strange notion for historical materialists. And these "binary histories" (Williams *et al.*, 1987) do indeed cause serious distortion. Thus, just-in-time systems are in many ways more rigid than just-in-case systems and better suited to mass production (on the

"Old Times" side of the table) than to small batch production. Adherents of the dualistic framework are obliged to deny this, either by ignoring it or by trying to argue, extraordinarily, that the likes of Toyota and Matshushita are no longer in mass production or that the Japanese are following the lead of the Italians (e.g., Friedman, 1988).

It is as well to keep in mind the extraordinary influence of such rigid forms of rhetoric. However, rhetoric is unavoidable and judging its value is an *a posteriori* matter (Sayer, 1989; Mäki, 1989), so we now turn to evaluate the descriptive and explanatory adequacy of these perspectives.

## Beyond Fordism

The first problem of talking about post-Fordism is that our uncertainty about it is not compensated for by any certainty about Fordism. The sense and reference of this term – chiefly associated with the French regulation school – is itself extraordinarily loose, particularly when used to denote a major epoch of capitalist industrialization (e.g., Aglietta, 1979).[5] Thus the current debate has not even got the easier side of its binary opposition firmly anchored. Let us consider some of the main senses and justifications of the term Fordism:

1   A labor process involving moving assembly line mass production.
2   A group of volume production sectors, including automobiles, steel, chemicals, dominant in terms of production of surplus value and propulsive in their effects on other sectors.
3   An allegedly hegemonic form of industrial organization consisting of large integrated corporations and factories, sectoral oligopolies, and characterized by the production methods of 1.
4   A "mode of regulation" (Aglietta, 1979), in which mass consumption absorbs the output of mass production, thanks to productivity-related wages for core workers in Fordist industry (sense 2).[6]

Fordism in the strict sense of 1 is not as large in absolute terms as one might imagine from all the attention it receives. One recent British estimate puts workers on such lines as only 700,000 of the total British workforce of 20.4 million (Littler, 1985). Adding workers employed in the same sectors, but not directly working on such lines, doubles the total. Adding continuous process production increases the employment a little more; however, few of the problems which allegedly now dog Fordism (e.g., line balancing and labor resistance) apply in these sectors, let alone others

[5] See Foster (1988), Williams *et al.* (1987), Brenner and Glick (1991) and Walker (1991) "the metahistory of Fordism" for critiques.

[6] Aglietta attempts to derive many other elements regarding money, the state and international relations, etc. We shall not address them since our focus is on production.

sometimes cavalierly lumped in with Fordism (cf. Meegan, 1988). Indeed, in practice, the alleged attributes of Fordism apply not even to the narrow definition in 1 above, but only to the subset which manufactures complex standardized products, such as cars and televisions. The many simple mass assembly processes, like that for ball point pens, are unlikely to face the same characteristics and problems.

Definition 2 is much more plausible, though lacking in clear support from economic statistics. In fact, the propulsive character of Fordism is questionable, since there is no evidence of a dramatic leap forward in manufacturing productivity in the US in the 1920s, as Aglietta claims (Dumenil and Levy, 1989). Sectors outside the mechanical assembly industry, such as agriculture and big construction (Freeman, 1982), have also shown remarkable increases in productivity and have been propulsive, though to call them Fordist would of course devalue the concept. Nevertheless 2 could be defended on the grounds that small batch and other non-Fordist forms of production are sustained by mass production (though this dependence works both ways), and non-Fordist producers (e.g., construction firms) can also sell to each other.

It is quite possible that Fordist forms of organization (3) have influenced the way in which non-mass production is organized, though again supporting evidence is needed. However, the association between mass production and large firms (and between small batch production and small firms) is itself questionable. Simple mass production can be done in small firms, and many large firms do small batch production. There is also a school of thought which argues that the growth of large firms is not strictly tied to the scale of production but is chiefly the result of financial factors (Marris, 1979; Nelson and Winter, 1982). Finally, calling such firms Fordist overlooks the fact that the multi-divisional corporation was explicitly a counter to Ford's obsession with mass production (Hounshell, 1984).

The idea of regulation (4) takes the concept of Fordism beyond production into macroeconomic relations.[7] The balance between mass production and mass consumption must occur across many sectors and many consumers. In particular, it does not depend solely on the minority of workers employed in mass production industries proper. Despite the much-cited $5 day, Henry Ford sold few cars to his own workers, for demand was plentiful from other classes and other workers, such as farmers, clerks, rentiers, and managers (Foster, 1988). In the United States, mass unionization, the productivity wage, and the state-funded social wage did not produce any discernible jump in mass consumption in the mid-20th century; wages and mass consumerism had expanded at a relatively steady pace for the preceding century or more (Dumenil and Levy, 1989; Brenner

---

[7] Indeed, in some versions, production virtually drops away from the definition of Fordism altogether, e.g., Jessop (1990).

and Glick, 1991). What mass unionization did unequivocally achieve was greater cyclical stability (i.e., dampened cycles) in the advanced capitalist countries during the post-war period, and a rising standard of living for large swathes of the working class, but these are lesser claims than the regulation school is making.

### Is Fordism in crisis?

The above criticisms do not amount to a firm rebuttal of the concept of Fordism, though they do expose its uncertain foundations. But let us suspend doubt about the concept and examine the proposition that Fordism is in crisis. The argument, developed by Aglietta and supported by many others, is that Fordism's limits appeared in the shape of line-balancing problems, long setup times, down-time, labor porosity, inventory build-up, labor disaffection. Others, particularly in the flexible specialization school, stress the significance of an alleged break-up of mass markets in undermining mass production. There are many problems with such claims.

First, aggregate empirical evidence on rates of output change and productivity growth in mass production industries would be useful in assessing these arguments – with the crucial caveat that Japanese mass production must be included so we do not mistake a local western decline for a universal change (Brenner and Glick, 1991). However, even in Britain, Meegan (1988) finds that evidence for a relative decline in these rates is inconclusive, with some classic mass production industries (e.g., television) growing rapidly while others have been declining more slowly than many non-Fordist industries (also Williams *et al.*, 1987).

Second, there is the simple question: if small batch production was, until recently, resistant to automation and rapid increases in productivity, why don't we speak of the crisis of small batch production? After all, in employment terms it is bigger than the mass production sector (Meegan, 1988). Moreover, an improvement in batch production is not the same as a decline in mass production; it may only reflect the inherent tendency of capital to revolutionize the forces of production.

Third, in principle, there is no reason why new mass production sectors which do not compete with old ones should not flourish using Fordist or even pre-Fordist methods. Thus, an alternate interpretation of the crisis would attribute it not to problems inherent in Fordism as a labor process, but to market saturation and the relative lack of new sectors and products.

Fourth, some of the supposed inherent limitations of Fordist mass production are doubtful. Fordism need not generate progressively worse working conditions for workers and hence build up labor resistance (not that the latter follows automatically from the former in any case) (contrast Sabel, 1982; Bowles *et al.*, 1983). Labor resistance did build up in a few Fordist sectors, particularly the automobile industry, but less so in others, and no more so than in many non-Fordist sectors. Arguably, the crescendo of disputes in the 1960s was more a function of the general politico-economic conjuncture than an inevitable consequence of Fordist labor processes. Line-balancing problems and other rigidities are more credible as limitations of Fordism, though their effect on productivity can be overridden by further automation and labor displacement. As we saw in the previous chapter, Japanese developments in managing mass production labor processes have in any case largely solved the problems of line balancing and rigidities; whether one calls the result neo- or post-Fordism or a non-Fordist kind of mass production is largely a semantic issue: the important point is that mass production still flourishes.

Fifth, it is not evident that the breakup of mass markets can be attributed to Fordist production. Declining mass markets may be due to exogenous features, such as crisis itself, sagging demand, penetration of national markets by the different products of foreign producers (consider food especially), and product innovation in general.

Sixth, the proposition that Fordism is in crisis may amount to nothing more than the following suspect syllogism: A. post-war capitalism can be characterized as Fordist; B. capitalism is currently in crisis; C. therefore Fordism is in crisis. Even if we accept A and B, it does not necessarily follow that Fordism is the cause of capitalism's crisis. Indeed, it may be just as reasonable to suppose that investment in plant and equipment simply ran ahead of the rates of growth of productivity and of addition of new labor to the industrial system, generating a fall in the rate of profit of the classic sort depicted by Marx. In fact, it is very difficult to offer any other explanation for the persistent fall of the rate of profit on corporate capital throughout the whole post-war period in all the advanced capitalist countries (see, for example, Devine, 1986).

More generally, arguments about the crisis of Fordism are unacceptable because they switch between concepts of Fordism without acknowledging their differences. Thus if one is going to use a broad version of the concept such as that of the regulationists, or a version which includes everything from continuous process industries to the entertainment industry, one cannot justify the claims regarding crisis at those levels by reference to the alleged problems of Fordism in the narrow sense of complex, mass, mechanical assembly. Of course, if Fordism is defined to cover everything one can argue that it causes everything, but there is no explanatory value in such a strategy.

We indicated above that any discussion of mass production must take account of Japan's experience. One cannot ignore Japan in discussing allegedly epochal changes in western capitalism, for it is quite likely that some changes in Europe and North America have not been independent of Japanese competition.[8] Even a cursory examination of new consumer

---

[8] In view of Lash and Urry's claims that capitalism is becoming more disorganized it is ironic that the main challenge to the industrial hegemony of the (western) countries they examine comes from a country whose capital (though not its labor) is organized with a vengeance.

products shows that mass production is not in decline but flourishing, particularly in Japan, and that the problems of Fordist mass production come not so much from within as from outside, in the shape of competition from Japan and the Newly Industrializing Countries. Productive efficiency is always a relative matter under capitalism, and although certain limitations may have been inherent in Fordism it took a different system to expose them. (Under capitalism, a problem shared is a problem halved – but once a competitor solves the problem, it becomes far worse for the others, until they find a solution too.) As we saw in the previous chapter, many of the Japanese "solutions", such as just-in-time systems, originated in the late 1940s and 1950s, long before the supposed crisis of Fordism emerged, and were developed as solutions to problems different from those later experienced outside Japan (Cusumano, 1985). However, once the innovations were developed and applied, they started to expose the weaknesses of western mass production.

## Beyond Flexibility

Buzz-words like "flexibility" are initially attractive because they offer the promise of escape from old concepts whose limitations are all too familiar and whose strengths bore rather than impress. They quickly die because their promise proves false, their unities and generalizations full of holes. As Pollert shows, many users of the term have deliberately or unintentionally overlooked its double-edged, value-laden character and its tendentious usage in management and political circles: flexibility sounds agreeable in the abstract, but not always when considered in the concrete, e.g., the debilitating effects of working alternating day and nighttime shifts. The resulting "discourse of flexibility" involves ideological slippage between description, prediction, and prescription which masks the vital political issue of the different interests at stake – whose flexibility, and in whose interest (Pollert, 1988; Standing, 1991)?

Even in a purely descriptive role, so much is expected of this concept that quite inflexible practices are labelled flexible and vice versa. It can be argued, for example, that in the present recession labor turnover is understandably low, and hence the labor market is more inflexible than during the post-war boom. Similarly, some observers argue that in the heyday of Fordism, inter-firm relations were highly flexible as they were generally short-term, arms-length market transactions which could be made or broken according to short-term market signals. By contrast, in recent years less flexible inter-firm linkages, involving lasting interventionist and collaborative relationships which ignore short-term price fluctuations, have supposedly become more popular. In some cases, these relationships have become more rigid where suppliers are linked into customers' proprietary information technology networks. However, it is quite possible that these rigidities actually facilitate greater flexibility in production.

What has to be recognized is that capitalist industry has always combined flexibilities and inflexibilities. What may be emerging now are new permutations of each rather than a simple trend towards greater flexibility alone. The variety of such combinations cannot be grasped by inflexible dualistic frameworks which counterpose the old as inflexible to the new as flexible. Indeed, mass production and flexible production need not even be considered as alternatives.[9] What is therefore needed is not an obsessive focus on flexibility but a broader awareness of new forms of the division of labor and new ways of organizing those divisions.

## Varieties of Flexibility

The paradoxes of "the flexible firm" are easily explained once we unpack the concept of flexibility. Several types, not all of them entirely distinct, suggest themselves:

- Flexibility in output volume (Piore and Sabel, 1984);
- Product flexibility, allowing the firm to change its product configuration over the short run without marked losses of efficiency (Storper and Scott, 1988). This can involve either permutations of an existing range of products (e.g., variants of a certain model of car) or limited design changes (e.g., new fashion clothes).
- Flexible employment (Atkinson, 1984; OECD, 1986; Boyer, 1988; Standing, 1991);
- Flexible working practices (absence of demarcation restrictions) (Atkinson, 1984);
- Flexible machinery (Kaplinsky, 1984);
- Flexibility in restructuring (Piore and Sabel, 1984; Saxenian, 1988).
- Flexible organizational forms, e.g., networks of specialist producers (Beccatini, 1978; Scott, 1988a; Storper and Christopherson, 1988; Sabel, 1989).

There has been much confusion among these in the literature. The most common examples are: to confuse product flexibility, which need involve no restructuring of capital within the firm, with flexibility in restructuring; failing to distinguish product and restructuring flexibility, which refer to individual firms, from flexible organizational forms, which refers to networks or industries; and the confusion of any of these with innovative capacity, or prowess in R&D and applying innovations, which cannot be usefully subsumed under "flexibility".

Note that while there are many complementarities between the above features, they can exist and generate benefits to capital separately; flexible

---

[9] Ironically, 20th century mass production was greatly facilitated by the highly flexible electrical technologies; compared to steam power, the electric motor was far more flexible (Hounshell, 1984).

technology need not be associated with flexible working practices (and vice versa), even though capital often obtains greater benefits when they are combined (capitals do not always act in their own best interest; Jaikumar, 1986). There are also many different routes to flexibility. For example, extensive use of external labor markets and greater use of temps, part-timers, and outside contractors or consultants can assist numerical flexibility (Harrison and Bluestone, 1988; Christopherson, 1989). On the other hand, task flexibility can be facilitated by greater use of internal labor markets, reducing labor turnover, increasing training in a range of skills (winning flexible working practices), and willingness to move within the firm in return for job security. (Ironically, flexibility can therefore mean that core workers' skills become more company-specific and less transferable between firms than more limited skills (Brown, 1984)). This kind of combination of rationales may be responsible for the increasing dualism of labor markets, which some have observed, with core workers more and more sharply differentiated from peripheral workers (Atkinson, 1984).

Further, many of the above kinds of flexibility can either facilitate the production of a fixed product range, even a single product, or they can facilitate product flexibility. It follows that some kinds of flexibility can be useful not only to so-called flexible specialization but to standardized mass production. We should be wary of assuming that mass production is synonymous with inflexibility, even though strict Fordist mass production was (cf. Hounshell, 1984; Williams et al., 1987). In other words, we challenge the very idea of treating mass production and flexible production as alternatives, the opposition which underpins the Fordist/post-Fordist debate. In addition, some kinds of flexibility may be most strongly associated with kinds of work which do not fit either – in Britain, some of the most publicized cases of numerical and employment flexibility have been in coal mining, railways, ferries, and public services (see also Boyer, 1988; Christopherson, 1989).

### Flexible machinery

This deserves special mention because it has been the subject of such an extraordinary degree of hype that some theorists were speaking of the widespread existence of a new kind of production organization – neo-Fordism, based on the use of microelectronics in manufacturing – before it had hardly begun. In 1985 an international survey by Ingersoll Engineering Consultants of "computer integrated manufacturing" (CIM) failed to turn up any examples (Riley, 1985). Localized, specific, and often atypical developments have been eagerly seized upon and used as the basis for wild extrapolations and generalizations in an unseemly scramble for theoretical novelty.

In general, the flexibility and power of the new technologies have been greatly exaggerated. Compared to human labor, even the most flexible machine is crude and rigid, and unintelligent when it comes to handling the unexpected. Robots are a case in point. Often seen as flexible, they are in fact extremely limited: a paint-spraying robot cannot insert chips in circuit boards and a pick-and-place robot cannot weld steel; each task is done by a special-purpose robot. Even reconfiguring them to do different tasks within their narrow range can be time-consuming. Significantly, Japanese manufacturers usually classify robots as dedicated machinery (Tidd, 1990).

No matter how flexible technology may be, it is still fixed capital and shares the inherent problems of that status (Harvey, 1982). Schonberger (1987) argues that flexible manufacturing systems tend to be extremely expensive and require round-the-clock operation to make them profitable (see also Schoenberger, 1990). This puts them out of reach of small firms (in contrast to the idea that flexible technology presents a new dawn for small firms) and requires user firms to push for both higher volumes and more product variety than the market may be able to bear. Dedicated machinery may be inflexible in one sense but it is not always big and costly. Small dedicated machines are more likely to be made in-house, and hence better suited to the firm's needs. They may also be relatively cheap and quick to show a return (Schonberger, 1987). Unfortunately, in stressing the negative quality of rigidity the recent literature has forgotten the virtues of dedicated machinery relative to the suboptimalities of general purpose machinery. Again, the flexible may turn out to be inflexible, and vice versa.

In addition to its flexibility, microelectronics-based automation excites interest because of its systemic scope, offering the potential of computerized integration of related tasks (Kaplinsky, 1984). But precisely because of its systemic nature, the upheavals its introduction entails, and hence the resistance which it generates, are all the greater (Freeman and Soete, 1985), as in the case of computer integrated manufacturing.

If systemic technology is introduced without profoundly altering the immediate environment it runs into problems. As the Ingersoll consultants commented, "... too often its purpose has been to automate existing complicated systems ... Manufacturers should not be thinking of bringing computers in to integrate their systems until they have taken a hard look at the way their factories are run without computers" (Riley, 1985). Similar comments have been made regarding office automation. The Japanese have looked hardest at their work organization, and enjoyed significant benefit, and their experience shows that exhaustive study of the work process and learning-by-doing are necessary to get the most out of new technology – and sometimes to show it is not needed. (Indeed, it is striking how frequently one reads of major Japanese firms choosing manual rather than mechanical transfer of parts, and verbal or paper communication over more expensive and sometimes less flexible information technology.) So although information technology can assist coordination, its success depends on the prior solution of organizational problems and on designing the information system as an integral part of the labor process and wider production system. It certainly does not automatically solve these problems on its own.

Outside Japan, the situation appears to be different. Jaikumar's research on flexible manufacturing systems (FMS) found that "with few exceptions, the . . . systems installed in the United States show an astonishing lack of flexibility" (Jaikumar, 1986, p. 69; Hayes and Jaikumar, 1988); most were used within a traditional Taylorist automation framework for cutting operating costs, not for broadening product **range** and raising the rate of product innovation. In Japan, however, FMS were used for this, and were more efficient, thanks again to flexible working and learning-by-doing.

The implications of this situation can be clarified, with the benefit of hindsight, by looking at the introduction of an earlier major technology – the electric motor.[10] When manufacturers first replaced steam power with electric power they assumed it was proper to apply it just as steam power had been used, with a single massive engine driving every process in the factory via a labyrinthine network of transmission belts. This was the easiest way of introducing electric power, because it minimized disruption to the existing system; however, it also perpetuated the problems of the old system with its inefficient production layout. Only later were the benefits of using multiple, decentralized power sources realized. We would suggest that we are largely in an early phase of flexible/systemic automation, and that the new kinds of production, whatever we wish to call them, have more to do with the organization of work, both in terms of labor processes and industrial organization, than with the hardware of new technology.

Even where flexible, systemic automation does exist, it does not offer the kind of flexibility signalled in Piore and Sabel's work. Microelectronics may facilitate limited short-run flexibility involving permutations of existing products, but it offers little for flexibility involving novelty, whether in restructuring or new product development. As Williams *et al.* (1987, p. 433) remark in response to Piore and Sabel, "New generations of computer-controlled equipment may deliver a more varied output but they do not restore an economic system based on redeployable productive resources and low fixed costs. That is a world we have lost". Moreover, flexible technology has only a limited impact on innovative capacity; it may assist in the representation, processing, and communication of new ideas, but it relies on people to solve novel problems and originate those ideas.

Finally, very little can be inferred from technology alone regarding the shape of industry. If flexible machinery rapidly increased productivity in small batch production, it could either expand the size of this sector or reduce it, depending on price and elasticity of demand. In short, there may

[10] We are indebted to Professor Chris Freeman, University of Sussex, UK, for this example.

well be a microelectronic revolution, but if does not play the role given to it in the literature on neo-Fordism, post-Fordism, or flexible specialization.

## Flexible product mix

A key claim of the flexible specialization thesis is that stable mass markets are breaking up into niches, which must be served by a greater variety of batch-produced products. Again, the aggregate data needed to substantiate this is conspicuously absent, but if it is increasing, how should we explain it?

The main, but widely-overlooked, part of the answer is that it is simply the general, structural tendency of capitalism to multiply the number of different kinds of commodities over time (and with it, increase the division of labor). The increased range and diversity of demand from all kinds of consumers is hardly new, for the changes wrought by capitalism, in production and ways of life, continually create new needs. There is therefore no need to enter into tiresome debates about which came first, changes in demand or changes in supply. Particular demand-side, or technological and organizational, changes might have been prominent (e.g., the ability of information technology to cope with complexity), but they invariably originate from, and operate within, the continuing dynamic interaction of production, distribution, and consumption (see chapter 2; Walker, 1985). Possibly, as noted earlier, market fragmentation (due to penetration by foreign producers selling different products) has been mistaken for product differentiation; that is, the number of products available in a given national market may have increased due to increased internationalization of markets and consequent penetration of foreign goods rather than because each firm is producing an increased range of goods.

This multiplication of products can take two forms: proliferation (i.e., whole new genera) or variation (i.e., speciation within existing genera). Often, instances of variation – for example the thousands of permutations of particular models of car offered by some manufacturers – have been conflated with proliferation,[11] but the economic effects of the two are likely to differ. When firms do find that the market for a particular product is saturated, it is rarely possible to relieve the condition through variation alone. The arrival of new varieties of bread is unlikely to make people eat any more than before, and hence the consequences of niche strategies are likely to be of a zero-sum nature at the level of whole industries. In contrast, proliferation (e.g., diversifying into pastries) might push up demand for the baked goods industry as a whole.

While the range of commodities continues to expand, from the consumer's point of view it is not clear that product variety is more important

---

[11] Microelectronic technology has probably enhanced the scope for variation more than proliferation.

than price and quality (Schonberger, 1987). The desire and need for lower costs, and for better quality and performance (use-value), are still fundamental to capitalist industrialization and still lie behind the drive for mass production. Nor is there any necessary inverse relationship between the standardization of products and their quality. Individual firms see increases in the breadth and flexibility of product mix not as ends in themselves but as means to survive in competition. Limiting variety and increasing output are attractive to all types of production, including small batch; there is no point in diversifying unless competitive survival demands it. Not surprisingly, firms generally prefer stable markets to unpredictable ones, and tend to achieve no more flexibility in product mix than is necessary to compete.

The issue is not, therefore, one of mass versus batch production, but the continued search for improvement in both. Scale of production and production type are significant, but not in the way described by the flexible specialization school. Economies of scale still matter, and not just in production. Even if optimum scales of production for individual products have fallen, there may still be important economies at the level of factories. Many kinds of capital investment still need to be very large and lumpy, beyond the reach of small firms. For example, digital public switching telecommunication systems require billions of dollars to develop and the R&D is not easily split up and subcontracted; moreover, huge markets – far bigger than any European national market – are needed to amortize such investments. What evidence is there that such products are becoming less important?

Niche marketing strategies are not new but as old as capitalism. They are also often defensive rather than offensive ways of competing. As Stone (cited in Meegan, 1988) points out, the announcement of niche market strategies is frequently a sign of imminent death, as it often presages the firm's eventual exclusion from the market by those able to live in the competitive world of volume production. British electronics firms have repeatedly capitulated to competition in this way and weakened their position (Morgan and Sayer, 1988). Moreover, by its very nature it is usually the standard product which is more resilient; overly specified or customized products are less likely to be adaptable.[12]

At the level of the economy as a whole, there is no advantage in a niche market/flexible product mix strategy (Solo, quoted in Gertler 1988, p. 427):

> No matter how flexible flexible specialization is it cannot wriggle out from under the descending force of a decline in aggregate demand ... Nor would

[12] Some British examples of niche strategies favored by flexible specialization advocates have proved to be short-lived. Thus in retail clothing, niche strategies apparently depended heavily on short-lived spending booms, when consumers could add a few luxury items to their wardrobes. When the boom subsided, supposedly "dated" strategies of selling standard clothes in department stores such as Marks and Spencer's proved to be more resilient.

future certainty be less debilitating for the flexible specialization technology than for that of mass production ... Investment in that technology will operate under the same constraints, and have the same need for future certainty, as investment for mass production

The flexible specialization school not only exaggerates the potential of niche strategies, but radically underestimates the difficulties firms experience in moving into new products and markets. Even when firms change product and market within the same industry, they face considerable problems. For example, electronics firms that have tried to move from specialized production to volume production have often encountered severe difficulties and had to abandon the attempt. In the computer industry, even the task of moving from producing computers for expert scientific or educational users to computers for novice users has proved beyond the marketing, technical, and organizational abilities of many firms (Morgan and Sayer, 1988).

In other words, far too little recognition has been given to the fact that:

- Firms accumulate and come to depend on product-specific and market-specific knowledge, often of a tacit nature. Different activities require radically different kinds of expertise and material support. Expertise still matters, even if it is largely concentrated among managers and technologists. Flexibility of the types discussed above does not overcome the difficulties of the learning process by which firms change their character. Information technology is not of much help here, as any change inevitably throws up problems and flexible technology does not usually have a learning and problem-solving capacity.
- Firms, together with their suppliers and customers, are to some degree embedded in and dependent on wider organizational relationships. Much of the difficulty of moving into new activities lies in the need to create new relationships, especially with suppliers and buyers. Products are not simply thrown onto the market; the market is an institution and must either be constructed, or, if it already exists, entered. Access to markets is rarely achieved without setting up these relationships and information exchanges, especially where complex custom products for big users are involved (e.g., business telecommunications networks), for here there are usually long-established buyer-seller arrangements in which the partners have evolved together and become dependent on one another's specific technologies.

If anything, there has been a move away from the kind of flexibility or resilience afforded by "conglomerization", though this is still popular, especially in Britain. The contrary trend, or "back-to-basics", involves firms shedding not only peripheral activities (like cleaning and catering) but recoiling from ventures into neighboring fields in which they lack expertise. This happened with the much-vaunted convergence of com-

puter and telecommunications firms: despite the common digital technology, differences in expertise, technology, and organization between the
two sectors proved a formidable barrier to flexibility in terms of changing
product range.

This kind of point is further reinforced from a different angle by the
rediscovery, in industrial geography, of the importance of local expertise in
specialist industrial agglomerations (Scott, 1986; Storper and Walker,
1989). Perhaps in practice as well as theory, western capitalists tempor-
arily came to believe that the material and intellectual bases of profit were
secondary; the "accounting syndrome" of the 1950s and 1960s epitomized
this. Now the back-to-basics movement, and the increase in collaborative
ventures between firms with different but complementary aptitudes, point
to a growing recognition of the importance of accumulated know-how in
competitive success (Morgan and Sayer, 1988).[13] As we shall see, Japanese
industry provides many lessons in this respect.

The above arguments directly contradict a common view of flexible
specialization. For example, Lash and Urry (1987, p. 200) argue:

> Just as there is a growing divergence *within* any market, so there is an
> increasing convergence of products across different markets. Companies are
> much less distinguishable by their previous participation in a given market. A
> particular market is much less controllable by a given set of firms since
> wholesale invasions can be mounted. In recent years a clear example has
> been the extraordinarily rapid capture of the market for watches by digital
> watch companies. Hirschhorn comments "The information capital content
> of digital watches did not develop from watchmaking production technology
> per se, and thus traditional watchmakers could not anticipate the scope and
> strength of their unanticipated competitors." The distinction between "com-
> munications companies" and "information-processing companies" has like-
> wise broken down ... This break down of clear-cut markets is associated
> with an increased flexibility of the capital stock.

Lash and Urry go on to associate this increase in flexibility with "cybernetic technology", i.e., flexible automation.

There are several problems with this kind of interpretation. Selective
evidence, of watches or whatever, is not enough to establish the validity of
general statements about the changing development of the social division
of labor. In the history of capitalism there have been many examples of
one sector and its technology displacing another; there is nothing particularly new about it. And the ease with which it can be done depends on,
*inter alia*, the type of product and market: with inexpensive consumer
products such as watches it is relatively easy, as marketing and producing
them involves no collaboration with the user. Matters are very different,
however, in specialized production for industrial and institutional users.

---

[13] Such collaborative ventures are common among firms as big as Philips and Sony, or General
Motors and Toyota – firms that do not fit the flexible specialization image.

Lash and Urry's example actually cuts both ways: it also demonstrates the difficulty traditional firms have in learning how to operate with entirely different forms of production and industrial organization. Conversely, it was easy for electronics firms to move into the market not because of some new kind of flexible means of production but because they had the accumulated expertise for making microelectronic products. Here, Lash and Urry confuse process technology (cybernetic technology) and product technology (digital watches), and so mistake the diffusion of basic technologies for industrial convergence. There has been no more convergence with other industries than with, say, the diffusion of the electric motor; the watch industry has been revolutionized, but it remains the watch industry. The machinery and organization needed to produce digital watches is not particularly flexible; it cannot make entirely different products. Certainly many types of chips can be flexible in the sense that they can be used for different purposes, but earlier technologies have also had this quality (e.g., electric motors), and flexibility in production and in the market depends upon much more than polyvalent equipment and products: it also depends on something which is all too often taken for granted as unproblematic – organization and marketing.

## Flexible production systems

So far we have concentrated on flexibility within individual firms, but the flexible specialization thesis emphasizes flexibility primarily as a property of networks of firms, often highly localized, in which each individual firm may be quite specialized and indeed inflexible. The members of such networks do not interact merely through market exchange but have well-established formal and informal relationships. This enables them to take advantage of synergies and to adapt continually. At the same time, since each firm is separately owned, each is disciplined by exchange value and sufficiently independent to allow adjustment to uneven market changes with less friction than in large, vertically-integrated firms. There is impressive evidence in the literature (e.g., Storper and Christopherson, 1987; Scott, 1988b), and from Japan, of cases where industries have agglomerated and experienced increases in both vertical disintegration and in the number of firms.[14] Yet their conformity to the model varies significantly, and explanation of that variation has proved elusive.

This tendency toward vertically disintegrated but strongly localized industries has been generally attributed to the need to adapt to greater uncertainty and more rapidly changing markets.[15] We see a rationale for

[14] Note that there could be increasing vertical disintegration and a progressive deepening of the social division of labor with a *stable* number of separate firms; existing firms could simply specialize more and trade.

[15] Transaction cost theory also sometimes treats vertical *integration* as a response to uncertainty, which reinforces doubts about its explanatory power. See chapter 3.

this type of configuration in industries where most firms produce specialized products for continually changing demands. It allows participants to
pool risk and maximize external economies of scale (Storper, 1988). Many
of the most frequently-cited examples of flexible specialization industrial
districts fit this rationale, such as the defense electronics industry of Orange
County, California (Scott, 1988a). However, as we shall see in the next
section, they can also under certain circumstances be favorable to standardized volume production.

A net shift towards vertical disintegration is likely to be in large part the
result of innovations in organization which, though prompted by recession,
will probably outlive it. These innovations could have had similar effects if
introduced earlier, but management simply was not aware of the possibilities. We cannot understand changes in industrial organization just in terms
of the timeless logic of choice among known alternative patterns of cost,
perhaps qualified by limited information or uncertainty, because such a
framework cannot hold the fact that individuals and organizations learn:
organizational forms are discoveries.[16] Equally, we suspect that the
emphasis on economies of scale and giantism in the post-war boom had
more to do with the discovery and application of a new form of organization than with permutations from a range of known alternatives in
response to shifting cost structures. Similarly the "small is manageable"
philosophy of the 1970s and 1980s, when workforces were split into
smaller units on separate sites, was a social innovation (though helped by
improvements in communications technology) whose cost effects were not
known beforehand.

The theoretical implications of such explanations deserve a closer look
for they are at odds with much of the so-called industrial organization
theory that has been invoked to interpret vertical disintegration and industrial agglomeration. We have already criticized the main variant of this
theory, transaction cost theory, *à la* Coase, Williamson, and Scott (see
chapter 3), but here we want to bring out two points:

- Organizational practices or modes of organization cannot be reduced
  to industrial ownership structure; they must be explained *sui generis*.
- We need an evolutionary perspective to avoid the causal ambiguity and
  ahistorical, aspatial character of industrial organization theory.

Each has several facets. Regarding the first point, the vertical integration/
disintegration distinction implicity emphasizes formal ownership, but
actual modes of organization are more important. A firm which looks

---

[16] Storper (1988) offers a variant of this argument closer to Piore and Sabel's concept of techno-
logical/organizational "branching points". Experimentation during a period of uncertainty establishes a new organizational form which, though not initially more efficient than others, becomes
dominant as increasing returns to scale – both within firms and within wider, vertically-disintegrated
complexes – give it advantages and eventually close off other choices.

vertically integrated, in terms of its range of divisions, may actually be very weakly organized; some of its subsidiaries or divisions may operate as merchant, or semi-captive, suppliers rather than as wholly captive suppliers (e.g., GEC of Britain has, *inter alia*, both electronic components and electronic systems products companies, but the links between them are minimal and potential synergies have been consistently ignored). Conversely, given the right internal organization, a large firm as a unit of ownership can embrace many strongly interacting product divisions and small workplaces, some with output that is wholly captive, others only partly so.[17] By the same token systems that are formally vertically disintegrated may have weak or strong inter-firm organization. Thus, as we shall see, there are critical differences between loosely organized, highly market-mediated industries, such as the Silicon Valley electronics complex, and the even more disintegrated, yet more strongly organized, complexes of Japanese industry. What matters is how the market is used, how far prices are allowed to govern relationships, how far market relations are embedded in non-market organizational relations, and what form the latter take within or between firms. So, while there is evidence of an increase in vertical disintegration, an increase in what might be termed "vertical organization" in both vertically integrated and vertically disintegrated cases could be of more significance.

Consider the vastly different organizational structures within the same industries in different countries. Whereas Japanese semiconductor producers are parts of large diversified electronics firms, a major portion of US semiconductor output comes from specialized merchant producers such as National Semiconductor. At the level of national economies, compare the vastly different size structures of Japanese industry with that of Britain, West Germany, and the United States: while the three western countries have only 16-18% of total employment in firms of less than 100 employees, in Japan the figure is 58%, including 30% in minuscule firms of one to four employees, including proprietor (implying that Japan is the place to study vertical disintegration) (McMillan, 1984; Patrick and Rohlen, 1988). Are we to believe these differences are the product of different patterns of transactions costs? In a sense they are, yet those patterns are largely the product of organizational forms and practices.[18]

Contrary to the impression given by industrial organization theory, the performance of any particular industrial structure, integrated or disintegrated, depends on specific practices. These practices (including "modes of governance") cannot be read off from industrial structure or patterns of

---

[17] Sabel (1989) now concedes this but perversely sees it as an imitation of the flexible specialization district. Such are the contortions necessary for preserving an inappropriate model.

[18] "Transaction costs have a well-deserved bad name as a theoretical device, because solutions to problems involving transaction costs are often sensitive to the assumed form of the costs, and because there is a suspicion that almost anything can be rationalized by invoking suitably specified transaction costs" (Fischer, quoted in Hodgson, 1988, p. 200).

transactions costs according to some timeless rationality – though it does not take much ingenuity to rationalize them after the fact as if no other behavior could have occurred. Organizational practices require recognition in their own right, for it is on them that economies of scope, transactions costs, and innovative capabilities depend.

Regarding the need for an evolutionary perspective, industrial structure and organization are consequences, partly intended, partly unintended, of many processes of development and competition, of the organizational evolution and learning that characterize capitalism. They are not uniquely the product of some timeless logic of rational choice; if they were, the increased uncertainty of recession allegedly responsible for accelerating the process of vertical disintegration would imply a *cyclical* rather than a secular change. Transaction cost theory would seem to suggest that there must have been a widespread shift in the pattern of costs making vertical disintegration more attractive for capital. But aside from the problem of independently measuring the transactions costs involved in alternative forms of organization, we must still identify the causes of changes in those costs – the answer might be organizational innovation itself. However, such changes are taken as exogenous; hence, in the manner for which neoclassical economics is notorious, transaction cost theory either ignores what it has to explain or assumes it to be given. (Like the economist who, needing to open a can but lacking a can opener, solved the problem by assuming the existence of one.)

The lack of an evolutionary perspective generates a systematic equivocation about causality in industrial organization theory.[19] In its theoretical exegeses, it offers a largely unidirectional explanation, moving from cost patterns associated with organizing transactions to disintegration or integration. Yet in its empirical application it has to be fitted to contexts where causality has moved in the opposite direction, and where industrial structure is also an unintended consequence of other processes. This is not to deny that cost structures related to unavoidable technical and market characteristics influence industrial structure: such phenomena obviously go a long way toward explaining the contrasting structures of, say, the clothing, steel, and tourist industries. And changing technical characteristics can prompt integration or disintegration as well. But cost patterns are also a product of organizational forms: organizational forms are to some extent the creator as much as the creature of the patterns of costs distributed across production systems. Scott acknowledges this, and Storper and Scott's (1988) emphasis on the importance of institutional structures implies the same point, but to resolve the ambiguity about causality we need an evolutionary approach.

[19] In the terms of realist philosophy, transaction cost theory is an example of an attempt to represent an open system, in which industrial organization is the partly intended, partly unintended outcome of many processes, as a closed system uniquely determined by rational choice (see Sayer, 1984).

The problem of causality applies equally to spatial organization. Certainly, some economic activities need more proximity to suppliers and buyers than others, but even this need can be a function of organizational form and practice rather than a cause of it; for example, it is the small size of some firms that makes them unable to market over long distances. Other firms may have spatially limited purchasing fields through accident rather than design. Transaction cost theorists might see this as a happy coincidence, one which produced economic benefits, but that would not explain its origins. The coincidence may come to influence the subsequent evolution of the industry so that it reproduces spatial agglomerations wherever it develops (cf. the evolutionary approaches presented in Hodgson (1988) and Storper and Walker (1989)). Thus it can be argued that the way in which Japanese, or for that matter any, industry is organized has been influenced by, *inter alia*, pre-existing spatial forms. Perhaps the extreme physical restrictions on space and the population density of Japan have been more congenial to agglomerated, highly organized, and disintegrated production than has the American context, in which giant, decentralized, multidivisional companies prospered.[20]

An evolutionary perspective recognizes the role of place and local happenstance in the development of new forms of industrial organization, not in the trivial sense that innovations have to come from somewhere but in the sense that their causes and conditions are highly time- and place-specific. The Third Italy is one of the best known illustrations of this (Brusco, 1982; Murray, 1983), but as we indicated in the previous chapter the innovations of Henry Ford and Toyota were also place-specific. This implies more possibilities than either the Fordist (global corporation) or flexible specialization theories acknowledge. It does not mean, however, that locally inspired innovations cannot spread by imitation, as we saw in the case of just-in-time.

Yet if organizational innovations cannot be grasped wholly by a timeless decision logic, neither can they be explained purely in terms of local idiosyncrasies. While there may be local stimuli – such as growth of new consumer markets (British clothing industry), or highly exploitable immigrant labor (Los Angeles garments), or rural labor (Japanese consumer

---

[20] Consider also the following arguments from Lash and Urry (1987, p. 47): "What thus appears to have happened in Britain is that because it was the first industrial nation, there was the emergence of extensive and systematic market mechanisms which made it unnecessary for any individual company to develop an enormously complex 'visible hand'. Companies could rely upon the 'invisible hand' which had served British companies so well during nineteenth century liberal capitalism. Why integrate forwards and purchase a number of sales outlets when existing commercial firms operated more efficiently via the market? Why integrate backwards into raw materials production when the market mechanism was already highly developed? . . .

Geographical distances made such an effective market system inoperable in the US, where vertical integration and hierarchy were the only viable strategies. But in places like Sheffield, where the costs of information and transaction were low, there seemed no need for vertical integration in metal working".

electronics), or anti-trust measures (Los Angeles films) – there are durable endogenous mechanisms in capitalist development, too. The possibilities presented at particular times and places (not only local, but national and global) are exploited within the framework of increasing division of labor and increasing production system complexity, combined with continual organizational innovation. This need imply neither the final victory of the large corporation nor its defeat (Harrison, 1989). Deepening of the social division of labor can be compatible not only with vertical disintegration but with increasing integration because multi-product firms can internalize parts of that division (see chapter 3).

We have argued that many of the key themes of the post-Fordist/flexible specialization theory are misconceived and too restricted in their perception of new developments in capitalist industry. In particular, it is necessary to look not just at ownership structure but at organization and its constituent practices. The Japanese experience is illuminating in both respects. It is time to rectify its astonishing neglect.

## Flexible Rigidities? Japanese Industrial Organization

While some features of Japanese industry resemble those of the flexible specialization model, there are others which either contradict it or are simply absent from it. Given the competitive success of Japanese manufacturing firms, neither capitalists nor academics elsewhere can afford to ignore these discrepancies. Crucial to Japanese industry's success has been its exceptionally organized character. In no other comparable capitalist country has the visible hand of organization and planning pushed back so far the invisible hand of market forces. Moreover, the degree of protectionism – formal and informal, intended and unintended – is exceptional. "Flexible rigidities", the oxymoronic title of Dore's (1987) analysis of industrial organization in Japan, aptly captures the paradoxes, paradoxes which confound many of the expectations of both orthodox economic theory and the post-Fordist debate itself.[21] In relationships within and between firms, and with respect to international trade, Japanese industry is characterized by rigidities such as obligated trading relationships, lifetime employment, and seniority-based payment systems, which would seem to impede the flexible working of market mechanisms. Yet far from inducing lethargy and sclerosis, Japanese companies have proved remarkably competitive and nimble in restructuring and innovating.

[21] The following account draws mainly upon Dore (1987), Aoki (1988, 1989), Yamamura and Yasuba (1987), and Florida and Kenney (1988b). It must be said that both western and Japanese commentators differ surprisingly among themselves over their characterizations of Japanese industry. We cannot say how far these differences stem from their having studied different sectors with different characteristics. For example, Dore's case studies in *Flexible Rigidities* are drawn mainly from textiles though he has considerable expertise on other sectors.

We shall look at the essentials of this "structured flexibility" (Kenney and Florida, 1988) in relation to the organization of industries, to subcontracting, and to labor markets.

## Organization of industries

As noted earlier, Japanese manufacturing is highly vertically disintegrated and has become more so since 1974 (Aoki, 1987; Mitsui, 1987; Nishiguchi, 1987). Yet it is also highly organized, not only within firms but at the inter-firm level. This is best understood in terms of industrial groups and relational contracting.

Regarding industrial groups, three, possibly four, major forms exist (see Figure 5.1):

1  *Gurupu* (or financial *keiretsu*), e.g., Sumitomo, Mitsui, Mitsubishi, are the remnants of the old *zaibatsu*, broken up during the US occupation. These are headless groups of major firms from different sectors (i.e., they are not conglomerates). "Each group has a bank, and a trading company, a steel firm, an automobile firm, a major chemicals firm, a shipbuilding and plant engineering firm and so on and, except by awkward accident, not more than one of each" (Dore 1987, p. 178). The members meet regularly and confer continuously over strategy (Sako, 1989). There is some limited overlapping ownership, and members give one another preferential treatment in trade and inter-firm relations (for example, a car firm is unlikely to buy steel from outside its *gurupu*). The *gurupu* effectively protect member firms from takeover and spread risks by giving each member an interest in the survival of the others. They ensure that investment is undertaken mainly by committed, knowledgeable investors with long-term interests in group survival rather than speculative investors out for short-term individual gain. The pattern of company finance reflects this protective role of the *gurupu*: heavy reliance on low interest loans from the *gurupu* bank, limited equity, low reserve ratios, high gearing, and evaluation in terms of long-term growth and stability rather than short-term profit. These characteristics help firms to move along the learning curve and price aggressively. However, it must be said that the *gurupu* are weakening (Murakami, 1987). Not all major firms belong to them and new industries such as information technology are poorly represented in them.

2  *Keiretsu* (or capital *keiretsu*) consist of major firms and their dependent supplier/subcontracter firms, organized into three or sometimes four tiers. For example, Matsushita has over 600 firms in its *keiretsu* (Aoki, 1984). A limited degree of mutual and superordinate/subordinate shareholding is common, so vertical disintegration, in terms of ownership, is not total, but most subcontractors are legally and practically distinguished from subsidiary companies (Mitsui, 1987). Generally, the higher the firm in the hierarchy of the *keiretsu*, the longer-term the transactional relations.

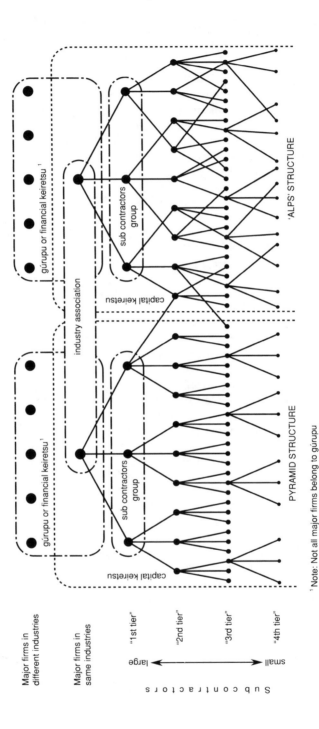

Major firms in
different industries

Major firms in
same industries

"1st tier"

"2nd tier"

"3rd tier"

"4th tier"

large ←——————→ small

S u b c o n t r a c t o r s

gúrupu or financial keiretsu [1]

industry association

sub contractors
group

capital keiretsu

'ALPS' STRUCTURE

PYRAMID STRUCTURE

[1] Note: Not all major firms belong to gúrupu

*Figure 5.1* Typology of industrial groups in Japan (based on Dore, 1986)

While the suppliers often sell to outside firms, the *keiretsu* are dominated by relational contracting in which many suppliers are dominated by a single buyer. Commentators differ over the degree of closure of these groups. Dore implies that relatively closed, pyramidal structures are dominant; Nishiguchi (1987), describing the car industry, emphasizes an "Alps" structure in which suppliers, especially in the higher tiers, serve several customer firms (see Figure 5.1). Dore notes that textile subcontractors generally do not want to diversify their customers (even where customers want them to). Nishiguchi emphasizes the development of flexible specialization-type relationships through the Alps structure; Aoki (1988) claims that between-group links are increasing. Despite these variations, the *keiretsu* are still hierarchical rather than matrix organizations.

3  *Industry associations of major firms in the same industry.* These tend to be strongest in early stages of development when the industry, often with the aid of MITI, can arrange collaborative research and strategy. Alternatively, they form as a defensive or recession cartel when an industry faces declining prospects. In a growth situation, an industrial association gives MITI an interface for fostering growth through organizing collaborative R&D and a division of labor in products, restricting competition and targeting products (Murakami, 1987). Things do not always work out so tidily, but the contrast with western, disorganized growth is still strong. In a situation of decline, the stable relational subcontracting relationships of the *keiretsu* and *gurupu* have the effect of facilitating cartelization because cartel members know their customers are unlikely to cut loose and divide (and rule) their suppliers.

4  *Cooperation groups (kyoryokukai).* These are emerging within *keiretsu*, as organizations linking small numbers of subcontractors operating in the same tier of the hierarchy, and involved in technically and functionally related work. "Major firms orchestrate the cooperation groups in the first tier of suppliers, institutionalizing both formal and informal interaction of people involved in the relational manufacturing process" (Nishiguchi, 1987, p. 13). Again, discussing the car industry, Nishiguchi writes (p. 13):

> Through an enormous number of meetings orchestrated by the institution – at virtually all the personnel levels from top executives down to manual workers – exchange of information takes place between relational manufacturing units. Through "study group" activities in particular involving both production engineers and shop-floor managers, better or best practices are discussed, engineered and experimented.

In some cases, first-tier suppliers have also replicated the model established by their "parent" firm by organizing cooperation groups for second-tier suppliers.

Effectively, this structure means that the Japanese economy's input-output matrix is partitioned in a way which limits and channels direct price competition. According to Dore, fierce inter-company price competition is limited largely to consumer markets which are expanding, though competitive pressures between large firms are always transmitted back to their suppliers – albeit often through non-price rather than price mechanisms, usually by tightening quality standards.

This form of organization also inhibits firms from buying into a range of activities as a way of becoming conglomerates. Takeovers are rare, and most firm growth comes through internal expansion or creation of subcontractors, a situation very different from that in most western firms. This is not simply a consequence of the inhibitions of group membership, but reflects management philosophy; Japanese directors, in contrast to the British, "... are apt ... to believe that running a company actually requires substantive expertise in the products and markets with which the company is involved" (Dore, 1987, p. 63). The success of Japanese firms in innovating without the "flexibility" in investment patterns that western firms enjoy may well have contributed to the growth of the "back-to-basics" movement in western management circles.

In situations of decline, the strong rooting of firms in particular sectors evidently has advantages as Japanese firms facing declining markets have proved better able to survive than their more diversified (and flexible?) western counterparts. In the textile industry, Dore shows how Japanese firms formed a cartel to spread capacity-cutting evenly, to regulate entry into the industry, and pursue a common strategy of moving up-market into higher quality fabrics which were less affected by competition from cheap labor countries. Imports have been limited by raising quality controls on imports. Larger firms pressed suppliers to innovate, so that relatively advanced technology is found even in small family firms. High levels of task flexibility within firms allowed further economies in increased work intensity and greater utilization of capital. The industry is likely to decline further in the face of Third World competition, but thanks to these developments, partly fostered by the cartel, it has withstood that competition significantly better than the textile industries of most advanced countries. A similar controlled reduction of the Japanese steel industry in the face of declining markets has also been effected (Shapira, 1989).

### Organization of subcontracting

Vertical disintegration is often associated in the literature with non-routine kinds of production, in which specialization reduces uncertainty and increases capacity utilization and flexibility.[22] Nevertheless, the incidence of

---

[22] Dore (1987, p. 190) attributes increasing vertical disintegration in textiles to the fact that, as production becomes more capital intensive, the balancing problems of large mills become more acute because they cannot fully utilize all their capacity simultaneously. Consequently, economies of scale can more easily be realized on a cross-enterprise basis.

relational contracting where single buyers dominate the output of suppliers suggests that specialization of function in response to balancing problems is not the dominant reason for the highly disintegrated structure. In their respective studies of Hitachi and the car industry, Ikeda (1987) and Nishiguchi (1987) offer explanations close to those in the post-Fordist literature: namely, that diversifying consumer demand has prompted a shift to smaller lot production by subcontractors. (Just how significant that diversification has been is unclear due to the ambiguities surrounding product diversification mentioned earlier.) Yet vertical disintegration is common in the immensely successful Japanese mass production industries, where subcontractors do the same kinds of repetitive work that might be carried out within integrated western firms. The rationale here is that volume production actually fits better with relational contracting than does small batch or custom production (Aoki, 1989; Sako, 1989). The latter might have some need for buyer-seller collaboration for a short period, or for a succession of projects, but in mass production there is more scope for a continuous relationship in which the parent intervenes in the subcontractors' internal processes as well as influencing the nature of their product/output.

Subcontracting relationships are highly organized and involve far more than simple market exchange. Collaboration with, and surveillance and domination of, suppliers facilitate the achievement of high quality levels. Indeed, according to Dore, big firms dominate their subcontractors more by tightening quality requirements than by threatening to turn to cheaper suppliers. As we saw in the last chapter, western industry has long underestimated how quality improvements not only make products more saleable but lead to considerable economies in production. Both of these aspects – coordination and quality – are key features of organization within and between Japanese firms.

Generally, parental companies "often support or assist subcontractors with funds, guaranteeing credits, lending production tools and machinery, giving technical advice for managerial improvements" (Dore, 1987). Large contractors forego short-term price advantages offered by alternative suppliers for the sake of maintaining close relations with, and control over, the subcontractors in their *keiretsu*. Management collaboration and intervention is strong, with major firms often "lending" managers to subcontractors and helping establish new suppliers. Although the *keiretsu* are dominated by major assembly firms, there is also considerable delegation so the top firm is "exempted from the burden of managing too many external manufacturing units." The top firm generally has to approve who suppliers subcontract to, so as to control quality, but the purchasing of second-tier products is entirely managed by the second-tier firms, and so on. Thus, while "in 1987 General Motors had some 6000 'buyers' managing on average 1500 suppliers per plant ... Toyota in 1985 had only 337 people in its centrally controlled Purchasing Department. The contrast becomes even more striking when one learns that Toyota's outsourcing percentage of manufacturing costs is 70% whereas GM's is 25%" (Nishiguchi, 1987, p. 15). Formal vertical disintegration is therefore complemented by strong, but highly stratified, vertical organization.

This kind of relational subcontracting has dynamic advantages in facilitating coordinated design and investment (Ikeda, 1987). For example, the design of TV components can be coordinated with that of TV chassis so as to facilitate both the performance of the final product and the efficiency with which it can be assembled.[23] Therefore, the product development process, from planning/conceptualization through designing and engineering to manufacturing, is one of overlapping, rather than discrete, stages. According to Nishiguchi, this interactive process helps to anticipate and avoid problems before they occur, so that it takes only four years to develop a new model of automobile, whereas the sequential process of US firms takes five to seven years (see also Aoki and Rosenberg, 1987). And Freeman (1987) argues that Japan's success in "reverse engineering" foreign products (i.e., analyzing and improving them in an integrated way) owes much to the horizontal and vertical communication fostered by the high degree of organization within and between firms.

Relational subcontracting, coupled with a high degree of intra-firm organization, therefore encourages joint learning-by-doing and the diffusion of innovation within industrial groups. The opposite tendency for diffusion between groups to be limited is presumably offset by the cross-cutting structure of different kinds of industrial groups – *gurupu*, *keiretsu* and industry associations (Figure 5.1). However, it is clear that the overall structure inhibits the diffusion of innovation to foreign competitors. The semiconductor industry is instructive. In the United States, a largely unorganized, vertically disintegrated industry, coupled with high employment mobility between firms, facilitated innovation, diffusion, and imitation by foreign companies. But the weakness of links between chip manufacturers and equipment makers, links which are essential for innovation in semiconductors, led to inferior performance in mass-produced chips. Moreover, innovations often took place in new (inevitably small) start-ups which were ill-equipped to follow through and maximize long-term gains from them (Stowsky, 1987; Florida and Kenney, 1990). Meanwhile, in the Japanese industry, locking in key employees through lifetime employment, and equipment suppliers through relational subcontracting, had the effect of locking out foreign imitation, creating a formidable non-tariff protectionist barrier. On top of this, the longer-term orientation of the financial environment in Japan meant that development programs were less often interrupted by the short-term fluctuations of the volatile semiconductor market. And the tendency to keep innovations within large firms or captive

---

[23] Even in vertically integrated British TV firms it was common for marketing departments to promise new designs without even consulting production engineering.

spin-offs facilitated follow-through. This example amply illustrates the advantages of organized over unorganized capital.[24]

Finally, although subcontractors that fail to meet requirements can be jettisoned, the system is relatively stable. This results in lower transactions costs for information search and marketing than those experienced by the more open, flexible, and unorganized inter-firm relationships common in other advanced industrial countries. As Ikeda (1987, p. 6) puts it, because of "the long-term, almost permanent relationship between subcontractors and finished product manufacturers, subcontractors feel no need to establish sales and other non-productive [sic] divisions."

### Organization of labor markets

The segmentation of labor markets corresponds to the stratified industrial structure. Firms tend to homogenize their workforces as much as possible – for control reasons, avoiding comparability claims. Generally, the lower the value added and the simpler the task, the further down the hierarchy the work is likely to be done, the lower the job security and pay, and the higher the proportion of female employees.

In terms of the Japanese labor market as a whole, contrary to the assumptions of many western observers, individual mobility is roughly the same as in other industrial capitalist countries; the much publicized phenomenon of lifetime employment covers only about 20% of the nation's workforce, chiefly key workers, invariably male, in major companies. Nevertheless, given the strategic importance of these workers and their immobility relative to western counterparts, lifetime employment, along with seniority-based payment systems, is still a "rigidity" worthy of attention. Again, these characteristics seem economically irrational, apparently discouraging individual effort and merit and inhibiting labor force numerical flexibility. Yet task flexibility is a condition of lifetime employment: having to change tasks need not entail losing one's job. Other workers are encouraged to develop multiple skills, too. There is also a general absence of notions of demarcation – a characteristic reinforced by the use of payment systems based on seniority rather than rate-for-the-job, by company unions, and by the weakness of labor organization in small firms. Security of employment is generally believed to promote lethargy, but it must be remembered that such workers cannot expect to find employment of the same quality in another firm. As Dore says, Nissan would no more employ a Toyota manager than the British army would have employed a member of the SS! For the individual lifetime employee, the only alternative to staying with the firm is the insecurity and inferior pay of the secondary labor market.

---

[24] This contradicts Lash and Urry's (1987) Eurocentric notion of increasingly "disorganized" capitalism.

But the adaptability of Japanese workforces cannot be attributed wholly to the formal features of labor markets. Informal modes of social organization play a major part, particularly through the exploitation of peer-group pressure and the groupishness widely noted in Japanese culture. For example, if one's colleagues have been willing to forego a holiday or accept a new way of working it is difficult to refuse to do the same, especially given the centrality of group norms and the sanctions against those who break them. Such features, coupled with the formidable sticks and carrots of Japanese management, are enough to produce an adaptable labor force in the primary sector.

In the long run, the lifetime employment commitment,[25] combined with the high degree of insulation from short-term profit pressures, effectively discourage firms from simply cutting back employment in the face of adversity, and encourage them to diversify and innovate their way out of trouble (Kenney and Florida, 1988). At the same time, the high level of job rotation and horizontal communication among key employees facilitate problem-solving and restructuring (Aoki, 1989). In other words, these short-term operational rigidities are exploited in such a way that long-term flexibility in restructuring is gained.

Numerical and output volume flexibility are not greatly limited by lifetime employment partly because this covers only a minority of the total workforce. Fluctuations in overtime, including overtime by lifetime employees, and changes in the number of female and temporary contract workers account for a large part of output variation. Supplier firms are apparently not used as shock absorbers as commonly as many observers expect, and what the subcontractors do absorb is mainly, again, through overtime fluctuations (Aoki, 1984; McCormick, 1985; Dore, 1987). As we have seen, big firms have an interest in maintaining stable relations with suppliers rather than driving them to the wall; using them as shock absorbers could be extremely costly in the long run. In any case, the high degree of insulation of firms from short-term market forces reduces the need to sacrifice suppliers in recessions.

As might be expected, these characteristics are also a response to local Japanese conditions (Dore, 1987; Mitsui, 1987). The extreme unevenness of economic development within Japan, with its heavily subsidized agricultural sector, and the extreme form of patriarchy, differentiate pay and working conditions along urban-rural and gender lines to a greater extent than in most industrial countries. This in itself makes vertical disintegration attractive. It is still common, for example, for subcontractors to employ workers who can supplement their income from agriculture, where high subsidies support continued rice cultivation (Ikeda, 1979). The extreme form of patriarchy that exists in Japan produces a workforce which is unusually differentiated along gender lines, with women's average earnings only 52% of men's, a low proportion for an advanced industrial

[25] Lifetime employment is not generally guaranteed by legal contract (McCormick, 1985).

society (Tasker, 1987). The rigidity of lifetime employment contracts for some men is complemented by the flexibility of female employment arising from the pressure on women to leave employment, at least in larger firms, at marriage or childbirth.[26] For a significant minority of women, the choice is not only between employment and unpaid domestic work, but also involves unpaid non-domestic work in the family firm (Dore, 1987; Patrick and Rohlen, 1988). For many workers, 50- to 60-hour weeks are common and the practice of foregoing vacations is widespread. From the point of view of capital, it could be argued that the degree of organizational work required both within and between firms (often of an informal, "after hours" nature) to secure coordinated development and innovation demands such hours. At the same time, long work weeks reinforce patriarchy, by making it difficult for married male workers to do anything in the home (cf. Phillips, 1984). Moreover, in order to facilitate the reproduction of key (male) workers, employers sometimes influence marriage decisions of key workers or provide bachelor flats for unmarried workers not living with their mothers.[27] Some of the same phenomena are common in the west of course, but not to the same degree and with so little resistance.

Given these circumstances, some commentators still wonder if the Japanese subcontracting system is a sign of backwardness rather than advancement, and criticize the institutional obstructions to the market mechanism as quasi-feudal relics. While the organization of Japanese industry does have some pre-capitalist roots, these are thriving, not anachronistic or relict features of the economy, because, as we saw in the last chapter, Japanese capital has made a virtue out of necessity, adapting developments from outside to local circumstances, and evolving forms which are often superior to those found elsewhere.

Important though these contextual conditions are, the success of highly vertically disintegrated systems depends on the way in which they are actually organized, on the specific practices involved. Provided there is sufficient organization, then, there are advantages in vertical disintegration. In particular, the *keiretsu* structure and its associated mode of governance have the effect of allowing degrees of organization between firms to approach those within firms, with the added advantage of financial independence of partial vertical disintegration of ownership.

## Conclusion

We have argued that recent industrial change cannot be forced into the frameworks of binary oppositions found in post-Fordist or flexible special-

[26] Female workers in big firms are expected to retire on marriage, usually at 25, though half the married women in Japan are employed, usually in small firms.
[27] Thanks to Karen Wigen, Department of Geography, University of California, Berkeley, for this point.

ization theory. On theoretical grounds, the unities and oppositions of Fordism and post-Fordism, mass production and flexible specialization, can be shown to be illusory; in practice, the key case of Japanese industry illustrates that these theories "do not compute", and it is the program, not the data, that is wrong. Unexpected combinations such as mass production with vertical disintegration and relational contracting show that, ironically, these theories are insufficiently flexible.[28]

One of the great dangers of theorizing about the development of industrial capital is to assume that it derives simply from a dynamic purely endogenous to industry, ignoring the influence of local conditions. The Japanese example makes a powerful case for giving greater consideration in studies of uneven development to the role of the wider context in which industrialization occurs. Of course, it is partly a combination of cultural distance and ethnocentrism which makes us notice this context in Japan but not in Europe or North America. There is no such thing as normal capitalism, and no reason to regard western industry as any less influenced by its local cultural and socio-political context than that of Japan. Perhaps taking more notice of Japan will enable us, like anthropologists returning home from the field, to see our own societies with new eyes. In this respect, as Storper and Scott (1988) advise, we need to study the social institutions associated with distinctive kinds of industrialization, extending this to what Freeman (1987) terms "national innovation systems". Indeed, we could profitably look further still to broader social and cultural characteristics (cf. Burawoy, 1985). The absence of studies of the context of industrialization has tended to lead to reductionist explanations of uneven development based upon selective studies of a firm or sector. What these do not explain, even when summed up, is why "in certain historical periods particular *countries* tend to do exceptionally well in export performance not just in one or two industrial sectors, but in many simultaneously, indeed sometimes in almost all of those sectors which are not dominated by natural resource availability or long-term traditional fashion-based factors" (Freeman, 1987, p. 96). The answers to this puzzle are to be found in areas such as education, the social and institutional form of capital, and state-capital relations, and – moving beyond the bounds of national innovation systems – in labor market characteristics, employment relations, and culture.

While the organizational forms of Japanese capital do have some features in common with flexible specialization, they also have characteristics which call into question not only the nature of flexibility, but also the central contrast between the alleged decline of mass production and rise of small batch production, and the implicit associations between vertical integration and mass production. Though not part of current concepts of post-

---

[28] e.g., see Friedman's (1988) attempt to interpret the Japanese phenomenon as flexible specialization involves numerous adjustments of the data.

Fordism, there are also other aspects, such as the relationships between industrial capital and financial capital, and the significance of accumulated product and market-specific expertise which – remembering the threat of Japanese capital to the west – are too important to ignore. Finally, the above quotation from Freeman neatly summarizes our principal objection to the mass production/flexible specialization contrast which underpins the post-Fordist debate: the superior industrial performance of Japan or West Germany over the United States and Britain is not so much a question of their relative commitments to mass production or flexible specialization but has to do with broader "environmental" characteristics which cut across labor process and production scale distinctions. In this respect comparative studies may serve us better than binary histories.

# 6

# Capitalism, Socialism, and The Social Division of Labor

"... (U)nder normal circumstances internally divided and fragmented labor is at the mercy not only of the ruling class and its state but also of the objective requirements of the prevailing division of labor". (Meszaros, 1986, p. 21)

These have been unsettling times for political economy, both as theory and as practice: we have witnessed the end of the post-war boom in the 1970s, the rise of the new right in the 1980s and, most dramatic of all, the disintegration of "already-existing socialism" in the USSR and Eastern Europe. Political economic theory has been slow to respond to the enormous challenge posed by these changes. In our view, all of the major problematics of political economy have structural flaws which limit or distort their understanding of the new social economy, in which, once again, division of labor figures prominently. In this chapter we shall focus mainly on the problems of Marxist and radical theory in understanding the micro-economic coordination of the division of labor. Contrary to a common view on the left, we believe that the crisis of "already-existing" or "disintegrating" socialism does indeed suggest fundamental problems in Marxist theory – problems which also have implications for its understanding of capitalism.

We want to argue two things: first, that Marxist and Marxist-informed theory on capitalist economic development has suffered from a pervasive neglect of problems relating to the integration of the social division of labor, and has tended to attribute effects to class which derive more directly from division of labor; and second, that this neglect is a primary reason why socialists can say so little about the transition to socialism, or about socialist economic policies. In consequence, Marxist theory of how capitalist economies work is curiously disengaged from socialist economics – i.e., from the theory of how to transcend capitalist and develop socialist economies. Such theory is of course important in its own right, but it is impossible to consider it without looking back at the workings of capitalism in a new light. We shall argue that instead of ignoring socialist economics, as Marxists from capitalist countries often do, we might fruitfully ask what lessons it has for our understanding of capitalism.

The opening quotation from the Marxist philosopher, Istvan Meszaros, alludes to a central point of the argument, and many readers may consider it unexceptional. Yet while it might seem obvious that divisions of labor divide labor, this is widely ignored on the left, being overshadowed by the attention given to labor's class position. Sometimes this pattern of emphasis is justified on the grounds that within-class divisions are a distraction from class struggle. However, this has deeply damaging political consequences for it supports the practice of underestimating the problems posed for socialist strategy by the enormous complexity of the social division of labor.

Modes of production contain not only relations of domination but mechanisms for integrating the division of labor. Any power structure which runs into difficulties in integrating the division of labor in society is itself endangered. Conversely, in any given mode of production, a challenge to prevailing relations of domination invariably has effects upon the regulation of the division of labor. This is most clear in the case of patriarchy but also applies, in a more complex way, for class. The extent to which any class or group can control its economic circumstances is not only a function of its ownership or non-ownership of the means of production, but of the extent to which it can control the division of labor. And this second aspect does not follow automatically from the first, as the history of attempts to socialize production show.

Analytically, this suggests that the effects of the division of labor and the particular ways in which it is integrated cannot be reduced wholly to effects of social relations of production. For example, since market co-ordination can co-exist not only with capitalist social relations but with petty commodity production, state ownership, or worker-owned firms, it cannot be regarded simply as a secondary characteristic of capital. On the other hand, to recognize that relations of production and modes of integration are relatively autonomous does not mean we can ignore the interaction between them, for sometimes it may be mutually reinforcing, sometimes mutually antagonistic. In other words, the methodological problem we face here, like that encountered in previous chapters, is the risk of misattributing causality when analyzing structures which are always in interaction.

The order of the chapter is as follows. We begin by defining the problem of the integration of the social division of labor in terms of its complexity and the mode by which it is coordinated. In the next section, we examine the treatment of these matters by the major problematics of political economy, paying special attention to Marxism. Then we show how Marxism repeatedly interprets effects of the division of labor and its integration in terms of class. A discussion of control, power, and the social division of labor provides a basis for a different, more materialist conceptualization of the social division of labor, one which can comprehend the failure of capitalism to pave the way for the socialization of production. We next examine the crisis and dilemmas of socialism, both in its state and market socialist forms, in relation to the division of labor. Then, in the light of these lessons we look at the implications for policies and stances adopted by socialists in capitalist countries. We conclude by bringing into relief the interacting structures of class and division of labor which underpin the range of issues covered in the chapter.

A word of warning: while we make many criticisms of Marxism and socialist economies in what follows, these are not in any way intended to idealize the theory and practice of the right. It should be possible in an academic context to criticize one system without being thought uncritical of the system opposed to it.

## The Complexity of the Division of Labor and the Problem of Integration

"The wealth of societies in which the capitalist mode of production prevails appears as an 'immense collection of commodities' ..." So begins Volume 1 of *Capital*. As many commentators have noted, Marx never came to terms with the implications of the complexity of the social division of labor in which that immense collection of commodities was produced; and of course, over a century later, the scope and complexity of the division of labor and the output of commodities has grown exponentially, so that now many millions of different kinds of commodities exist, not only within capitalist countries but also within already-existing socialist countries.

If one is to understand the problems of integrating the division of labor, it would be disastrous to underestimate the extent of what has to be controlled. Even looking at the extraordinary range of products and separate producers in a product directory for a single sector – say the construction industry – is a salutary experience. And at the level of individual firms, although it is common to talk as if each produced only a handful of different commodities, it is not unusual to find firms with product ranges of hundreds or even thousands of different commodities.

Moreover, the division of labor does not exist outside space and time. (How much simpler the problem of integration would be if it did.) Instead it is stretched across the globe, so that the connections between production and consumption, between effort and effect, are extraordinarily indirect and hidden. The development of divisions of labor is also temporally open-ended.[1] We have only a rough idea of our future needs, and insofar as they depend on what others will do, they are not simply uncertain but unknowable, for people have to act in the knowledge that they cannot reconcile actions beforehand. Consequently, the idea of producers and consumers

---

[1] Giddens (1984) refers to this as "time-space distanciation".

in a complex division of labor jointly agreeing in advance what should be produced and consumed is far-fetched, to say the least; perfect information and foresight are out of the question. It is better to start by imagining a vast input-output matrix, disaggregated to the level of individual kinds of commodities and spatially disaggregated as well.

Some readers might argue that complexity, space, and time can be ignored because theory always has to abstract from detail, and indeed this is what makes it so powerful. In a sense this is right, but abstractions give us a general grasp of one kind of complexity by abstracting from another. They cut into the connective tissue of the world at different angles, and if we have too few abstractions over too narrow a range of angles we miss important things. What the complexity and the spatial form consist of is usually an empirical question, but we should at least expect our theories to alert us to the fact that complexity, spatial form, and historical time will make a significant difference. For the problems they pose are not only conceptual but practical, forming a hard material base with which any form of economic organization must engage: quantity, form, and time have qualitative effects.

There are several aspects to the problem of integration itself. First, complex systems of communication and material distribution must connect producers and users. However this is done, specific producers in specific places need to establish contact with specific potential users of their products, again in specific places. This does not happen spontaneously, and a vast amount of labor and resources is needed to effect coordination in addition to the activities of producing and consuming. This applies to any advanced division of labor, whether in capitalist or socialist society. (Hence our neutral, historically unspecific terminology of users and producers and products at this stage of the argument, rather than capitals and commodities.) As we saw in chapter 3, the growth of these connections is in itself one of the major accomplishments of economic development, both effect and cause of the growth of the social division of labor. Yet, as we shall see, all the major problematics of political economy neglect it.

A second aspect of integration is that material processes of coordination also presuppose, and help provide a way of coping with, the division of knowledge associated with the division of labor. Hayek (1949) has provided the clearest statement of this issue, which he regards as *the* economic problem: that of socially mobilizing a vast and changing array of fragmented knowledge held by producers and consumers. If no one can possibly know all the particular capabilities and needs of producers and consumers, and all the relative costs and scarcities, how is any coordination achieved? Closely related to this is the problem of motivation: what moves the myriads of producers across the social division of labor to invest, produce, and consume so some coordination is achieved?

It is essential to recognize the particularity of production and consumption capabilities and needs here. When a computer breaks down, it usually requires a particular component, and no other, and one must find not just a producer of electronic components, but a producer of that particular component out of the hundreds of thousands available. This is the level at which microeconomic coordination takes place. Multiplied across the millions of products made and used in an advanced economy, this indicates the hugeness of the practical problem of micro-economic integration. As the lessons of central planning tell us, it is not a matter of simply coordinating general categories of use-values, nor is it the same as the macroeconomic problem of realization, dealt with by a long range of theorists from Marx to the French regulation school, which concerns aggregate levels of output and demand.

A third aspect is more economic in character, and concerns the means by which equivalence in exchange is established, by which the different preferences and powers of users, and the different capabilities of producers, are registered and evaluated in the context of ineradicable scarcity. In other words, there must be allocational mechanisms, be they of force (legitimate or illegitimate), convention, negotiation, or exchange value. Allocational questions are never simply ones of substantive rationality, of assessing ultimate needs, for the costs (in the broadest sense, including expenditure of human and natural resources) of providing the means to those ends are usually significant and involve trade-offs or opportunity costs. There are always relative scarcities: how many of each type of computer should be allocated to different users; how much diesel fuel should go to the food industry, how much to shipping, how much to the different kinds of shipping, and so on.

This aspect of the coordination problem would be immediately visible to a planner in a centrally planned economy, but it is no less present in capitalism. The fact that the "solution" in capitalism is an unintended consequence of spending patterns, capital accumulation, and competition among many separate producers does not mean that it is of only secondary importance, for capital accumulation and economic development also recursively depend on past solutions to the problem.

Radicals may connect the issue of the allocation of scarce resources with introductory textbooks on neoclassical economics, where it is usually associated with discussions of markets and the operation of the price system, with market-clearing solutions to static allocational problems, and with a deafening silence on class and exploitation. But relative scarcity and allocational efficiency are too important to be thrown out with the bath water of static equilibrium and neoclassical apologetics for capitalism. Precisely because socialists want an economy in which people are the controllers rather than the victims of economic processes, it is especially important that they – more than anyone else – take microeconomic integration and allocational efficiency seriously. As we shall see the expropriation of capital does not solve the problem but merely poses it in a different form.

The best-known modes of integration are markets and central planning, but there are others, too, as we saw in chapter 3. Divisions of labor may be coordinated by various combinations of custom, authority, coercion, and democratic negotiation, through networks, states, or markets.[2] None of these can operate in isolation. Contrary to a common view in economics, markets are not self-sufficient, but presuppose acceptance of a certain moral order (custom, authority), rely on the coercive threat of the state, and involve plenty of negotiation. Planning has similar preconditions. The converse is not true, however: custom, authority, coercion, and negotiation can integrate some divisions of labor without markets or central planning. These means of coordination are not rendered anachronistic by markets or planning; they are widespread in contemporary societies, both supporting markets and planning and operating independently of them. Each provides a way of dealing with aspects of connection, information, motivation, and regulation, at scales ranging from the interpersonal to the internal organization of institutions, and to national and international levels.

Modes of integration differ in their strengths and limitations, and each is limited by the nature of the activities, and the form of the division of labor it attempts to coordinate, and by existing social relations of production and other relations of domination. In the following comments we shall outline salient characteristics of four of these modes, concentrating on the most commonly cited, but most misunderstood – markets.[3]

### Markets

As the most pervasive mode of integration in capitalism, markets and the price mechanism require special consideration. Their conceptualization is far less straightforward than commonly supposed, and much confusion derives from the increasing tendency in both popular and academic discourse to mythologize and essentialize the market. Worst of all is the pervasive use of "the market" or "market system" as a synonym for capitalism. Markets are a mode of integration which can co-exist with several different kinds of social relations of production.

Economists tend to consider markets purely in terms of the logic of the price mechanism or the "invisible hand", abstracting from the institutional or social context and producing an undersocialized view of exchange; sociologists tend to invert this, producing an oversocialized view.[4] Neither reduction will do, for in real markets the logic of the price mechanism interacts with the institutional or social context. As Tomlinson (1990, p.

---

[2] See Lindblom (1977) for discussions of alternative classifications.

[3] These are a subset of those discussed in chapter 3.

[4] Parker (1982) argues that contemporary economists have overlooked the fact that Adam Smith did not use the term "invisible hand" exclusively to refer to market exchange but identified it with the more general phenomenon of the constraining effect of the unintended consequences of actions.

43) puts it "There is no such thing as 'the market', but only markets, whose effects are determined by the agents who operate in them, their forms of calculation, their relations to other agents and the legal and moral ... framework in which they operate". Markets vary considerably according to how well-organized and well-informed consumers and producers are about the production and distribution process, implying different degrees of power of producers relative to consumers (Elson, 1988). The responses of producers and consumers vary according to degrees of trust, whether they can postpone consumption, depend on profits, etc. Such differences bring into question the notion of an invariant logic of choice and behavior.

Markets presuppose the regulation of economic activity according to exchange-value, and are generally understood to involve not merely isolated exchanges but routinized, regular exchange of commodities for money (Hodgson, 1988). Since buyers have some choice over how they spend their money and since many products are substitutes – even those produced by monopolies – markets are characterized by varying degrees of competition. When generalized, markets provide a decentralized form of control and coordination of the social division of labor, though they do not do this unassisted.

Contrary to the impression given by many economists, markets are not a kind of ether but are embedded in varying degrees in social relations (Granovetter, 1985; Tomlinson, 1986; Hodgson, 1988). Nor does the hidden hand work unassisted: considerable amounts of labor are involved in creating and sustaining market exchange, and that work is by no means reducible to the act of exchange plus mere advertizing.[5] Neglecting both the social embeddedness of exchange and the labor of market-making gives the impression that commodities sell themselves or that all markets involve spot trading and arms-length relationships between participants.[6]

Nevertheless, it remains true that the precise results of the operation of the market are largely unintended consequences of countless individual actions. Even where there are few direct transactions or markets between separate sectors, as with mining and tourism, prices still perform an allocative role in coordinating the social division of labor. It is therefore important not to switch straight from an undersocialized view of markets to an oversocialized one. There are face-to-face social relations, but this should not obscure the central role of the interplay of unintended consequences of actions through the price mechanism, the cost calculus, and economic logic. Relational contracting and other deeply embedded forms of exchange are still strongly affected by exchange-value signals in terms of

[5] Granted, Marxism acknowledges realization problems but these are invariably considered abstractly rather than as the failure of concrete processes of marketing.
[6] This makes it difficult to understand why there are so many non-production workers; in many computer firms, only a minority of the workforce are in hardware production, while the majority are involved in administration, development, marketing, sales, and servicing.

costs, rates of profit, and rates of accumulation, albeit over a longer period than in arms-length trading; exchange-value remains the final arbiter. Such practices therefore represent not an abandonment of this kind of economic logic but a more indirect yet sophisticated way of competing within it (see chapters 3 and 5).

Real markets are spatially fragmented and operate without perfect information and foresight, and they link production systems and consumption patterns which are often highly inflexible; as a result, the price mechanism is far less effective than in textbook models.[7] Most markets have relatively "sticky" prices and adjustment may take place more through actors withholding supply or demand than by price change. Contrary to the impression given by some neoclassical texts, volatile prices suggest poor coordination of supply and demand. In some sectors (e.g., agricultural crops), instability of supply and hence price are unavoidable, but generally industrial capital finds stable prices more conducive to long-term development and risk-taking, and it goes to considerable trouble to organize markets so as to stabilize its environment. As Chandler's (1977) history of the growth of the large corporation in the United States shows, firms had little incentive to improve their internal organization in an erratic economic environment. This unpredictability, in turn, was due to the weakly organized nature of markets and, more generally, of distribution and production.

Prices primarily reflect costs of production, but this does not mean that demand has no effect on what is produced (Walker, 1988a). Even if prices are fixed regardless of demand, over time the effects on revenue will tend to ration production in accordance with demand patterns, as well as cost. Markets do not merely establish an equivalence of productivity among competitors (average socially necessary labor time) and relate production and exchange to the relative costs of producing different commodities (the social valuation process); they also adjust production of different commodities according to demand levels, providing a continuing social validation of existing and new products.[8]

As Tomlinson suggests, economic behavior cannot be read off simply from descriptions of markets, e.g., by counting the number of competitors (Auerbach, 1988). A market in which many firms sell is not necessarily more competitive than one with few sellers, and exposure to competition through markets does not guarantee that firms will compete effectively, as liberals often suppose. Indeed, competitors often succeed by insulating themselves from price competition. Contrary to economic models of price competition, with their static equilibrium assumptions, the most significant

[7] For the purpose of understanding real markets, the concept of perfect competition is not a harmless abstraction but a gross obfuscation (Weeks, 1981).

[8] Contrary to the impression given by the labor theory of value, this implies that social validation (socially-necessary products) cannot be reduced to the matter of socially-necessary labor time involved in their production.

kind of competition in capitalism – strong competition – is relatively insensitive to current prices since it derives from technical and organizational changes which will alter the whole basis of pricing and create further disequilibrium (Storper and Walker, 1989). Nevertheless, *a posteriori*, price and exchange value considerations remain the final arbiter.

The chief advantage of markets over other modes is their ability to facilitate allocational efficiency while minimizing the information costs of microeconomic coordination. Markets allow decentralized coordination of individual consumers' and producers' actions, and the price mechanism makes the allocation of resources responsive to relative scarcities and consumer demand, insofar as that demand is backed by purchasing power. At the limit, suppliers and buyers need know little or nothing of each others' situations and motives, for feedback in terms of exchange-value – or more particularly price – can prompt appropriate responses without expensive surveys, endless committee meetings or cumbersome political agreements, which can never produce more than approximate and inflexible assessments of relative scarcities and demands.

Like all modes of integration, markets have well-known failings. The regulation of production and consumption by markets can lead to major injustices, economic irrationality, and ecological damage (Buchanan, 1985). Two drawbacks are most relevant to the discussion of micro-economic coordination. First, the price mechanism does not communicate all the information required to coordinate actions. Where certain transactions are of critical importance to agents, such as the purchase by a firm of some vital piece of equipment, extensive extra-market information exchange and negotiation may be needed to enable the market transaction to go ahead. In such cases – and they are common – minimizing information costs is not the issue, though given the dominant role of exchange-value in commodity production, price is still important, both to the firm and for allocation of resources in the economy at large. Second, although markets provide information, firms may not be able to act on it in a way which coordinates supply and demand owing to prisoners' dilemma situations (O'Neill, 1989). Thus competitors facing a declining market generally require some extra-market mechanism, such as a cartel, to coordinate scaling-down of production so that no one can take advantage of the cuts of the others. This suggests that in some cases it is decentralized control rather than exchange-value which is the problem.

Without delimiting particular activities or contexts, we cannot say *a priori* whether market prices are good or bad (Amsden, 1989). The same goes for state-regulated prices. Subsidies may enable the child of poor parents to be less disadvantaged than otherwise, or they may cause overproduction and wasteful consumption. The main point is that it is absurd to condemn or celebrate all markets on the basis of selected successes or failures, without looking at their contingent forms and without comparing the efficiency and equity properties of market coordination with those of other feasible modes of integration.[9]

While these qualifications are important, as are the equity implications of allowing widespread market coordination, it is essential to evaluate alternative modes of integration with equal assiduousness. Unfortunately this is often overlooked in radical circles. How do the alternatives compare in terms of their ability to respond to needs (including needs as defined by individuals), to use resources in economical and ecologically benign ways, to coordinate activities efficiently and fairly (including the work involved in coordination), and to motivate workers to work to high standards?

### Electoral democracy[10]

Democratic control is not just a means to an end. It has intrinsic virtues as an end in itself, but these may not compensate for its defects. Its scope is heavily limited by the size of the constituency, and the number and complexity of the issues at hand. While democratic control can be made less costly by delegation and representative democracy, the costs may still outweigh the benefits (Buchanan, 1985). The scope for electoral democracy is limited largely to voting either for simple single issues or for whole packages of issues – disaggregation is extremely costly. Where interests are diverse and the tasks complex, the degree of control can only be loose, not least because of the problem of the division of knowledge. Millions of users obviously cannot vote for millions of different products from thousands of different enterprises. Nor can "an elected assembly decide by 115 votes to 73 where to allocate ten tonnes of leather, or whether to produce another 100 tonnes of sulfuric acid" (Nove, 1983, p. 77). In sum, and in the light of socialists' enthusiasm for democratic control, one is reminded of Oscar Wilde's jibe, "socialism takes too many evenings."

In addition to these disadvantages, a free-rider problem weakens the incentive to vote (my vote won't make any difference). A different free-rider problem arises when people vote for things instead of paying for them individually; they may vote for more than the economy can deliver, squeeze out other activities, and generate inflation. Democratic control, unassisted by price signals, lacks a strong mechanism for responding to relative scarcities. Finally, electoral democracy has to confront a classic dilemma: to implement and enforce democratic decisions, a central authority – usually the state – is required, and this creates opportunities for anti-democratic forces to develop.

[9] Precisely because modes of integration are ways of coping with complexity, it is vital that any alternatives are specified in detail equal to that provided for the systems being criticized.

[10] Electoral democracy was not considered as a mode of organization of integration in chapter 3 because it so rarely occurs in the economic affairs of capitalism. We treat it here because of its importance in the political theory of capitalist societies and socialist economies.

*Central planning*[11]

Many segments of the division of labor are directly coordinated *a priori* by planning. Planning must be authoritarian and centralized to a significant degree. Democratic refereeing may be possible, but design and implementation must override local or sectional opposition, or the plan will be ineffective. When attempted at the scale of a national economy, coordinating vast input-output matrices of complex divisions of labor by central planning involves calculations that outstrip the capacity of any foreseeable level of computer technology. Inevitably, it is necessary to decentralize detailed decisions to a significant degree, and to aggregate and approximate needs and production capabilities. This produces inefficiencies in microeconomic coordination – just what kinds of electronic components, or drugs, are needed? Moreover, in view of the problems of democratic control just noted, the state must either decide what people should consume, and allocate it as use-values, or allow them to "vote" with money by buying commodities. The latter course predominates because it is very difficult to register relative scarcities and costs when the economy is coordinated through use-value output targets. Hence, exchange-value is retained, though prices are generally not market prices and do not influence the economy in the same way.

In no other situation is the complexity of the division of labor so apparent; unlike the hidden workings of the market mechanism, centralized allocation of resources among competing ends is transparent and highly politicized. As we shall see, it is not so much the class, bureaucratic, or undemocratic character of central planning which causes its distortions and inefficiencies, but the enormous task of planning a complex economy which makes bureaucratic structures and authoritarian decision-making inevitable (Nove, 1983).

*Networks*

Even in a highly advanced social division of labor there are areas in which closely related activities can be coordinated directly through various blends of collaboration, negotiation, and domination. These usually involve small numbers of actors with common interests, well-defined projects, and the possibility of direct interaction. Networks consist of interlinked units exchanging information and services for mutual benefit. These exchanges need not include cash transactions but often do; they are deeply embedded. Unlike markets, networks can only function with relatively small numbers and in relation to relatively simple goals. They are often important in linking particular segments of the social division of labor, most obviously production systems or supply chains, in which members' activities are

---

[11] We shall develop this topic more fully in the discussion of state socialism below.

technically interdependent. But unlike market coordination, they are of no help in regulating weakly-related activities.

We have defined our problem of the micro-economic coordination of the division of labor and introduced various modes of coordination. We shall now examine how these subjects are treated – or distorted – in political economic theory.

## Approaches to the Division of Labor and the Problem of Integration

Division of labor and the problem of micro-economic integration receive contrasting treatments in the major problematics of political economy: neoclassical, Austrian, and Marxist. None of these is satisfactory, but we shall dwell longest upon the silences of Marxism because of their influence upon the problems faced by the disintegrating socialist economies.

It has become a commonplace that the respective preoccupations of Marxist and neoclassical economics are mirror opposites, the former examining production and class with unrivaled detail and insight while largely ignoring exchange and allocational efficiency, the latter reversing these priorities. Neoclassical economics focuses on the achievement of equilibrium and allocational efficiency through maximizing behavior and the logic of choice in exchange in crude terms; it is more concerned with economizing than with the nature of the economy (Barratt-Brown, 1970).

When it does occasionally address production, neoclassical economics continues to use an exchange model in which production is merely a matter of combining inputs in the right proportions, rather than one of serial organization and transforming inputs into outputs (Leijonhufvud, 1986). This exchange model holds for many who consider themselves to have broken with the neoclassical tradition, as for example in Williamson's (1985) extraordinary attempt to reduce production to "transactions" (see chapter 3).

The Austrian school, in which Hayek is pre-eminent, shares the neoclassical blindness to class and exploitation, and a disinterest in the nature of production. Both are as overwhelmingly apologetic to capitalism as Marxism is critical of it, and both are explicitly hostile to socialism, nowhere more than in the work of Friedman and Hayek. Both emphasize division of labor, but largely as a prelude to an exclusive concern with (and eulogization of) markets and the price mechanism, a focus which relegates other modes of integration of the division of labor to the status of anomalies.

However, from the point of view of our problem of integration there is a significant difference between the Austrian and neoclassical schools. The former takes a robustly dynamic view of the coordination of the division of labor, one which recognizes the impossibility of perfect knowledge or foresight and looks at processes in historical, open-ended time where there is no possibility of reconciling demand and supply before exchange takes place (Arnold, 1989; Shapiro, 1989). This is quite different from neo-classical economics, which focuses mainly on the allocation of given scarce resources among known competing ends.

In contrast to these problematics, the great strength of Marxist economics is its analysis of production and the recognition of its social character. But no matter how highly these qualities are rated, it does not follow that Marxism's weakness regarding exchange and allocational efficiency is insignificant. Nor can it be excused on the grounds that Marxism is just not interested in these issues; the fact that a certain group of theorists is not interested in a problem does not make that problem unimportant in practice, as radical students of socialist economies have found to their chagrin.

Marxism's treatment of the problem of integration can be faulted on several counts. First, regarding complexity (Charles Taylor, quoted by Nove 1983, p. 65):

> Marx seems to have been oblivious of the inescapable opacity and indirect-ness of communication and decision in large bodies of men [sic], even in small and simple societies, let alone those organized around a large and complex productive system. ... [this] prevented Marx from seeing com-munism as a social predicament with its own characteristic limits ... [thus] Marx held a terribly unreal notion of freedom, in which the opacity, division and cross-purposes of social life were quite overcome.[12]

We shall examine Marx's own thinking on the social division of labor later, but this tendency to underestimate the complexity of the social division is evident in modern Marxism – even where other sources of oppression are fully recognized. Second, with regard to the problem of how producers and consumers are materially connected, Marxism is just as guilty as mainstream economics of implying that commodities, once produced, sell themselves without any marketing labor.[13] Third, the allo-cational problems and associated forms of economic calculation used to evaluate alternatives are largely foreign to Marxism.[14] In contemporary Marxist theory, as in Marx, "the question of economic calculation (in its most general sense as the criteria and methods for choosing between different uses of the available means (resources) for the achievement of optimum economic results) was never, or hardly ever, the subject of consideration" (Brus, 1972, p. 28).

---

[12] Although, as we shall see, Marx was aware of the implications of complexity within the technical division of labor of large-scale production, he was not where the social division is concerned.

[13] The work of Baran and Sweezy (1966) is a notable exception. See also chapter 2.

[14] In their useful critique of the Marxist and liberal problematics in political theory, Bowles and Gintis (1986) argue that Marxism lacks a fundamental theoretical vocabulary to represent con-ditions of choice and individual liberty, and hence cannot address the problem of despotism (see also Tomlinson, 1982).

The dominant impression in Hayek's problematic is of an economy atomized into individual economic agents, but miraculously coordinated by markets and the price mechanism. By contrast, although Marxism characterizes the social division of labor as "anarchic", it provides a remarkably holistic and integrated view of the economy – too integrated in fact, for it obscures the problems of coordination. Where neoclassicists and Austrians treat production (supply), distribution, exchange, and consumption (demand) as separate elements, Marxists (following Marx in the *Grundrisse*) tend to integrate these as "moments" in a dialectical movement dominated by production. Thus, distribution is first of all distribution of the means of production and hence not something subsequent to production, consumption is not autonomous for consumers are also workers and consume to reproduce their labor power for production, production involves productive consumption, etc. (Marx, 1973).

This form of abstraction does have the virtue of showing the interconnectedness of these moments, and of confronting the way in which economic systems are reproduced, which the neoclassical and Austrian problematics largely miss. But it achieves these insights at the cost of burying the problem of the coordination of the division of labor at the level of particular commodities and their producers and consumers. To acknowledge that consumption is not entirely autonomous is not to say it makes no difference of its own. We are not saying that Marxism takes the reproduction of the system to be automatic; indeed the risks of noncompletion of the circuits are often noted (e.g., the value of the commodity may not be realized), but this still skirts the problems of allocation and coordination.[15] Nor does Marxism wholly ignore issues of efficiency, but it does not address questions of allocative efficiency across substitutable commodities when choices must be made between alternatives.

We noted that neoclassicists retain an exchange model in their occasional ventures into production. The converse is true with Marxists. When they look beyond production within individual units to the level of the social division of labor, where the problems of the integration are concentrated, they tend to do so in ways which retain a production optic. Thus, as we saw in chapter 5, radical researchers (including ourselves) have recently become more interested in supply chains within production systems. This concerns sections of the social division of labor which have taken on a quasi-technical aspect, with exchange and allocation between separate firms subjugated to the requirements of production in the chain as a whole. Naturally, these parts of the social division of labor are more susceptible to Marxism's production orientation. Useful though this line of research has been it distracts attention away from the general problems of microeconomic integration and allocational efficiency as they occur not

---

[15] Marx's circuits of capital address some problems of microeconomic organization of capitalism, reminding us of the precarious nature of accumulation. However, they concern the situation of individual capitals rather than the coordination of many capitals.

just within production systems but between quite different sectors, such as smelting and optics. How is the allocation of resources (e.g., energy, labor power) between these determined?

The dominant modes of abstraction in Marxist economics which deal with the coordination or regulation of the system as a whole are macro-economic rather than microeconomic. In particular, the reproduction schema, as developed by Marx in Volume 2 of *Capital*, establish the macroeconomic preconditions for the reproduction of capitalism. They reduce the social division of labor to the division of the economy into two or three departments. This was an important theoretical innovation. How-ever, in taking the equation of output with consumption (both final and intermediate, productive consumption) as given, the economic problems of making them relate at the level of individual commodities, and hence of integrating the hugely complex division of labor of capitalism, get short shrift. Contrary to appearances, equilibrium models, whether Marxist, neoclassical, or Ricardian, do not show how coordination problems are solved; rather they illustrate equilibria and the conditions for their exist-ence without saying how they are reached (or not reached). They exclude all the problems of matching supply with demand at the level of individual commodities (Nove, 1984).

Reinforcing this blindness to the problem of coordination is the com-mon conflation of workers and consumers. The Marxist and Keynesian approaches make us appreciate that all waged workers are consumers (though the contrary is not true) and that wages – contrary to the divisive rhetoric of the right – are a source of revenue as well as a cost in the economy. But few individual workers produce even a tiny fraction of the range of commodities that they and others consume. Hence the conflation of people as workers and people as consumers, effectively defining them only as wage-workers, has the disastrous effect of obscuring the social division of labor again and creating the illusion that producer and con-sumer interests are always in harmony.[16] Workers' actions then appear to be in the best interest of consumers, and if they are not we are tempted to assume that it must be capital which divides them rather than the division of labor.

By such means, the Marxist problematic has evaded issues of micro-economic coordination. We now show that where effects of coordination problems have been exposed, they have had to be attributed to the social relations of production.

## Class Reductionism and the Social Division of Labor

Marxism's strength regarding production and class, and its weakness re-garding division of labor and integration, make it prone to attribute the

---

[16] It also obscures the division – strongly gendered – between paid and unpaid workers.

effects of division of labor to class relations. This class reductionism is evident in Marxist explanations of a wide range of features of capitalism. To help explain this we use a model which highlights the form of the social division of labor in relation to capital-labor relations and technical divisions of labor, as shown in Figure 6.1.

Production units contain capital-labor relations and technical divisions of labor. The relationships between production units constitute the social division of labor, assuming for simplicity that most units are owned by separate capitals.[17] The repetition of these relationships, thousand of times over, is indicated by the horizontal rows of commas, mimicking mathematical notation. Some capitals may be competitors, some may relate as suppliers and customers, but the social division of labor also comprises firms working in weakly related sectors such as air transport and mineral extraction. Even though these are relatively independent of one another in terms of exchange of commodities, they are still crucially, if invisibly, related by the price mechanism governing the allocation of capital and labor; their relative sizes still depend upon comparative costs and demand patterns. As the diagram indicates, some parts of the social division of labor involving supply chains are hierarchically structured, but the overall pattern is far more messy and complex, and is best thought of as a matrix. Some flows of products move up the diagram from primary producer to final consumers; the metaphors of "upstream" and "downstream" and "supply chains" conceal the fact that upstream producers can be using downstream products. Not represented are labor markets or flows of money and capital, weaving in and out of the structure, which would form an even denser pattern.

Compared to the usual Marxian reproduction schema, Figure 6.1 is inelegant because it includes both a repetition of relationships and a failure to close the circuits of the economy between output and investment, and consumption via profits and wages. This inelegance has a point, however: to problematize the consequences of repetition in real economies for their microeconomic regulation. When these issues are not problematized, there is a tendency to read most eventualities in capitalist development as outcomes of the changing relationship and struggle between capital and labor. We shall demonstrate this by looking at the treatment of competition, restructuring, spatial divisions of labor, management and discipline, workers and localities, and class struggle.

## Competition

Insofar as Marxist accounts deal with inter-capitalist competition, they do so overwhelmingly in terms of the capital-labor relation and the exchange-

---

[17] The distinction between production units or workplaces and firms, and the qualifications regarding the distinctions between technical and social division of labor made in chapter 3, are not relevant to our argument at this point.

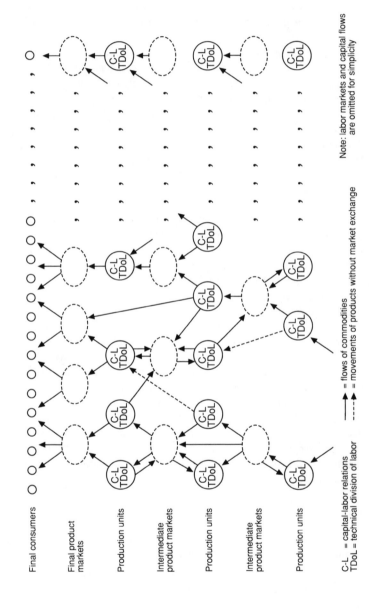

Final consumers

Final product
markets

Production units

Intermediate
product markets

Production units

Intermediate
product markets

Production units

C-L = capital-labor relations
TDoL = technical division of labor

⟶ = flows of commodities
⤑ = movements of products without market exchange

Note: labor markets and capital flows
are omitted for simplicity

*Figure 6.1*  Key relationships in the social division of labor under capitalism

value of the commodities in competition. So, competition is seen both as the whip hand forcing individual firms to extract more output from their workers, and as a consequence of firms succeeding in doing just that. We can readily agree that this is an important, perhaps the most important, process in the interaction between, on the one hand, competition among capitals, and the capital-labor relation on the other, especially in volume production industries such as consumer electronics.

Nonetheless, competition acts upon both aspects of the nature of commodities, not only on their exchange-value but on their use-value, and some successes or failures of individual firms have more to do with the latter than the former. Capitals survive not only by increasing exploitation, whether by lengthening the working day (absolute surplus value) or by increasing productivity (relative surplus value), but by developing new commodities (which might be termed "extended surplus value"; see chapter 1). This product innovation is often obligatory for survival in a way that is largely independent of the state of firms' capital-labor relations, though of course it may have a major impact on labor. Even where developments in capital-labor relations, especially regarding professional and managerial workers, are directed principally towards facilitating continual product improvement, this is rather different from trying to push up labor productivity in producing a fixed range of products, the case the left traditionally focuses upon (cf. Sayer, 1985).

The relationship between conditions inside a firm and its market position is not a simple one. Firms which are dominant in the market may have relatively favorable industrial relations, but the link is not inevitable; market domination can be built on worse than average working conditions. Likewise while "sweatshop" is a by-word for both marginal market position and oppressive work conditions, the connection between marginality and internal oppression is not universal; democratically run co-ops often have a marginal market position.

We can illustrate the relative autonomy of the use-value side of commodities in competition by looking at IBM. This firm has an unparalleled degree of monopoly in the electronics industry, with 70% of sales in mainframes; it is the largest computer supplier in every advanced capitalist country except (recently) in Japan, and generally overshadows even its biggest rivals. Yet in the late 1980s it was losing market share and (discreetly) shedding labor. How could so dominant a firm get into such trouble? The reasons cannot be found within the sphere of capital-labor relations and the immediate organization of production: it has invested heavily in process technology, spending $10 billion on factory automation in the early 1980s; its internal use of information technology to link global operations is among the world's most advanced; its annual R&D budget is bigger than the US space program; and its labor-management relations are the envy of companies all over the world. The basic explanation of its problems simply was that it had the wrong product strategy. It tried to continue to lock customers in to its products when the rate of growth and diversification of information technology products had outstripped even the capacity of IBM, and made users want products compatible with those of other firms (Morgan and Sayer, 1988; *Guardian*, June 30, 1988).[18]

This shows that strength in the capital-labor/production process sphere is not sufficient to guarantee success, and that even the largest firms cannot fully control technological and market trajectories. It also shows how hopeless it is to reduce the problems that capitals face in maintaining a place in their markets to the problem of minimizing costs and labor time directly involved in producing a familiar, standardized commodity whose marketing is unproblematic.

Some may object to this conclusion by arguing that product innovation is a response to the need of user firms to introduce process innovations, perhaps in some cases as a way to control labor better. Some accounts tend to use this as an excuse for reducing the situation of the supplier firm to that of the user firm, so that again technological change can be seen purely in terms of the capital-labor relation. Thus, the challenge of information technology is sometimes presented as a matter of merely increasing control over labor (e.g., Locksley, 1986). Not all product innovations involve process innovations (e.g., videos), but our point is that firms must be seen as technology producers as well as technology users: the characteristics deriving from the former cannot be read off from those deriving from the latter as simple consequences of the dynamics of the capital-labor relation.

If we ignore product innovation and growth in the total number of commodities of different kinds, then the prospects of capital continuing to win surplus value look dimmer than they actually are. Looking only within production and prioritizing the labor process, we learn much about how things are produced and about the forces determining the exchange-value of the commodity, but we learn little about what is produced, its use value, and its fortunes in the product market.

### Restructuring

Stories of industrial restructuring undertaken to improve process efficiency and control over labor have entered the conventional wisdom of the left. The political implications of this approach are generally expressed in terms of the need to strengthen labor against capital and its management, so that capital cannot pit one group of workers against another, to the detriment of both (e.g., Gough, 1986). Yet where this does occur, capital divides and rules because it is itself divided and in continual internecine conflict. We

---

[18] The converse, in which the success of a company has more to do with its product than process efficiency, is the case of Levi's, with its 501 jeans.

do not deny that such restructuring occurs, or reject the political goal, but we stress that this is far from the only process by which capital accumulation proceeds. However, as we shall see later, it is because of the one-sidedness of the orthodox view of restructuring that its proponents can say so little about what a strengthened labor might do with industry.

A preoccupation with the capital-labor relation makes it easy to overlook the fact that firms face *other* problems besides those concerning their relationship with labor, in particular the need to respond to changing product technologies and markets, and these, in turn, cannot simply be reduced to consequences of successes or failures in disciplining labor (Tomlinson, 1982; Morgan and Sayer, 1988). To be sure, any kind of restructuring affects the capital-labor relation, but it is often begun for other reasons. In the 1970s, when firms producing telephone exchanges shifted from electrical to electronic equipment, it caused thousands of job losses in Europe and the United States, but this was not the reason it was done. Although it created the opportunity for changing to new, green workforces, the restructuring would have been necessary even without this opportunity (contrast Bannon and Thompson, 1985). This illustrates that the relations between capitals and their product markets are relatively autonomous from capital-labor relations and production.

## Spatial divisions of labor

Prevailing treatments of spatial divisions of labor at both the international and intra-national scales echo the problems already noted in three respects (Massey, 1984). First, spatial patterns tend to be seen too little in terms of the social division of labor and too much in terms of technical divisions of labor, which of course relate closely to capital-labor relations within firms; indeed spatial relations of class, rather than divisions of labor, are often the primary focus. Inter-regional differences within some countries are now marked more by distinctions in corporate, technical, hierarchical divisions of labor than by distinctions in the social division of labor (i.e., distinctive sectoral bases), but it does not follow that spatial divisions of labor are no longer social divisions of labor, only that the latter are more jumbled spatially, at least at the inter-regional level. Similarly, the growth of international corporate hierarchies should not make us overlook the fact that the international division of labor is still overwhelmingly a social division – between different sectors and between competitors within the same sectors – and uneven development at this scale is still heavily structured by what the neoclassical tradition terms comparative advantage (Morgan and Sayer, 1988; Storper and Walker, 1989).

Second, technical spatial divisions of labor are seen primarily as reflections of capitals' relationships with different kinds of labor rather than as possible by-products of market strategies. For example, in the English M4 corridor many electronics firms have developed the elite nature of the area's labor markets, but this has often been a side effect of location decisions based mainly on reasons of market access (particularly access to large specialized users); hence, the resulting spatial division of labor is to a great extent an unintended consequence of investment decisions. Again, labor is not necessarily capital's main problem, so in explaining uneven development we should be wary of putting too much weight upon concepts of spatial divisions of labor: spatial differentiation of markets is still important in the geography of capitalist production (Morgan and Sayer, 1985).

Similarly, in analyzing the rise of the Newly Industrializing Countries it is not enough to attribute development to state repression, superexploitation of cheap labor, and the position of branch plants within multinational technical divisions of labor. These conditions certainly need to be exposed, but they do not amount to a sufficient condition for the rise of the NIC's, for many countries with cheap, super-exploited labor show little sign of catching up with more developed countries. What matters is not only the state of capital-labor relations, but the position of capital in the former countries with respect to technological and market trajectories, the selection or targeting of particular products and sectors, backed by long-term state support, and the relationship of indigenous capital to foreign capital in terms of subcontracting, licensing, joint ventures, technology transfer, and so on; the Korean case illustrates this perfectly. Success depends on strategies with respect to use-values and market, too (Morgan and Sayer, 1988). A third problem in thinking about spatial divisions of labor is a tendency to read spatial relations between employers and employees within the same firm, but in different places, purely in class terms, rather than as between occupants of different positions within technical divisions of labor.

### Management and discipline

The above points lead to important implications for understanding capitalist management. Management has been seriously neglected, and its activities largely reduced to those of controlling labor and enhancing productivity (Tomlinson, 1982), a tendency reinforced by the lingering influence of Braverman (1974) and of Marglin's (1974) widely-cited essay on the origins of the factory. But control of labor is not necessarily the main problem facing firms – indeed, material production as a whole may not be the main problem either (Morgan and Sayer, 1988). Product and marketing strategy, distribution and administration, the reorganization of management and financial restructuring are frequently more important. These emphases are not necessarily irrational, for in many companies direct production is not the major source of costs. These are considerations that new worker-controlled enterprises must quickly learn to appreciate, but radical analyses of capitalism generally fail to warn them.

Even to the extent that management is about controlling the technical division of labor there are deficiencies. Marx himself recognized two sides to the capitalist production process: that dictated by its capitalist form and that dictated by its use-value, or material, technical form. Similarly, in analyzing the social relation between capitalist and workers he acknowledges that certain tendencies derive simply from the size of organizations and their technical possibilities (Marx, 1863, pp. 448–9; cf. Nove, 1983, p. 50):

> All directly social or communal labor on a large scale requires, to a greater or lesser degree, a directing authority, in order to secure the harmonious cooperation of the activities of individuals, and to perform the general functions that have their origin in the motion of the total productive organism, as distinguished from the motion of its separate organs. A single violin player is his own conductor: an orchestra requires a separate one.

However, as Rattansi (1982) notes, what seems to have happened following Braverman is that Marxists came to see the management of labor, and the existence of hierarchical control of the technical division of labor, almost entirely as products of the class character of capitalist production, not as partial products of its scale and complexity. Similarly, technical difficulties are often read primarily as problems of social control. Thus, for example, the left has tended to interpret the software productivity bottleneck in the information technology industries as a problem of skilled workers keeping control over their work in opposition to management, not as a problem deriving largely from the inherent technical difficulty of working with software (CSE Microelectronics Group, 1980; Morgan and Sayer, 1988).

### Workers and localities: determinants of weakness

The implication of these points for particular groups of workers is that their insecurity and lack of power derives from several sources (Selucky, 1973):

- First, as Marxism amply recognizes, from being propertyless.
- Second, from their position within a technical division of labor, which both divides them and restricts their experience and expertise, consigning the majority to positions where they are cut off from matters of strategic planning and competitive survival in product markets. This naturally reinforces the difficulties workers have in determining their destiny.
- Third, by belonging to an organization competing within, and dependent upon, a wider social division of labor. The latter in turn is governed by the law of value and mediated through the largely

uncontrolled and uncontrollable drift of technological and market trajectories.

The second and third sources of insecurity are important in their own right and by no means reducible to the first. They are also likely to occur in some form in any economy with large-scale production and an advanced social division of labor.

At one level, it may often be correct to attribute job losses through restructuring to employers' failure to keep up with changing product markets and technology (although not all can keep up simultaneously, and there may be job losses even when they do choose the "right" strategy). But while the point is generally acknowledged, it is often ignored in radical theoretical and empirical analyses of industry, or marginalized by a preoccupation with relative and absolute surplus value.

These determinants of the weakness of labor also illuminate the power relationship between firms and localities. Regional characteristics, especially the nature and prior socialization of labor, affect the competitiveness of firms within those regions, not just in terms of the cost of inputs – particularly labor power – but in terms of how those inputs can be used. Yet there is also a deep divide between the interests of the firm and those of the worker in the region or regions in which it is situated. Unless the region consists of an integrated complex of firms, the interdependence between firm and region is highly asymmetric, for there are critical determinants of company performance which have nothing to do with the region. In other words, the region usually depends on the firm more than the firm depends on it. Consequently, each region is at the mercy of wider changes in product markets and international competition beyond its control; a region and its workers can hardly be blamed if a superior product from Japan threatens its local electronics plant or for redundancies caused by the introduction of new assembly technology. Again, at one level these points are widely recognized on the left, but if we read competitiveness entirely as a matter of process efficiency, there is a risk that we may actually overestimate the importance of local variations in workforces and their productivity, simply because we underestimate the weight of non-labor factors. In fact, the left may inadvertently provide ammunition for the right in blaming regions for their own problems. For labor, the asymmetry between firm (or plant) and region can be seen in terms of the three determinants of workers' insecurity already noted.

- The class relation may involve ownership and control by capitalists outside the region.
- A spatial technical division of labor which consigns local people to a lowly position within the corporate hierarchy gives them little knowledge of, or interest in, the arcane details of their firm's changing technologies and markets, even though their livelihoods may be strongly affected by the firm's performance.

- Where the region or locality has a highly specialized role within the wider social division of labor, it is vulnerable to the latter's opacity, changeability, and uncontrollability.

It is, of course, the social division of labor rather than class which isolates particular groups, making them dependent on distant, invisible producers and consumers. Benefits and costs of actions do not coincide in space or time and therefore tend, intentionally or unintentionally, to be unequally distributed and unaccountable (e.g., a decision to invest in the long-term development of one sector often implies hidden costs and benefits and opportunities foregone at different times and places).

## Class struggle and the division of labor

It might seem perverse to say that class struggle itself has often been interpreted in class reductionist terms, but our point is not that there are no class struggles, only that Marxism characteristically underestimates the fact that they are invariably combined with struggles over the changing social division of labor.

To illustrate this, we shall take a classic example of class struggle – the British miners' strike of 1984-85. There can be no doubt that this was indeed a major instance. Most radically, it involved a struggle over who makes investment and disinvestment decisions and what criteria they use, and need was counterposed to profit (though as, we shall see, a significantly limited concept of need). One of the arguments used by the National Coal Board in the battle for public opinion was that the miners' actions and their demands were contrary to the interests of other workers, both in the energy supply industry and in the rest of the economy. This is a familiar line for capital in strike situations, and as always the argument is disingenuous because capitalists have no material interest in the welfare of other workers, both because of their class position and, where workers in other sectors are concerned, because those workers are under the control of rival capitalists.[19] Such arguments are put forward by capitalists not because they believe them to be correct (though they may indeed do so), but to mask their class interest and to weaken opposition to it.

However proper the response of the left was in defending the interests of workers – all workers, not just the miners – as a class, it does not follow that the capitalists' argument was wrong. The point relates again directly to our opening quotation – "internally divided and fragmented labor" and "the objective requirements of the prevailing division of labor". What one set of workers does has implications for workers in rival indus-

---

[19] Although publicly-owned, the NCB behaved as a capitalist enterprise, trying to force itself into profit. The appointment of one of the most ruthless US capitalists, Ian MacGregor, as chair reinforces this point.

tries, most clearly in this case those in the rest of the energy sector. Some on the left recognized this, and argued that it wasn't enough simply to defend coal mining in isolation and to hope workers in the largely competing sectors of gas, oil, electricity, and nuclear power would support it, too. What was needed was an alternative strategy spanning all the energy sectors (cf. Howells, 1986; in retrospect, we might add ecological considerations). Even if such a strategy were primarily driven by need, there would still be competition between rival or alternative needs, and it would be necessary to evaluate the relative costs of alternative inputs, so exchange-value considerations would not be eliminated. Furthermore, such an energy strategy would confront the problem of its relation to the rest of the social division of labor, both in terms of the demand for its products and the distribution of income. Bottom-up planning by different sectors inevitably involves some competition, both for investment and the distribution of income. While this is most pressing in the short run, where the situation approaches a zero-sum game, competition – whether in the market or for the priorities of a central plan – could only be eliminated if scarcity were eliminated, a situation which Marx himself abandoned as a possibility in his later works (Rattansi, 1982). Although it is possible for wages to rise at the expense of capital's profits, it is virtually impossible to stop selective increases in real wages producing selective decreases elsewhere.[20]

We noted that profit was counterposed to need in the miners' strike but, as is often the case, the concept of need was not examined too closely. While the producer needs of the mining communities, in terms of employment, were clear enough, the case regarding needs on the consumption side, whether short-run (how to dispose of huge coal stocks) or long-run (restructuring the wider energy supply industry), was more difficult. Even in coal mining communities many households didn't use coal-fired heating.

If all this seems controversial let us rephrase it in the unexceptional terms of our initial "obvious points". The restructuring of capital always has distributional effects and frequently has distributional causes. If workers are obviously divided and fragmented by the social division of labor, how could there not be conflicts of interest between different workers? Trade unions are usually first and foremost concerned with securing their members' place in the social division of labor, though this is not unconnected to class struggle, or indeed to gender struggle. At every meeting between trade unions the truth of our opening quotation is driven home – class struggle and distributional struggle between different parts of the social division of labor are simultaneously present, and however much the left may want the latter to give way to the former, the problems of the

---

[20] This is not at all an argument against strikes, for apart from the point that a successful strike by the strong may advance the general class interests of labor, it may be in the interests of justice that some workers do indeed raise their income at the expense of others.

division of labor will never be very far away. This is in no way intended to imply that common class interests are unimportant. But just as it is now firmly established that the pursuit of class interests, per se, has often ignored gender divisions, it should also be clear that class has obscured the problem of the social division of the labor, its uneven development, and its integration.

## Control, Power, and the Social Division of Labor

Marxism may underestimate the complexity of the social division of labor and attribute its effects, and those of modes of integration, to class, but it is certainly not uncritical of the division of labor. Though overshadowed by the opposition to class, this older strain of criticism goes back to the early works of Marx. While these have had little influence on economic thought, it is highly instructive to consider how Marx's thought evolved regarding division of labor for it illuminates the antipathy, still common among radicals, to market coordination and a lack of understanding of the crisis of socialism (Rattansi, 1982).

From our contemporary position it comes as a surprise to find that in Marx's early work alienation is attributed more to division of labor than to class, though the two concepts are not strongly distinguished. As the famous (or infamous) romance of the "hunter, fisher, cattle-herd and critic" suggests, emancipation apparently depends on transcending the social division of labor. In his later work this theme is muted (though not absent), and class and capital are more sharply distinguished and become the main enemies. Although Marx no longer calls for the abolition of the social division of labor, he implies it must be drastically simplified (with each individual still expected to acquire a variety of capacities; Rattansi, 1982, p. 177). Production might still be done for others but it should be done as an end in itself, rather than for exchange-value, and without the cleavage between effort and consumption inherent in commodity exchange (Moore, 1980). He therefore regarded communism as incompatible with regulation by exchange-value.[21] The establishment of a classless exchange economy, socialism, was taken to be only a transitional stage to communism rather than a final goal.[22] Curiously (though not surprisingly in terms of his earlier thinking), Marx seems not to have considered the possibility of a market socialism of commodity-producing worker-owned

---

[21] There are some curious paradoxes in this for Marx saw market exchange *per se* as a realm of equality and recognized that it frees people from direct political dependence on others (Selucky, 1979).

[22] As Rattansi (1982, p. 193) points out, the tendency to underestimate the role of market forces under socialism is prefigured by "too close an identification between commodity production and private ownership, a tendency reflected in some of the confusions in the concept of social division of labor" (also Selucky, 1979, especially chapters 1 and 2).

enterprises (Selucky, 1979; Moore, 1980; A. Wright, 1980; Buchanan, 1985).

Further evidence of Marx's continued antipathy to an advanced division of labor and market exchange can be found in his critique of commodity fetishism in Volume I of *Capital*. This is virtually the mirror image of the liberal celebration of market coordination. Whereas for the likes of Hayek, the anonymity, functionality, and information-saving character of markets are prime virtues, for Marx they are dehumanizing and irrational, replacing direct relations between people producing use-values for visible ends with relations between people and things (commodities), in which they are the victims of unintended consequences of past economic actions.

Marx's critique of commodity fetishism has had a remarkably uncritical reception on the left, despite its extraordinary implications for the division of labor (e.g., Harvey, 1989). It is understandable that we should want to know that what we buy is not harmful or the product of super-exploitation, whether of people, animals, or of the environment. And there may be some point in knowing a bit about production, producers, consumption, and consumers in general terms in order to be well-informed. But why should we waste time checking up on matters which are irrelevant to social welfare, justice, and ecology? Why should individuals have to get involved in decisions regarding production or consumption of specific products about which they know nothing and could never know enough, however much time they devoted to the task? If we want the benefits of division of labor – and they are of historic importance in economic development – then we have to accept much of what goes with it, including the division of knowledge and the anonymity and functionality of social relations between producers and consumers. This does not mean total submission to market forces, for market exchange is always supplemented by other forms of coordination and control. National and international economies can only function on a largely *gesellschaft* basis, not as a local community (Holton and Turner, 1989; Miller, 1989). The anonymity and opacity of the integration of a complex social division of labor are far from unmitigated evils; to abolish them we would have to sacrifice enormous benefits in terms of economic development,[23] and liberation of time and interest.

Marx believed that capital grows out of commodity (value) circulation, to become the leading power over the whole system. Unless there are restrictions on hiring of wage-labor and private ownership and control of the means of production, so that profits of commodity producers cannot be used to establish capitalist social relations of production, capital fetishism inevitably grows out of commodity fetishism. But in pointing to the power of capital Marx stopped pointing at the division of labor as a power. Many Marxists have parked in the middle: they focus on the power of markets, underestimating *both* capital and the division of labor.

---

[23] Or accept comprehensive central planning.

Marx (1863, p. 929) considered that capitalism itself would overcome the social division of labor through the centralization of capital: "Centralisation of the means of production and socialisation of labor at least reach a point where they become incompatible with their capitalist integument. This integument is burst asunder. The knell of capitalist private property sounds. The expropriators are expropriated". Apparently, the social division of labor is supposed to be transformed into a technical division on a social scale in which the whole economy functions as one big factory – an idea endorsed quite explicitly by Engels, Lenin and Kautsky.

This idea, so central to Marxist thinking on the transition from capitalism to socialism, has caused immense damage in terms of socialist practice since it legitimizes extending the despotism of the technical division of labor to the economy as a whole. It is profoundly mistaken, both in its description of the development of capitalism and in its conceptualization of division of labor. As Auerbach et al. (1988) have made clear, centralization of capital does not replace the "anarchy" of the social division of labor with the "despotism" of the technical division. This is not only because of the reversal or slowing of the process of centralization, it is also because growing firms create new markets as well as replace old markets by internal organization. Here, as in questions of industrial organization, a zero-sum view of markets versus hierarchies (firms) just will not do. Any reduction in the apparent anarchy of the social division of labor since Marx's time is not so much because the social division has been partly supplanted by corporate technical divisions but because markets themselves and their preconditions (communication and distribution systems) have become more institutionalized and routinized, improving the coordination of demand and supply.

Moreover, insofar as large firms plan the development of their sections of the division of labor, we should not imagine this "planning for the market" is a direct correlate of central planning. Planning in relation to a particular target (maximizing accumulation) is not at all like planning for a whole society with all its divergent interests and needs.[24]

The failure of the centralization of capital to overcome the social division of labor further implies that development of the forces of production, far from eliminating the need for market exchange, will increase it, and that division of labor should itself be regarded as an aspect of the forces of production.[25] It is therefore not the backwardness of the forces of production that prevents the abolition of exchange, but their degree of advancement.

There is a more serious misunderstanding of the nature of control in an advanced division of labor, namely the view that the so-called anarchy of the social division derives from the fact that it is governed by capitalists,

[24] It is significant that the big problem Soviet planners faced in considering the use of programming models for planning was the choice of objective function. What should they maximize?
[25] Selucky (1979) has pointed out that in the German Ideology Marx and Engels appear to accept that the social division increases with the development of the productive forces.

each individually seeking to accumulate capital. But to say that the control of the social division of labor happens to be fragmented under capitalism, and hence the development of the economy tends to be anarchic, is to put things thoroughly backwards. The fragmentation results not so much from the individualism of competing capitals (though capitalist social relations increase it), but from the inherent difficulty of controlling an advanced division of labor, which, by default, provides scope for individualistic control. Given the expansionary dynamic of capitalism, individual firms would grow and swallow up more of the social division of labor if they could, as Marx supposed. They do not because they cannot. Although the modes of organization deployed by capital are becoming ever more effective, the burgeoning division of labor continues to outstrip them, proving too intractable for centralized control.

Now in *Capital* Marx defines the social division of labor under capitalism primarily in terms of ownership – it consists of separately-owned capitals, the relations between them being controlled *a posteriori* through the market.[26] But it also has a material dimension, for the divisions correspond to material differences between the various kinds of commodities – food, steel, transport, energy, videos, etc. The range of these different kinds of production, and the qualitative differences between them, limit the opportunities for continued centralization toward control by a single entity.

Overcoming the decentralization of control that defines the social division of labor therefore implies a solution to these material (simultaneously technical and political) problems. One could, of course, try central planning, but its history largely confirms our point: the division of labor between radically different activities is not simply contingently uncontrolled by virtue of fragmented ownership, but is materially unruly whatever the relations of production. Centralization[27] of activities under a single administration may work for some sectors, but at the level of society as a whole it is pulled apart and subverted by the intractability of the "inevitable cross-purposes of social life" in an advanced economy. Division of labor is therefore at least as much a barrier to the socialization of production as private ownership of the means of production.[28]

A social division of labor cannot function as a technical division of labor on a social scale (as "one big factory") just through a change in ownership. Within a particular technical division of labor, say for assembling airplanes, the proportions of different kinds of labor and materials are largely

---

[26] There is an exception where he defines it more abstractly as the "totality of heterogeneous forms of useful labor, which differ in order, genus, species and variety" (Vol. 1, p. 132).

[27] Centralization is equivalent to internalization in the terminology of chapter 3.

[28] "Socialization of production" is a notoriously ambiguous term, it being unclear whether it means internalization of formerly separate technical divisions of labor under a single administration as the centralization of capital proceeds or whether it also includes the increasing interdependence between individuals via the social division of labor (the Durkheimian sense).

a reflection of technical characteristics, the number of wings per plane, for example. Within the social division, the relative sizes of aerospace and newspaper production are not influenced in the same way but reflect relative production costs and demand levels. Consumers can demand more newspapers and fewer air flights, but they cannot have planes with three wings. Putting aerospace, newspapers, and all the rest of the social division of labor under social ownership does not annul these differences between the determinants of the technical and social divisions. As Selucky (1979, p. 33) puts it, "No legal, political or organizational change is able to abolish the social division of labor or to turn the division of labor into that within the enterprise."

The same point applies to arguments for centralization and public ownership as a way to internalize major externalities in the use of a particular good or service; e.g., roads and sewage. Although the arguments against decentralized and competitive provision are strong in these cases, centralization is not necessarily effective. Mere central control of a range of activities does not necessarily solve the problems of coordination (Nove, 1989a, p. 249):

> It is apparent from Soviet experience that one cannot internalize *all* externalities, i.e., consider everything in the context of everything, because of informational overload. Tasks must be subdivided. The centre itself becomes a loose federation of semi-autonomous ministries and departments presenting not only major burdens of coordination but also creating administrative boundary lines. These in turn reproduce the externality problem: anything that is not within the purview of a given official or department is, for him or her, an externality.

This obviously applies to large organizations, private as well as public, in any mode of production. Externalities are not purely functions of private property, of the form of ownership; rather, they also arise through material divisions, separations, and interdependencies in production at a social level. As such they require material organization, not merely formal ownership, to internalize them and include them in economic calculation and decision. Control, organization – and the material barriers to them – are the decisive forces.

We therefore need a much more robust view of the materiality of a complex social division of labor, one that interprets its fragmentation in terms of the scope it affords for effective control, rather than in terms of the contingent form of ownership of the fragments. This is not to suggest that the form of divisions of labor have a purely technical character; we can still acknowledge that they also reflect power and class and other interests. For example, private capital restricts workplace size partly for reasons of social control, to avoid giving labor too much leverage, and this influences the form of the division of labor. But we do need to keep in mind the more technical aspect, that divisions of labor are not inert but have their own powers and constraints.

A further consequence of these arguments is that the problem of industrial organization merges into the problem of the socialization of production. Economies and diseconomies of scope continue to contribute their own influence to the pattern of organization across the division of labor, aside from the question of ownership.

Radicals have always regarded ownership of property in the means of production as the primary source of power in capitalist society and with justification, both in terms of individual capitals and the general interests of capital. Yet property rights are never absolute (Macpherson, 1978). Capitalists are free to use and dispose of their capital as they wish, but they can only maintain their power if they conform to the disciplines of their economic environment, producing commodities not only with the average socially necessary labor time but commodities deemed socially necessary by their customers. When capitalists try to expand their power by expansion and acquisition, and indeed by collaborative ventures, they sooner or later run up against constraints in conforming to those disciplines.

Similar problems of control apply to labor. We have already noted the non-class sources of workers' vulnerability. If they were to expropriate their employers, and set themselves up in self-managed firms, their ability to use and dispose of their assets would still depend on finding users of what they produced, be they commodities or use-values. This is particularly evident when worker co-ops take over a capitalist firm that has gone bankrupt. The workers gain potential power through becoming owners of the means of production, but at first they lack power in terms of a secure place in the social division of labor; their productive capabilities are not socially validated. Capital is willing to let go of the means of production in these circumstances precisely because they no longer afford the necessary power with respect to division of labor. It is therefore no sacrifice to capital, and a most inauspicious position for labor in which to begin worker-control.

If, on the other hand, an industry or all enterprises were nationalized, with ownership formally belonging to the people, then real control would belong, as Aganbegyan (1988) says of Soviet industry, to everyone and to no one, since individual "owners" would have no significant influence on their "property". Real divisions would remain.

In this section, we have reconsidered the properties of the social division of labor in relation to control, coordination, and power and have found Marx's own contribution seriously flawed. Throughout these arguments, we have seen the dual influence of the intersecting structures of the social relations of production and division of labor on the distribution of power and control. This duality conditions all advanced political economic systems and must be taken into account in any evaluation of alternatives. As we shall see in the following discussion of socialist systems, examining alternatives illuminates these structures still more clearly.

## The Social Division of Labor and Socialism

To a greater extent than commonly realized on the left, the current crisis of the socialist economies is partly one of failure to control and transform the division of labor. Our main purpose here is not to evaluate different variants of socialism normatively, but to relate their characteristics to the problem of micro-economic integration and to illustrate the relative autonomy and mutual interaction of social relations of production and division of labor.

### Centrally planned systems

Under centrally planned or state socialism (these terms will be used interchangeably), capitalists are replaced by those in charge of distributing the surplus, the means of access to this power being the Party. As long as there is a complex division of labor and scarcity, there will be differences of interests among producers, and between producers and consumers – even in the absence of such divisions as ethnicity, gender, or nationality. These differences are not removed simply by instituting central planning, but are transformed into party, antagonism between direct producers and the planners, and administrators. Although the planning authority does not own the surplus product, as do capitalists, it controls its distribution. In view of this, Szelenyi (1986) and many others have argued that this gives rise to a basic class distinction around the processes of production and expropriation and the allocation of surplus.

This has been associated with a view of state socialism which sees its deficiencies almost entirely in class terms, and implies that workers must liberate themselves from the power of the bureaucratic and political class, and take direct control as a free association of producers, abolishing the state. But while the class or quasi-class character of state socialism is clear, the conclusions drawn from this are hopelessly awry; indeed, this is the mistake we encountered before of supposing that all workers in capitalism need do to gain power over their situation is to dispossess capital.

The bureaucratic character of planning is not merely an unfortunate contingent feature. It is inevitable that a complex economy which is run centrally should be despotic, even if the planners are democratically elected (Nove, 1983, p.77). Either individual units must be subordinated to the discipline of the plan or there is no plan, though certain matters can be devolved to intermediate levels. A free association of workers is impossible, not because planning happens to be centralized and despotic but because, "(C)entralism is the inevitable price which must always be paid for the abolition of the market without the prior eradication of its preconditions" (Selucky, 1979, p. 34). The quasi-class character of state socialism derives from the need to have a vertical division of labor of control for centrally coordinating a complex horizontal social division. As Bahro (1978) has demonstrated with considerable force, already-existing socialism has not removed the division between the labor of strategic planning and control of many activities on one hand, and the subaltern labor of executing particular tasks on the other.

Alienation therefore worsens, rather than eases, in this situation. The token nature of public ownership, compared to that of property over which people actually have some power (e.g., private agricultural plots), invites free rider problems and a lack of a sense of responsibility for public property (Aganbegyan, 1988). Work is no more meaningful when driven by plan targets than by profits and exchange-value criteria. Workers are still alienated from the product and from their wages as well, since producers do not have to respond to spending patterns but to the plan (Selucky, 1979).

In a highly complex economy, replacing market coordination by *a priori* planning does not increase the transparency of the relationship between production and consumption. If anything it has the opposite effect, for *a priori* control of such an economy has to be centralized, so the plan is interposed between producers and users, direct feedback from the users through prices is cut, and producers become accountable to the central planners. Producers are monopolists in a sellers' market, with centrally-determined prices, and central (vertical) control of supply chains supplants their horizontal relations, making low quality and poor coordination almost inevitable. Further inefficiencies stem from the costs of gathering vast amounts of information (remember the production of millions of different kinds of commodities has to be planned), and as a consequence of the inevitable abstraction of complex needs to simple statistical data (Nove, 1983).

In other words production is neither for profit nor need, but for the plan. Likewise production is valorized not by users through exchange but by the plan. Yet consumption is still atomized, as under capitalism and other market-coordinated systems, and follows a different logic from that of production: consumers attempt to spend their money according to their assessment of the use-value of products, but they have little choice; where they can say yes or no with their money, there is little direct incentive for producers to respond (Nove, 1983). In this schizophrenic form of coordination, workers "are left with the interest of maximizing their incomes, while bureaucrats function as a ruling class having an interest in accumulation . . ." (Selucky, 1979, p. 36).

Since millions of users cannot value millions of products by voting, and central planners cannot value them on their behalf (in terms of use-value), except in the crudest terms, exchange-value and money have not been eliminated in centrally planned economies. Rather, market prices determined by competition and the interplay of supply and demand have been replaced by centrally determined prices (Arnold, 1989). One does not have to idealize market prices to recognize that these regulated prices are unlikely to reflect relative scarcities and so act to ration consumption and production accordingly. They produce distortions and disequilibrium on the supply side, and can encourage inefficiency in consumption; for example, in the USSR bread is so cheap that it is widely used as animal fodder. Furthermore, disequilibrium and the absence of competition invite black markets and corruption. In the Soviet Union and some Eastern European countries, cooperatives have benefited from these disequilibria, often making large profits denied to those in state enterprises. Ironically, like black-marketeers in wartime, this has made cooperatives objects of contempt rather than hope.

Alternatives to market prices are problematic. Fixing prices purely according to costs ignores use values, demand, and opportunity costs. The average socially necessary labor time for producing a pair of shoes or a pair of skis tells us nothing about the relative use-values or social necessity of shoes and skis. On the other hand, making relative need the sole criterion ignores questions of allocational efficiency in meeting those needs. Allocation in kind is feasible where there is little differentiation of the product and consumption (e.g., water), but as differentiation increases so does the need for decentralization of decision-making (Kornai, 1986). Lastly, setting plan targets in output terms rewards waste compared to allowing demand to regulate output.[29]

In practice, the weak concern for allocational efficiency and the lack of competition mean that central planning tends to be heavily influenced by political considerations, often at the expense of economic efficiency. Leante (1989) claims that in this respect it combines the worst of centralism and localism. One paradox is that the very impossibility of comprehensive planning makes centrally planned economies susceptible to influence by powerful local or sectional interests, such as that of heavy industry in the Soviet Union. Conflicts between interests of the center and particular production units are inevitable – as it is often in the latter's interest to deceive the center. Not surprisingly, enterprises develop contradictory relationships to the center, wanting both more autonomy and more paternalistic favors from the center. At the same time they barter and make deals with suppliers and user enterprises through informal networks. The inadequacies of central planning have been so great that these practices have often taken over as the dominant mode of integration (Kornai, 1986).[30]

A crucial element of this combination of security without autonomy is the policy of prohibiting unemployment and the extreme rarity of bank-

[29] See Nove (1983) for a full discussion of these conundrums.

[30] Market systems are also very susceptible to political influence by large or well-organized groups of producers (e.g., the oil industry), so this is not uniquely a problem of central planning.

ruptcy at the enterprise level. Workers are therefore shielded from conflicts inherent in the uneven development of the social division of labor, though they indirectly suffer the costs of this fettering of the development of the forces of production. This allows situations where hardly any enterprises show losses, yet the national economy is in deep trouble (Kornai, 1986). In effect, while centrally planned socialist economies still use exchange-value, it is not the arbiter of economic survival: all budgets are "soft".

The clearest indicators of the coordination problems of centrally-planned economies are economic shortages.[31] Explanations of these differ (e.g., Nove, 1983; Kornai, 1986; Aganbegyan, 1988; Burawoy and Lukacs, 1989), but there is a consensus on the negative point: it is certainly not a result of deliberate restrictions on output, but rather the contrary. As to possible cause, the following circumstances are thought to encourage inefficiency in the use of resources and so produce shortages:

- it is in the interests of individual enterprises to restrict output, and to avoid declaring their maximum possible capacity to central planners so as to avoid being stretched (remember they have no competitors);
- poor horizontal coordination between complementary activities through their vertical subordination to the plan invites bottlenecks (note that shortages in just a few products can hold up whole systems of production);
- once investment projects have begun, the government will always give extra funds for completion. The demand for investment goods therefore tends to be insatiable, creating shortages;
- given investment inertia and soft budget constraints (the state will invariably bail out enterprises in trouble), there is very little incentive for economizing, or enhancing efficiency;
- the shortage syndrome is self-reinforcing, since the enterprises' (and consumers') natural response to it is to hoard resources.

By contrast, "hard" budget constraints (dependence on revenue from customers, competition, exogenously determined prices, no tax exemptions, limited credit) are more likely to induce internal accumulation and efficiency savings, as under such conditions enterprise efforts are directed more to improving internal operations than to exploiting a soft environment.

Two further aspects of centrally planned economies have retarded their development with regard to industrial organization and innovation.

- Industrial organization is characterized by large scale irrespective of diseconomies, e.g., the average size of the workforce in USSR enter-

---

[31] Shortages have not been uniformly severe throughout state socialism, however; East Germany, Czechoslovakia, and (less surprisingly in view of its greater use of market coordination) Hungary have suffered less.

prises is 800, and industrial concentration in Hungary is greater than in Sweden (Aganbegyan, 1988). Large size does make for economies of scale in that it is easier for the center to control a few large units than many small ones, but this does not necessarily correspond with the most efficient size distribution of enterprises or workplaces (Horvat, 1982) – even aside from the question of whether or not there should be competition and horizontal disintegration. What is certain is that it is absurd to impose a single organizational model on "industry" whether it is making steel, producing clothing, retailing food, or raising cattle.

- Finally, there is the matter of innovation. Under central planning, this is limited first and foremost by the inherent difficulty of regulating an advanced social division of labor. Innovation tends to be imposed from above, there being little incentive from below; indeed, independent action from below is often resisted from above because it disrupts the established order. If enterprises cannot raise prices, must maintain employment levels regardless of technological change, have no competitors, respond to plan targets rather than user demand, and if the horizontal links so vital to learning-by-doing are absent, then there is little incentive to raise quality or to introduce product innovations (Leante, 1989; Nove, 1989a; Burawoy and Lukacs, 1985, 1989). Moreover, under a centrally planned regime, those likely to be negatively affected by innovation have more power to resist than under capitalism, precisely because the allocation process is *a priori* rather than *a posteriori*, centralized, and more politicized. Consequently, although central control may be particularly effective for pushing through innovation in some sectors, it would generally seem to fetter the development of the forces of production – though such "fettering" is not necessarily a bad thing.[32]

Many of these problems are prefigured, so to speak, by the silences and emphases of Marxist theory. It is significant that Marxists schooled in the critique of capitalism are often surprised to discover that socialist economists tend to be preoccupied with the problems of coordination, rather than with property relations. There may not seem much that is socialist in this, but in the light of the experience of centrally planned economies they have a point. The main lesson is that if the quasi-class of the state bureaucracy is to be undermined, what is needed is not expropriation of the means of production – for it doesn't belong to anyone – but decentralization of the control of the coordination and division of labor. In a complex economy this must involve recourse to markets, though that is not all there is to it.

---

[32] These remarks do not apply to public sponsored innovation under capitalism.

*Market socialism*

Let us consider an ideal type of market socialism in which all enterprises are worker-controlled and owned, and produce commodities for sale in competitive unregulated markets, and in which economic power is highly decentralized. Yugoslavia is often considered the only example of a national economy to come anywhere near this model, though the differences are great enough to cast severe doubts on treating it as a test case. Closer fits, far more modest in scale, are to be found in cooperatives such as those of the Mondragon group in the Basque country (Horvat, 1982; Thomas and Logan, 1982).

For workers, the alternatives of "pure" central planning and "pure" market socialism present a simple trade-off. Under market socialism, as we have defined it, they have strong control over their own immediate work, but their economic survival is precarious since they are subject to competition and the vicissitudes of *a posteriori* market coordination. Under state socialism, the environment is more stable with protection from bankruptcy and unemployment but the price is lack of control over their own work. In addition, the two systems bring differing efficiencies and standards of living, particularly in terms of the benefits and costs of competition, and differing degrees of uneven development. But what the masses cannot have is *a priori* control over their economic environment *and* local autonomy over their own workplaces: an advanced division of labor makes this a structural impossibility.

Existing socialist systems are always a blend of several different kinds of control, so in practice things do not work out so tidily, and for good reasons. As we argued above, varying patterns of economies of scope and scale are always likely to encourage mixtures of centralization and decentralization, and the material divisions and interdependencies of technical and social divisions of labor always afford additional scope for local differences in power, of which certain strategically placed groups may take advantage (Burawoy and Lukacs, 1989). Nevertheless, as a starting point it is helpful to abstract from these effects and consider a heuristic model of pure market socialism as a way to highlight the effects of major economic structures.

The fact that both are dominated by a market mode of coordination means there are many similarities between market socialism and capitalism. Macroeconomic problems are likely to be similar. Although income differentials within cooperatives are likely to be less than those in capitalist firms, they may diverge just as much between firms according to the firms' competitive positions, unless there is an income policy at a social level. Competition is still likely to produce losers as well as winners, and non-class exploitation between enterprises is no less probable (Roemer, 1986). Under either system, innovating firms are able to escape any negative effects of their innovations (unemployment, obsolescence) which fall elsewhere in the economy. Markets may afford decentralized control but that in no way guarantees equality of power among enterprises; in fact, they have strong tendencies toward concentration of power. Laissez-faire and monopoly have essentially the same deficiencies under both systems, and the arguments for centralization of control of certain activities under capitalism tend to apply to market socialism, e.g., natural monopolies like electricity generation, the need to internalize externalities in activities like urban transport, or to ensure universal access regardless of income (Nove, 1989b). Clearly, ending private property in the means of production does not end uneven development or render markets a benign mode of coordination – they still have strong tendencies which run counter to socialist objectives.

But there are also significant differences in economic behavior between market socialism and capitalism (Horvat, 1982; Arnold, 1987; Schweikart, 1987; Miller, 1989). For example, abolishing the capital-labor relation simultaneously abolishes the enormous disparity in mobility – and consequently power – between capital and labor, which contributes so much to uneven development (Harvey, 1982). Capital may decide to close a plant and employ a new workforce on the other side of the world, but this is not an option for the members of a co-op. Nor, quite clearly, can cooperators gain, as individual capitalists can, by cutting wages. Other differences between market socialist and capitalist enterprises have been suggested as evidence of the formers' alleged inferiority. One of the most puzzling features of market socialism in general, and cooperatives or worker-managed firms in particular, is their rarity, which many take to indicate endemic weaknesses. Some have argued that worker-owned firms tend to degenerate into capitalist or quasi-capitalist organization (e.g., Arnold, 1987). The main processes are alleged to be:

- vertical technical divisions of labor in firms tend to permit the rise of quasi-capitalist social relations of production between a managerial elite, which has the information and skills needed to control production and exact high incomes for itself, and a mass of nominal worker-owners who cede control to them;
- a lack of interest among workers in the responsibilities of co-ownership and management;
- worker-owners are tempted to restrict additions to the workforce in order to maintain a high share of income; alternatively they are tempted to turn the fruits of competitive success, in terms of money, into capital by hiring wage-labor, thus re-establishing a capital-labor relation and diluting the cooperative principle;
- the difficulty of obtaining credit without ceding control to outsiders and a corollary tendency to prefer current consumption to investment;
- finally, a tendency to be out-competed by capitalist firms unencumbered by these problems.

While there is evidence for some of these deficiencies, it is debatable whether they are intrinsic to worker-ownership or are effects of contingent factors (e.g., lack of worker involvement in management may be due to the injuries of class inherited from years of capitalist control). But there is also evidence which suggests they can be effectively countered by institutional arrangements, and that cooperatives can be at least as efficient as capitalist firms (Schweikart, 1987; Elster, 1989; Miller, 1989). For example, the success of the Mondragon co-ops owes much to the creation of networks of supporting institutions that provide education and training, management services, research and development, and special credit provisions (Thomas and Logan, 1982). These links include some market transactions and also go far beyond them.

Our ideal-type of market socialism cuts many corners, ignoring especially the particular form of ownership and control as regards workers, creditors, and consumers, and the unavoidable need for central state influence. But it highlights the extent to which it is possible to separate ownership of the means of production and the mode of microeconomic coordination. The ideal-type takes market coordination as a necessary condition of decentralized control, but it does not establish that it is sufficient – either as a form of decentralized control or in meeting socialist objectives. Other, supporting forms of horizontal control, such as relational contracting or industry associations, are just as likely to be needed, as the case of Mondragon shows. In any case there will inevitably be activities which need to be centralized under the state. From a normative point of view, market integration cannot be treated simply as a benign accompaniment to worker ownership.

### Mixed ways and third ways

There are no perfect solutions that remove the contradictions and trade-offs in any mode of production e.g., centralization v decentralization, security v efficiency, equality and solidarity vs differential rewards for work according to its social contribution, need vs ecological constraint, etc. (Kornai, 1986; Nove, 1983). Real economies will always have some variety in types of ownership, types of industrial organization, and modes of coordination (Hodgson, 1988). As is clear in the Soviet bloc, normative proposals are bound to offer mixed solutions, with the eventual combination depending on how the various trade-offs are assessed in relation to the material possibilities afforded by social and technical division of labor, and issues of equity and social justice (e.g., Davies, 1990).

Some radicals have responded to the deficiencies of capitalism and already-existing socialism by seeking a "third way" which uses neither the market nor the plan. Others – including some of those who propose mixed solutions – deny that there is a third way, in the sense of an alternative which can replace markets and planning (a denial that is often put in Latin

– *"tertium non datur"* – to give it added authority). Much confusion is found on both sides of this argument.

First, it must be remembered that economic systems always contain modes of integration other than markets or planning, though often in a supporting rather than an alternative role. Both planning and markets are, moreover, embedded in relations involving other forms of integration through custom, negotiation, voting, coercion, or authority, including direct control within institutions, and behavior is a product of the interaction of these relations. In state socialist economies, informal deals between enterprises supplements the directives of central planning; in capitalism, relational contracting supplements market coordination. "Pure" markets and market behavior are therefore an abstraction from the variable influences always present in real markets. And it is possible, under market socialism as under capitalism, to modify the operation of markets through taxation, regulation, force, cooperation, government ownership, bureaucracy, moral authority, etc.

The mythology of the pure market is all too apparent in commentaries on the rise of "the market" in the Soviet Union and Eastern Europe, even where – as with McDonald's – it is manifestly the case that what is involved is markets in the highly organized form of relational contracting. Multinationals are hardly likely to set up in these countries and then fish around for whatever is going in some embryonic market; rather they recruit, train, and organize suppliers so their subsequent market transactions achieve the quality and reliability they need.

In a complex economy, these additional modes are in one sense pervasive; little happens without them. But in another sense, they provide coordination mostly at local rather than society-wide levels: in households or in single-purpose associations, in the deliberations of local government or institutional management, in supply chains and industrial groups or associations, rather than in a comprehensive manner across extensive and diverse parts of the division of labor, as do planning and especially markets. Those who expect to find a third way which can replace markets and central planning are wildly overestimating the ability of the alternatives to cope with the complexity and differentiation of an advanced economy. This does not mean that alternatives as supplementary modes cannot play an expanded role in certain areas, nor that new combinations of markets and planning might not offer many benefits.

Finally, impure, mixed forms of economic organization must be acknowledged in a sense that stems from the situation of socialist economies within a capitalist-dominated world economy. Socialist planners or worker-managers cannot escape the world system and its exigencies. There is not only a matter of markets, international trade and finance, but of comparisons (by citizens, leaders, and other nation-states) of standards of living, technologies, and military capabilities. Inevitably these must influence socialist economic development.

## Socialist Policies in Capitalism

When we learn about the dilemmas facing socialist organization, it is hard not to look back at capitalism in a different light. Our explorations of the dual basis for economic control and power, deriving from class and division of labor, make it easier to illuminate the limitations of some common features of socialist thinking and policy in capitalist countries.

We can begin quite concretely with the left-controlled Greater London Council's Industrial Strategy (GLC, 1985). This was undoubtedly a landmark in political economic praxis, embodying bold and imaginative socialist policies. It was unavoidably limited by its geographical scope and budget, but it also illustrated the persisting problem of the relationship between the social division of labor and class. Its analyses of London's industry were not only fully cognizant of capital-labor relations, but were authoritative studies of industry's ability to survive in the social division of labor, both in terms of the internal organization of firms (productivity, costs) and in terms of the prospects for their products; such issues as the state of play of competition, and the short-term horizons within which firms are obliged to operate, were authoritatively assessed. But as regards policy, the emphasis was largely on the capital-labor relationship, particularly on democratizing workplaces and opening them up to formerly excluded groups of workers. While these are of course proper socialist concerns, the policies proposed with respect to the relationship of the firms to the users of their products were less clear. Where the division of labor is sufficiently simple and localized (e.g., household labor or public transport), the Strategy could deal with the issue of the relationship between producers and users. But where firms produce commodities for wider markets under competitive conditions this unity is absent and the socialist content is harder to specify. In other words – and this is not intended as a cynical point – the Strategy faced the same constraints as more traditional socialist strategies in capitalist countries, namely the need to combine an industrial policy (enabling industries to remain competitive in the division of labor), which does not seem particularly socialist, with policies regarding labor and its role within production, which are more recognizably socialist but which have only a minor influence on industrial performance. This, again, provides a practical demonstration of the partial autonomy of division of labor and modes of coordination from the social relations of production.

The GLC recognized more clearly than most that a viable socialist strategy can never be restricted to relationships within workplaces, purely in terms of peoples' lives in paid work. But it still faced a dilemma in confronting the coordination of the division of labor, not only for the practical reason that it covered only a small corner of the national and international economy, but because the available intellectual resources in socialist theory had consistently ignored these issues or treated them as minor technicalities. This shows just how far what is perceived as socialist in the capitalist countries suffers from the failure to consider questions of microeconomic coordination and allocative efficiency.

Two concepts extensively used in the GLC strategy illuminate the problems further: "restructuring for labor" (as opposed to restructuring for capital) and "socially useful production". In the first of these "the rhetorical contrast . . . – 'for labor' and 'for capital' – mystifies the trade-offs that would have to be made" (Rustin, 1986, p. 83) and gives the impression that there can be no conflict between producer and consumer interests across the division of labor, or between present and future consumption. Reinforcing this failure to confront the dual grounding of power and economic control is an understandable, yet unreasonable, reluctance to admit that increased productivity is a necessary (though not sufficient) condition for economic development and rising real wages, and that in many cases this is bound to mean reductions in the workforce required for producing any given commodity or use-value (though not necessarily a reduction in employment as a whole).[33]

The concept of socially useful production is still less clearly defined, reflecting a weakly examined antipathy towards markets, a preference for public spending over private spending and for collective over private consumption. But these preferences are pitched at too abstract a level to distinguish what is or is not beneficial. For example, "it can no more be assumed that 'public' decisions will benefit the public than that 'free enterprise' will benefit the consumer" (Rustin, 1986, p. 80). Such assumptions fail to come to terms with why markets and privatized consumption are often popular with the working class.

Concepts of restructuring for labor and socially useful production are related to an older duality, opposing production for profit against production for need. It would be hard to find a clearer illustration of the blindness to the problems posed by a complex division of labor than this. Intending to produce for need does not guarantee that needs are met more fully than they might be by intending to produce for profit. To suppose that it can do so automatically, without specifying mechanisms, is to ignore completely problems of division of knowledge (establishing needs and production capabilities), and of choosing priorities among competing needs under conditions of relative scarcity.

The antipathy towards profit itself merits further examination. Profits have a dual aspect, corresponding to ownership or class and to the integration of the division of labor. Regarding the former, the prime question is who has a right to appropriate profits; regarding the latter, the function of profit as an indicator of production performance must be evaluated.[34] The first question should not obscure the second, nor should positive responses

---

[33] This reluctance is of course ironic given the left's theoretical awareness of the significance of labor time.

[34] *Pace* Schweikart (1987), we do not think any harm is done here by including surpluses obtained by worker-owned firms as "profit".

to the second question be allowed to suppress the first. But this is usually what happens: on the left the critique of capitalist control and appropriation of profits is often taken to disqualify arguments in favor of the motivational and regulatory qualities of profits (and losses); on the right, these latter qualities are disingenuously used to conceal capital's class interest. Note, however, that profit-takers need not convert their profits into capital: this is a crucial distinction. Worker-owners need not use their profits to employ wage laborers and hence become capitalists; consequently, production for profit need not be capitalist. This is not the place to evaluate the efficiency and equity implications of profit and loss as performance indicators and incentives, but we do want to insist that profit has more than a class dimension, and that it does have some effect in encouraging adjustments to the division of labor in accordance with supply and demand, albeit imperfectly and less directly than commonly supposed (cf. Walker, 1988).

A similar connection to structures of class and division of labor can be established in relation to the left's traditional antipathy toward competition. There are indeed many good reasons for this antipathy (e.g., the corrosive effect of competition on opposition to capital, its effects on individual development and social relations) and some bad ones, too (e.g., the assumption that firms usually compete by cutting wages, or that cooperation is necessarily harmonious rather than conflictive), but there is also a tendency to underestimate its benefits. Although competition doesn't always guarantee increased consumer choice and greater efficiency and more innovation, there are many cases where it does just that. (Recall that "consumers" includes intermediate as well as final consumers.) Cooperation may seem healthier, and likely to encourage greater unity among workers, but posing cooperation as the simple opposite of competition is naive since it only considers relationships among producers. As long as producers are not identical to consumers (i.e., as long as producers consume things other than what they produce) then the opposite of competition is also monopoly.[35] And the dangers of monopoly lie not just in extortionate pricing and the dominance of producer over consumer, but in the lack of a strong need to consider the consumers' voice. Again, the arguments are far more complex than this, but our point is to note that the naiveté owes much to a gross underestimation of the division of labor, and an unwillingness to recognize the difference between producer and consumer interests.

One of the biggest challenges to socialist policy in capitalist countries concerns the state. In some countries the public sector has become increasingly unpopular, even where (as in the cases of nationalization and the welfare state) it was originally a consequence of working class pressure. Marxism has made a major contribution to theories of the state in capital-

---

[35] Nove (1983) is wrong to deny that cooperation is not *also* the opposite of competition.

ist society by theorizing its relation to capital and class, and more recently by demonstrating the patriarchal and racist character of state institutions. These are important considerations, but they fail to disclose another aspect of the state which relates to the social division of labor. For the state is both part of the division of labor in society and an organizer of certain sections of that division, and many of its characteristics stem from those facts. Neglect of this aspect has led to numerous misattributions of causality, particularly with respect to responsibility for failings of planning in providing services.[36] Problems such as the lack of a sense of responsibility for public property, free riders, failure to make producers responsive to user needs, failure to coordinate diverse but related activities, failure to improve efficiency and quality, cannot all be simply laid at the door of the class character of the state. To a significant extent, these are all likely consequences of the difficulty of coordinating those parts of the division of the labor for which the state is responsible and of the modes of coordination which it uses.[37]

The stereotypes of tax-financed public sector service and nationalized industry monopolies versus revenue-financed, market-coordinated, competitive private capital producing for profit, conceal this relative autonomy and limit alternatives. The ills attributed by left and right to these stereotypes usually inaccurately assign credit or blame, either indiscriminately to the wholes or to the wrong elements: each has its own kind of misplaced essentialism in explanation. Yet the stereotypes can be burst open and their elements combined in quite different ways.[38] Not just any combination of these elements is possible of course, but the range of permutations is still far greater than the stereotypes suggest. There is a desperate need for creativity in exploring these in practice.

In short, socialist policy has paid too little attention to overcoming problems rooted in the division and integration of labor and has assumed too much that class responses will resolve all problems. The separation of producers from the means of production may be resolved, but the separations between producers, and between producers and consumers, will not go away. Micro- and macro-processes of power do not derive wholly from capitalist organization. Wherever there is scarcity and division of labor – whether it is organized hierarchically, competitively, or cooperatively – there is ample scope for conflict, unequal power, domination, and some-

[36] The British experience is not typical, and the public sector and the welfare state have functioned better and without stigma in other countries (e.g. in Scandinavia). This supports our view that the performance characteristics of public or private institutions do not follow simply from their ownership.

[37] In response to these failures, advocates of privatization on the right in Britain have mischievously claimed that privatization involves "giving industry back to the people". That this is disingenuous is obvious; more significant are the negative experiences of public ownership that allowed the right to even think they might be able to get away with such an outrageous claim.

[38] For a particularly effective critique of both left and right essentialist views with respect to health services provision, see Maynard and Williams, 1988.

times exploitation, often along racist and sexist lines. This is the lesson to be drawn from the similarities in the internal hierarchies and power structures of capitalist and state non-capitalist organizations. The capitalist environment in which the latter operate and their relationships to capital are only secondary influences.

## Conclusion

"When the question of ownership has been settled the question of administration remains for solution" (R.H. Tawney, in Murray, 1987, p. 88). Ownership is not the root of all economic power. We quite agree that non-class divisions and relations of domination, such as those of ethnicity and gender, have been underestimated by Marxism, but even when these faults have been remedied, there remain major problems on Marxism's traditional home ground – the political economy of capitalism.

Economic power derives also from our positions in the technical and social divisions of labor, from the opportunities and constraints they and the existing modes of integration or coordination allow us. How much influence do we have over the labor of others in other parts of the economy, and how much do others have over us? What mechanisms are there to ensure that production will be efficient, effective, non-exploitative, and ecologically benign, that consumption will not be wasteful? Ownership makes a difference to these things, but so do modes of coordination. Both interact: domination deriving from the form and coordination of the division of labor, can feed off, reinforce, or undermine domination deriving from ownership.

Even fortified by an understanding of patriarchy and race, Marxism still fails to come to terms with the division of labor, except for selective aspects such as its internationalization. The apparent expectation that the social division can be transcended, the underestimation of its complexity and opacity, of its materiality and the intractability of its differentiations, the limited awareness of problems of allocational efficiency, the class reductionism of its explanations of power and conflict, the critique of commodity fetishism, and the disregard of the benefits of exchange-value – all of these stand in the way of an improved understanding of capitalism and the already-disintegrating socialist economies.

If Marxism can grasp the mechanisms which perpetuate capitalism, then it ought to be of use to socialist economists trying to understand how to block and replace those mechanisms with superior forms of organization. The fact that it does not – and that it has actually been misleading for that purpose – does indeed reflect badly upon its critical theory of capitalism. Marxism's understanding of the dynamics and social relations of production at the level of individual capitals may still be unsurpassed, but its failings beyond this account for the gulf between its critical theory of capitalism and theories of socialist economies.

Meanwhile, mainstream economics persists in overextending its atomized, undersocialized, exchange view of the economy and remains hopelessly blind to class and other relations of domination. Marxism's class focus and liberalism's traditional concerns with the individual and society, and with the problem of reconciling sectional and general interests, cannot continue to be seen as mutually exclusive nor can one be reduced to the other (Bowles and Gintis, 1986). Liberalism's concerns are intelligible in the light of the centrifugal and centripetal tendencies of division of labor and modes of integration, but they need not be coupled to the usual evasions of relations of domination and the collective interests of class.[39] In short, it may seem rather dated to call for a new paradigm or problematic in political economy, but that is what is required.

Any division of labor introduces lags in distance and/or time between effort and effect, and this in turn creates problems of control, responsibility, and accountability, as well as efficiency and equity. The scope for each mode of integration, and its effects when in place, depend on the characteristics of the parts of the division of labor to which they are applied and the social relations of production with which they are combined. The same mode may be empowering in some cases, disempowering in others. Networking among related capitals or co-ops, or among political pressure groups with overlapping interests, may make them more successful; networking among large numbers of people having little in common is unlikely to work. It is by underestimating the significance of an advanced division of labor that many socialists are able to see only the disadvantages of *gesellschaft* relations epitomized by market exchange, ignoring their liberating qualities, and to entertain utopian hopes for a *gemeinschaft* economy, in which modes of social organization, feasible within small, knowable communities, are imagined to be possible at much larger scales (Holton and Turner, 1989; Miller, 1989).

Central planning is despotic and suppresses decentralized horizontal feedback between fragmented producers and users. Coordination through markets is more likely to encourage feedback and response, though often this needs to be supplemented by extra-market interactions – relational contracting and networks, market research, pressure groups and consumer associations, state regulation. Market coordination may be equitable among producers at similar levels of development but highly regressive when they are not. The powerless may, in some situations, be empowered by gaining rights of democratic control, in others by gaining more money to spend.[40]

In other words modes of integration cannot be wholly endorsed or condemned irrespective of the context and the nature of the activities being coordinated, be they the provision of education, electricity, food, books, broadcasting, childcare, or holidays.

---

[39] As recent writings show, it is possible to theorize class and class action in a way which recognizes the tensions of individualism and class interests (e.g. Przeworski, 1985).

[40] Recall the old slogan, "If voting changed anything they'd ban it".

Understanding the interactions of class, division of labor, and modes of integration is essential to grasp the constraints and opportunities for socialism in the face of the new social economy. Despite the speed of the revolutions of 1989, old modes of coordination and associated structures of power and interest are slow to dislodge; things will inevitably get worse before they get better. Arguments within those countries about new economic forms are distorted not only by these structures and by the advice and intrusion of capitalist interests, but by the limited understandings – theirs and ours – of the implications and possibilities of modes of integration. The Eastern Europeans are as much prisoners of the market versus plan orthodoxy as anyone else: only the Hungarians and Yugoslavs have any substantial experience in trying to overcome this (Horvat, 1982; Kornai, 1986). And much the same applies to reform in capitalist countries.

Yet options for economic reform go far beyond new combinations of social relations of production and modes of integration. There is also the whole question of the relationship to the domestic mode of production, the possibilities for re-evaluating the division between paid and unpaid work, and for decoupling rights to income from the need to work, as in proposals for a social wage (*Theory and Society*, 1989). In addition, we must introduce ecological considerations into the evaluation of all the possibilities. We note these things now to acknowledge the limits of our present concerns. For our part, work on them will have to wait until Monday morning; in the meantime, there is still unfinished work to do on the division of labor.

# Bibliography

Abercrombie, N. and Urry, J. 1983. *Capital, Labour and the Middle Classes.* London: Allen and Unwin.

Abernathy, W., Clark, D. and Kantrow, E. 1983. *Industrial Renaissance: Producing a Competitive Future for America.* New York: Basic Books.

Adler, P. 1985. Technology and us. *Socialist Review.* 85: 67-98.

Aganbegyan, A. 1988. *The Challenge: Economics of Perestroika.* London: Hutchinson.

Aggarwal, S. 1985. MRP, JIT, OPT, FMS? Making sense of production operations systems. *Harvard Business Review.* Sept-Oct. 8-12.

Aglietta, M. 1979. *A Theory of Capitalist Regulation.* London: New Left Books.

Agnew, J-C. 1986. *Worlds Apart: The Market and the Theatre in Anglo-American Thought, 1550-1750.* New York: Cambridge University Press.

Alchian, A. and Demsetz, H. 1972. Production, information costs, and economic organization. *American Economic Review.* 62: 777-95.

Alexander, S. 1976. Women's work in 19th century London. In: Oakley, A. and Mitchell, J. (eds.) *The Rights and Wrongs of Women.* Harmondsworth: Penguin. 59-111.

Allen, J. 1988. Service industries: uneven development and uneven knowledge. *Area.* 20(1): 15-22.

Altshuler, A., Anderson, A., Jones, D., Roos, D. and Womack, J. 1984. *The Future of the Automobile.* Cambridge, Mass.: MIT Press.

Amman, R. and Cooper, J. (eds.) 1987. *Technical Progress and Soviet Economic Development.* Cambridge, Mass.: Basil Blackwell.

Amsden, A. 1989. *Asia's Next Giant: South Korea and Late Industrialization.* New York: Oxford and Cambridge University Press.

Anderson, B. 1983. *Imagined Communities: Reflections on the Origin and Spread of Nationalism.* London: Verso.

Anderson, P. 1987. The figures of descent. *New Left Review.* 161: 20-77.

Antonelli, C. 1988. The emergence of the network firm. In: Antonelli, C. *New Information Technology and Industrial Change: The Italian Case.* Boston: Kluwer. 13-32.

Aoki, M. 1984. Aspects of the Japanese firm. In: Aoki, M. (ed.) *The Economic Analysis of the Japanese Firm.* Dordrecht: North Holland. 3-46.

Aoki, M. 1986. Horizontal versus vertical information structure of the firm. *American Economic Review.* 76(5): 971-83.

Aoki, M. 1987. The Japanese firm in transition. In: Yamamura, K. and Yasuba, Y. (eds). *The Political Economy of Japan.* Stanford: Stanford University Press. 263-88.

Aoki, M. 1988. *Information, Incentives and Bargaining in the Japanese Economy*. Cambridge: Cambridge University Press.

Aoki, M. and Rosenberg, N. 1989. The Japanese firm as an innovative institution. In: Shiraishi, P. and Tsuru, S. (eds.) *Economic Institutions in a Dynamic Society*. London: Macmillan. 137-54.

Appelbaum, E. 1987. Restructuring work: temporary, part-time and at-home employment. In: Hartmann, H. (ed.) *Computer Chips and Paper Clips: Technology and Women's Employment*. Volume I. Washington, D.C.: National Academy Press. 268-310.

Armstrong, J. 1973. *The European Administrative Elite*. Princeton: Princeton University Press.

Arnold, S. 1987. Marx and disequilibrium in market socialist relations of production. *Economics and Philosophy*. 3: 23-48.

Arnold, S. 1989. Marx, central planning and utopian socialism. *Social Philosophy and Policy*. 6(2): 160-99.

Arrow, K. 1962. The economic implications of learning by doing. *Review of Economic Studies*. 29: 154-74.

Arrow, K. 1969. The organization of economic activity: issues pertinent to the choice of market versus nonmarket allocation. In: US Congress, Joint Economic Committee. *The Analysis and Evaluation of Public Expenditures: The PPB System*. Washington, D.C.: US Government Printing Office. I: 59-73.

Arrow, K. 1974. *The Limits of Organization*. New York: Norton.

Arrow, K. 1985. The informational structure of the firm. *American Economic Review*. 75: 303-07.

Atkinson, J. 1984. *Flexibility, Uncertainty and Manpower Management*. Institute of Manpower Studies, University of Sussex. Report No. 89.

Auerbach, P. 1988. *Competition*. Oxford and Cambridge, Mass.: Basil Blackwell.

Auerbach, P., Desai, M. and Shamsavari, A. 1988. The transition from actually existing capitalism. *New Left Review*. 170: 61-79.

Auerbach, P. and Skott, P. 1991. Concentration, competition and distribution – a critique of theories of monopoly capital. *International Review of Applied Economics*. (Forthcoming)

Averitt, J. 1968. *The Dual Economy*. New York: Norton.

Badaracco, J. 1988. Changing forms of the corporation. In: Meyer, J. and Gustafson, J. (eds). *The US Business Corporation: An Institution in Transition*. Cambridge: Ballinger. 67-91.

Bagdikian, B. 1990. *The Media Monopoly*. Third edition. Boston: Beacon Press.

Bahro, R. 1978. *The Alternative in Eastern Europe*. London: New Left Books.

Bain, J. 1956. *Barriers to New Competition*. Cambridge, Mass.: Harvard University Press.

Ballance, R. and Sinclair, S. 1983. *Collapse and Survival: Industry Strategies in a Changing World*. London: George Allen and Unwin.

Bannon, E. and Thompson, P. 1985. *Working the System*. London: Pluto Press.

Baran, B. 1987. The technological transformation of white-collar work: a case study of the insurance industry. In: Hartmann, H. (ed.) *Computer Chips and Paper Clips: Technology and Women's Employment*. Volume II. Washington, D.C.: National Academy Press. 25-62.

Baran, P. and Sweezy, P. 1966. *Monopoly Capital*. New York: Monthly Review Press.

Barbalet, J. 1987. The "labor aristocracy" in context. *Science and Society.* 51(2): 133-53.

Barnard, C. 1938. *The Functions of the Executive.* Cambridge, Mass.: Harvard University Press.

Barratt-Brown, M. 1970. *What Economics Is About.* London: Weidenfeld.

Barrett, M. 1980. *Women's Oppression Today.* London: Verso.

Baudrillard, J. 1988. Consumer society. In: *Jean Baudrillard: Selected Writings.* Cambridge: Polity Press. 28-55.

Becattini, G. 1978. The development of light industry in Tuscany: an interpretation. *Economic Notes.* 3: 107-23.

Becker, G. 1964. *Human Capital.* New York: National Bureau of Economic Research, Columbia University Press.

Beechey, V. 1987. *Unequal Work.* London: Verso.

Belec, J., Holmes, J. and Rutherford, T. 1987. The rise of Fordism and the transformation of consumption norms: mass consumption and housing in Canada, 1930-45. In: Harris, R. and Pratt, G. (eds). *Social Class and Housing Tenure.* Gavle, Sweden: National Swedish Institute for Building Research. 187-237.

Bell, D. 1973. *The Coming of Post-Industrial Society.* New York: Basic Books.

Bell, D. 1980. The social framework of the information society. In: Forester, T. (ed.) *The Microelectronics Revolution.* Cambridge, Mass.: MIT Press.

Bello, W. and Rosenfeld, S. 1990. *Dragons in Distress.* San Francisco: Institute for Food and Development Policy.

Belussi, F. 1987. Benetton: information technology in production and distribution. Occasional Paper No. 25. Science Policy Research Unit, University of Sussex, Brighton, U.K.

Beneria, L. and Roldan, M. 1987. *The Crossroads of Class and Gender.* Chicago: University of Chicago Press.

Beniger, J. 1986. *The Control Revolution: Technological and Economic Origins of the Information Society.* Cambridge, Mass.: Harvard University Press.

Berger, J. 1973. *Ways of Seeing.* New York: Viking.

Berle, A. and Means, G. 1932. *The Modern Corporation and Private Property.* New York: Macmillan.

Berman, M. 1982. *All That Is Solid Melts Into Air.* New York: Simon and Schuster.

Bertrand, O. and Noyelle, T. 1988. *Human Resources and Corporate Strategy: Technological Change in Banks and Insurance Companies.* Paris: OECD.

Bhaskar, R. 1989. *Reclaiming Reality.* London: Verso.

Bianchi, P. and Bellini, N. 1991. Public policies for local networks of innovators. *Research Policy.* (forthcoming)

Blades, D. 1987. *Goods and Services in the OECD Economies.* Paris: OECD.

Blair, J. 1972. *Economic Concentration.* New York: Harcourt, Brace, Jovanovich.

Block, F. 1977. *The Origins of International Economic Disorder.* Berkeley and Los Angeles: University of California Press.

Block, F., Coward, R., Ehrenreich, B. and Piven, F. 1987. *The Mean Season: The Attack on the Welfare State.* New York: Pantheon.

Blois, K. 1972. Vertical quasi-integration. *Journal of Industrial Economics.* 20(3): 253-72.

Bluestone, B. and Harrison, B. 1982. *The Deindustrialization of America.* New

York: Basic Books.

Bois, G. 1984. *The Crisis of Feudalism*. Cambridge: Cambridge University Press.

Boserup, E. 1970. *Women's Role in Economic Development*. New York: St. Martin's Press.

Bott, E. 1971. *Family and Social Networks*. New York: Free Press.

Bourdieu, P. 1977. *Outline of a Theory of Practice*. Cambridge: Cambridge University Press.

Bourdieu, P. 1979. *La Distinction*. English edition 1984. London: Routledge and Kegan Paul.

Bowles, S. and Gintis, H. 1976. *Schooling in Capitalist America*. New York: Basic Books.

Bowles, S. and Gintis, H. 1986. *Democracy and Capitalism*. New York: Basic Books.

Bowles, S., Gordon, D. and Weisskopf, T. 1983. *Beyond the Waste Land: A Democratic Alternative to Economic Decline*. Garden City, NY: Anchor Press.

Boyer, R. (ed.) 1986. *The Search for Labor Market Flexibility*. Oxford: Clarendon Press.

Braudel, F. 1979. *Civilization and Capitalism: The 15th to 18th Centuries*. 3 volumes. New York: Harper and Row.

Braverman, H. 1974. *Labor and Monopoly Capital*. New York: Monthly Review Press.

Brenner, J. and Ramas, M. 1984. Rethinking women's oppression. *New Left Review*. 144: 33-71.

Brenner, R. 1985. Agrarian class structure and economic development in pre-industrial Europe. In: Ashton, T. and Philpin, C. (eds). *The Brenner Debate: Agrarian Class Structure and Economic Development in Pre-Industrial Europe*. New York: Cambridge University Press. 10-63.

Brenner, R. 1986. The social basis of economic development. In: Roemer, J. (ed.) *Analytical Marxism*. New York: Cambridge University Press. 23-53.

Brenner, R. and Glick, M. 1991. The regulation approach: Theory and history. *New Left Review*. 188: 45-120.

Brint, S. 1984. New class and cumulative trend explanation of the liberal political attitudes of professionals. *American Journal of Sociology*. 90: 30-71.

Britten, N. and Heath, A. 1983. Women, men and social class. In: Gamarnikow, E., Morgan, D., Purvis, J. and Taylorson, D. (eds). *Gender, Class and Work*. London: Heinemann. 46-60.

Britton, S. 1990. The role of services in production. *Progress in Human Geography*. 14(4): 529-46.

Brown, A. 1984. The emergence of enterprise unionism. Address to the Institute of Personnel Management, Harrogate.

Brown, C. and Reich, M. 1989. When does union-management cooperation work? A look at NUMMI and GM-Van Nuys. *California Management Review*. 31(4): 26-44.

Brown, R. 1979. *Rockefeller Medicine Men*. Berkeley: University of California Press.

Browning, H. and Singelmann, J. 1978. The transformation of the US labor force: the interaction of industry and occupation. *Politics and Society*. 8: 481-509.

Brownmiller, S. 1975. *Against Our Will: Men, Women and Rape*. New York: Simon and Schuster.

Brus, W. 1972. *The Market in a Socialist Economy*. London: Routledge and Kegan Paul.

Brusco, S. 1982. The Emilian model: productive decentralisation and social integration. *Cambridge Journal of Economics*. 6(2): 167-84.

Brusco, S. and Sabel, C. 1983. Artisanal production and economic growth. In: Wilkinson, F. (ed.) *The Dynamics of Labor Market Segmentation*. London: Academic Press. 99-113.

Buchanan, A. 1985. *Ethics, Efficiency and the Market*. London: Clarendon.

Buckley, P. and Casson, M. 1976. *The Future of the Multinational Enterprise*. London: Macmillan.

Burawoy, M. 1976. The functions and reproduction of migrant labor: comparative material from Southern Africa and the United States. *American Journal of Sociology*. 81: 1050-87.

Burawoy, M. 1979. *Manufacturing Consent*. Chicago: University of Chicago Press.

Burawoy, M. 1985. *The Politics of Production*. London: Verso.

Burawoy, M. 1989. The limits to Wright's analytical Marxism and an alternative. In: Wright, E. (ed.) *The Debate on Classes*. London: Verso. 78-99.

Burawoy, M. and Lukacs, J. 1985. Mythologies of work: a comparison of firms in state socialism and advanced capitalism. *American Sociological Review*. 50: 723-37.

Burawoy, M. and Lukacs, J. 1989. What is socialist about socialist production? Autonomy and control in a Hungarian steel mill. In: Wood, S. (ed.) *The Transformation of Work?* London: Unwin Hyman. 295-316.

Burnham, J. 1941. *The Managerial Revolution*. New York: John Day.

Burris, V. 1987. The neo-Marxist synthesis of Marx and Weber on class. In: Wiley, N. (ed.) *The Marx-Weber Debate*. Newbury Park, CA: Sage. 67-90.

Business Week. 1986. *The Hollow Corporation*. Special Report. March 3: 57-85.

Business Week, 1990. *The Global 1000*. Special Report. July 16: 111-36.

Cagan, L., Albert, M., Chomsky, N., Hahnel, R., King, M., Sargent, L. and Sklar, H. 1986. *Liberating Theory*. Boston: South End Press.

Cain, M. 1986. Realism, feminism, methodology and the law. *International Journal of the Sociology of Law*. 14: 255-67.

Callinicos, A. and Harman, C. 1987. *The Changing Working Class*. London: Bookmarks.

Camagni, R. (ed.) 1990. *Innovation Networks: A Spatial Perspective*. London: Pinter.

Carchedi, G. 1977. *On the Economic Identification of Social Classes*. London: Routledge and Kegan Paul.

Castells, M. 1990. *The Informational City*. Oxford and Cambridge, Mass.: Basil Blackwell.

Cavendish, R. 1982. *Women on Line*. London: Routldege and Kegan Paul.

Caves, R. 1982. *Multinational Enterprise and Economic Analysis*. Cambridge: The University Press.

Chalmin, P. 1985. *Traders and Merchants: The Panorama of International Commodity Trading*. English edition 1987. Chur, Switz: Harwood.

Chamberlin, E. 1933. *The Theory of Monopolistic Competition*. Cambridge, Mass.: Harvard University Press.

Chandler, A. 1962. *Strategy and Structure*. Cambridge, Mass.: MIT Press.

Chandler, A. 1977. *The Visible Hand.* Cambridge, Mass.: Harvard University Press.

Chandler, A. 1990. *Scale and Scope.* Cambridge, Mass.: Harvard University Press.

Chisholm, D. 1989. *Coordination Without Hierarchy.* Berkeley: University of California Press.

Cho, S. 1988. *How Cheap is Cheap Labor? The Dilemmas of Export-led Industrialization.* Unpublished doctoral dissertation, Department of Sociology, University of California, Berkeley.

Chodorow, N. 1978. *The Reproduction of Mothering.* Berkeley: University of California Press.

Christopherson, S. 1983. Households and class formation. *Society and Space.* 1(3): 323-38.

Christopherson, S. 1988. Overworked and underemployed: the redistribution of work in the US economy. Unpublished manuscript. Department of City and Regional Planning, Cornell University, Ithaca, NY.

Christopherson, S. 1989. Flexibility in the US service economy and the emerging spatial division of labor. *Transactions of the Institute of British Geographers.* 14: 131-43.

Christopherson, S. and Storper, M. 1988. New forms of labor segmentation and production politics in flexibly specialized industries. *Industrial and Labor Relations Review.* 42(3): 331-47.

Clark, C. 1940. *The Conditions of Economic Progress.* London: Macmillan.

Clark, G. 1989. *Unions and Communities Under Siege: American Communities and the Crisis of Organized Labor.* Cambridge: Cambridge University Press.

Clark, P. 1987. *Anglo-American Innovation.* New York: de Gruyter.

Clawson, M. 1989. *Constructing Brotherhood: Class, Gender and Fraternalism.* Princeton: Princeton University Press.

Coase, R. 1937. The nature of the firm. *Economica.* 4: 386-405.

Cockburn, C. 1983. *Brothers: Male Dominance and Technological Change.* London: Pluto Press.

Cockburn, C. 1985. *Machinery of Dominance.* London: Pluto Press.

Cohen, G. 1979. *Karl Marx's Theory of History: A Defense.* Princeton: Princeton University Press.

Cohen, S. and Zysman, J. 1987. *Manufacturing Matters: the Myth of the Post-Industrial Economy.* New York: Basic Books.

Cole, R. 1971. *Japanese Blue Collar.* Berkeley: University of California Press.

Collins, R. 1979. *The Credential Society.* New York: The Academic Press.

Conference of Socialist Economists Microelectronics Group. 1980. *Microelectronics: Capitalist Technology and the Working Class.* London: CSE Books.

Connell, R. 1987. *Gender and Power.* Oxford: Polity Press.

Contractor, F. and Lorange, P. (eds). 1988. *Cooperative Strategies in International Business.* Lexington: Heath.

Cooke, P. 1984. Region, class and gender: a European comparison. *Progress in Planning.* 22(2): 87-146.

Cooke, P. 1988. Flexible integration, scope economies and strategic alliances. *Society and Space.* 6(3): 281-300.

Coriat, B. 1983. *La Robotique.* Paris: La Découverte/Maspero.

Crompton, R. and Jones, G. 1984. *White Collar Proletariat: Deskilling and Gender in Clerical Work.* Philadelphia: Temple University Press.

Cumings, B. 1984. The origins and development of the Northeast Asian political economy. *International Organization*. 38(1): 1-40.

Cusumano, M. 1985. *The Japanese Automobile Industry*. Cambridge, Mass.: Harvard University Press.

Cyert, R. and March, J. 1963. *A Behavioral Theory of the Firm*. Englewood Cliffs, NJ: Prentice-Hall.

Dahrendorf, R. 1959. *Class and Class Conflict in Industrial Society*. Stanford: Stanford University Press.

Daly, M. 1978. *Gyn/Ecology*. Boston: Beacon Press.

Davies, R. 1989. *Soviet History in the Gorbachev Revolution*. Bloomington: Indiana University Press.

Davies, R. 1990. Gorbachev's socialism in historical perspective. *New Left Review*. 179: 5-28.

Davis, M. 1985. Urban renaissance and the spirit of post-Modernism. *New Left Review*. 151: 106-13.

Davis, M. 1986. *Prisoners of the American Dream*. London: Verso.

DeBeauvoir, S. 1954. *The Second Sex*. New York: Knopf.

DeBresson, C. and Amesse, F. 1991. Networks of innovators: a review and introduction. *Research Policy*. 20 (forthcoming)

DeBresson, C. and Walker, R. 1991. Networks of Innovators. Special issue *Research Policy*. 20 (forthcoming)

Delauney, J-C. and Gadrey, J. 1987. *Les Enjeux de la Société de Service*. Paris: Presses de la Fondation Nationale des Sciences Politiques.

Delphy, C. 1984. *Close to Home: A Materialist Analysis of Women's Oppression*. Amherst: University of Massachusetts Press.

Devine, J. (ed.) 1986. Empirical work in Marxian crisis theory. Special issue *Review of Radical Political Economics*. 18(1&2): 1-260.

Devine, J. 1989. The utility of value: the new solution, unequal exchange and crisis. *Research in Political Economy*. 12: 21-39.

Dicken, P. 1986. *Global Shift*. London: Harper and Row.

Dietrich, M. 1986. Organizational requirements of a socialist economy: theoretical and practical suggestions. *Cambridge Journal of Economics*. 10: 319-32.

Dinnerstein, D. 1976. *The Mermaid and the Minotaur*. New York: Harper and Row.

Dobb, M. 1947. *Studies in the Development of Capitalism*. New York: International Publishers.

Domhoff, W. 1970. *The Higher Circles: The Governing Class in America*. New York: Vintage.

Domhoff, W. 1974. *The Bohemian Grove: A Study in Ruling Class Cohesiveness*. New York: Harper and Row.

Domhoff, W. 1979. *The Powers That Be*. New York: Vintage.

Dore, R. 1987. *Flexible Rigidities: Industrial Policy and Structural Adjustment in the Japanese Economy, 1970-1980*. London: Athlone Press.

Dosi, G., Teece, D. 1991. In Winter, S. (ed.) "Towards a Theory of Corporate Coherence." *Journal of Economics Behavior and Organization*. (forthcoming)

Downs, A. 1967. *Inside Bureaucracy*. Boston: Little, Brown.

Doyal, L. 1979. *The Political Economy of Health*. London: Pluto Press.

Drucker, P. 1968. *The Age of Discontinuity*. New York: Harper and Row.

Drucker, P. 1986. *The Frontiers of Management*. New York: Dutton.

Drucker, P. 1988. The coming of the new organization. *Harvard Business Review.* 88(1): 45-53.

Dumenil, G. and Levy, D. 1989. Theory and facts: what can we learn from a century of U.S. economic history? Unpublished manuscript. Paris: CEPREMAP.

Dunford, M. 1990. Theories of regulation. *Society and Space.* 8(3): 297-322.

Durkheim, E. 1893. *The Division of Labor in Society.* English edition, 1984. London: Macmillan.

Eatwell, J. 1971. Growth, profitability and size: the empirical evidence. In: Marris, R. and Wood, A. (eds). *The Corporate Economy.* Cambridge: Cambridge University Press. 379-422.

Eccles, R. 1981. The quasifirm in the construction industry. *Journal of Economic Behavior and Organization.* 2: 335-57.

Eccles, R. 1985. *The Transfer Pricing Problem.* Lexington, Mass.: Lexington Books.

Edholm, F., Harris, O. and Young, K. 1977. Conceptualizing women. *Critique of Anthropology.* 3: 101-30.

Edwards, R. 1979. *Contested Terrain.* New York: Basic Books.

Eggertson, T. 1990. The role of transaction costs and property rights in economic analysis. *European Economic Review.* 34(2-3): 450-57.

Ehrenreich, B. 1983. *The Hearts of Men.* Garden City: Anchor/Doubleday.

Ehrenreich, B. 1989. *Fear of Falling: The Inner Life of the Middle Class.* New York: Pantheon.

Ehrenreich, B. and Ehrenreich, J. 1979. The professional-managerial class. In: Walker, P. (ed.) *Between Labor and Capital.* Boston: South End Press. 5-43.

Ehrenreich, B. and English, D. 1979. *For Her Own Good: 150 Years of the Experts Advice to Women.* London: Pluto Press.

Eisenstein, Z. (ed.) 1979. *Capitalist Patriarchy and the Case for Socialist Feminism.* New York: Monthly Review Press.

Eisenstein, Z. 1984. *Feminism and Sexual Equality: Crisis in Liberal America.* New York: Monthly Review Press.

Elfring, T. 1988. *Service Sector Employment in the Advanced Economies.* Aldershot: Gower.

Elfring, T. and Klosterman, R. 1989. The Dutch job machine: the fast growth of low-wage jobs in services, 1979-86. Unpublished manuscript. Rotterdam School of Management, Erasmus University, Rotterdam.

Elson, D. 1988. Market socialism or socialization of the market? *New Left Review.* 170: 3-45.

Elster, J. 1989. From here to there; or, if cooperative ownership is so desirable, why are there so few coops? *Social Philosophy and Policy.* 6(2): 93-111.

Englander, E. 1988. Technology and Oliver Williamson's transactions cost economics. *Journal of Economic Behavior and Organization.* 10: 339-53.

Estall, R. 1985. Stock control in manufacturing: the just-in-time system and its locational implications. *Area.* 17(2): 129-133.

Ewen, S. 1976. *Captains of Consciousness.* New York: McGraw-Hill.

Ewen, S. 1988. *All Consuming Images: The Politics of Style in Contemporary Culture.* New York: Basic Books.

Fanon, F. 1965. *The Wretched of the Earth.* English edition. New York: Grove Press.

Farmer, M. and Matthews, C. 1991. Cultural differences and subjective rationality: where sociology connects with the economics of technological choice. In: Hodgson, G. *Rethinking Economics*. (forthcoming)

Ferguson, A. 1989. *Blood At The Root*. London: Pandora Press.

Fernbach, D. 1981. *The Spiral Path*. London: Gay Men's Press.

Firestone, S. 1970. *The Dialectics of Sex*. New York: Bantam.

Fisher, A. 1939. Production: primary, secondary, tertiary. *The Economic Record*. 15(June): 24-38.

Flandrin, J. 1979. *Families in Former Times*. Cambridge: Cambridge University Press.

Florida, R. 1991. The new industrial revolution. Working paper 91-07. School of Urban and Public Affairs, Carnegie-Mellon University. Pittsburgh, Pa.

Florida, R. and Kenney, M. 1988a. Venture capital, high technolgy and regional development. *Regional Studies*. 22(1): 33-48.

Florida, R. and Kenney, M. 1988b. High technology restructuring in the USA and Japan: flexible specialization versus structured flexibility. *Environment and Planning A*. 22: 233-52.

Florida, R. and Kenney, M. 1990. *The Breakthrough Illusion*. New York: Basic Books.

Foray, D. 1991. The secrets of industry are in the air: industrial cooperation and the organizational dynamics of the innovative firm. *Research Policy*. 20 (forthcoming)

Forester, T. (ed.) 1980. *The Microelectronics Revolution*. Cambridge, Mass.: MIT Press.

Forester, T. (ed.) 1987. *The Information Technology Revolution*. Cambridge, Mass.: MIT Press.

Forty, A. 1986. *Objects of Desire: Design and Society, 1750-1980*. London: Thames and Hudson/Cameron.

Foster, J. 1974. *Class Struggle and the Industrial Revolution*. London: Weidenfeld and Nicholson.

Foster, J. 1988. The fetish of Fordism. *Monthly Review*. 39(9): 14-33.

Foucault, M. 1979. *Discipline and Punish*. English edition. New York: Vintage Books.

Fox, S. 1984. *The Mirror Makers: A History of American Advertizing and Its Creators*. New York: Morrow.

Frank, D. 1989. Labor's decline. *Monthly Review*. 41(5): 48-55.

Fraser, W. 1981. *The Coming of the Mass Market, 1850-1914*. London: Macmillan.

Freedman, C. 1976. *Labor Markets: Segments and Shelters*. Montclair, NJ: Allanheld, Osmun.

Freeman, C. 1982. *The Economics of Industrial Innovation*. Second edition. London: Frances Pinter.

Freeman, C. (ed.) 1984. *Long Waves in the World Economy*. London: Frances Pinter.

Freeman, C. 1987. *Technology Policy and Economic Performance: Lessons from Japan*. London: Frances Pinter.

Freeman, C. 1991. Networks of innovators: a synthesis of research issues. *Reseach Policy*. 20 (forthcoming)

Freiberger, P. and Swaine, M. 1984. *Fire in the Valley: The Making of the Personal Computer*. Berkeley: Osborne/McGraw-Hill.

Friar, J. and Horwitch, M. 1985. The emergence of technology strategy: a new dimension of strategic management. *Technology in Society*. 7(2-3): 143-78.

Friedan, B. 1963. *The Feminine Mystique*. New York: Norton.

Friedman, D. 1988. *The Misunderstood Miracle: Politics and Economic Decentralization in Japan*. Ithaca: Cornell University Press.

Friedmann, H. 1978. Simple commodity production and wage labour in the American plains. *The Journal of Peasant Studies*. 6(1): 71-100.

Fuchs, V. 1968. *The Service Economy*. New York: Columbia University Press.

Gadrey, J. 1987. The double dynamic of services. *The Service Industries Journal*. 7(4): 125-38.

Galambos, L. 1966. *Competition and Cooperation: The Emergence of a National Trade Association*. Baltimore: Johns Hopkins University Press.

Galbraith, J. 1967. *The New Industrial State*. Boston: Houghton Mifflin.

Gallman, R. and Weiss, T. 1970. The service industries in the 19th century. In: *Studies in Income and Wealth*. New York: National Bureau of Economic Research. 34: 287-353.

Gardiner, J. 1975. Women's domestic labour. *New Left Review*. 89: 47-58.

Gardner, C. and Sheppard, J. 1989. *Consuming Passion: The Rise of Retail Culture*. London: Unwin Hyman.

Garnsey, E. 1981. The rediscovery of the division of labor. *Theory and Society*. 10: 325-36.

Garnsey, E. 1982. Women's work and theories of class stratification. *Sociology*. 12: 223-43.

Garrahan, P. 1986. Nissan in the north-east of England. *Capital and Class*. 27: 5-13.

Gerschenkron, A. 1962. *Economic Backwardness in Historical Perspective*. Cambridge, Mass.: Harvard/Belknap.

Gershuny, J. 1978. *After Industrial Society?* Atlantic Highlands NJ: Humanities Press.

Gershuny, J. and Miles, I. 1983. *The New Service Economy*. London: Frances Pinter.

Gertler, M. 1988. The limits to flexibility: comments on the post-Fordist vision of production and its geography. *Transactions of the Institute of British Geographers*. 13: 419-32.

Giddens, A. 1979. *Central Problems in Social Theory*. Berkeley: University of California Press.

Giddens, A. 1980. *The Class Structure of the Advanced Societies*. Second edition. London: Hutchinson.

Giddens, A. 1984. *The Constitution of Society*. Berkeley: University of California Press.

Glasmeier, A. 1989. The role of merchant wholesalers in industrial agglomeration formation. Working Paper #17, Community and Regional Planning, University of Texas, Austin, TX.

Gleicher, D. 1985-86. The ontology of labor values. *Science and Society*. 49(4): 463-71.

Goldthorpe, J. 1982. On the service class, its formation and future. In: Giddens, A. and MacKenzie, G. (eds). *Social Class and the Division of Labor*. Cambridge: Cambridge University Press. 162-85.

Goldthorpe, J. 1983. Women and class analysis: a defence of the conventional view. *Sociology*. 17: 465-78.

Goldthorpe, J., Lockwood, D., Bechhofer, G. and Platt, J. 1969. *The Affluent Worker in the Class Structure*. Cambridge: Cambridge University Press.

Gomes-Casseres, B. 1988. Joint venture cycles: the evolution of ownership strategies of US MNEs, 1945-75. In: Contractor, F. and Lorange, P. (eds). *Cooperative Strategies in International Business*. Lexington, Mass.: Heath. 110-28.

Goodman, E., Bamford, J. and Saynor, P. (eds). 1989. *Small Firms and Industrial Districts in Italy*. London: Routledge.

Goody, J. 1976. *Production and Reproduction: A Comparative Study of the Domestic Domain*. Cambridge: Cambridge University Press.

Gordon, A. 1985. *The Evolution of Labor Relations in Japan: Heavy Industry, 1853-1955*. Cambridge, Mass.: Harvard University Press.

Gordon, D. 1988. The global economy: new edifice or crumbling foundation? *New Left Review*. 168: 24-65.

Gordon, L. 1976. *Woman's Body, Woman's Right: Social History of Birth Control in America*. New York: Grossman.

Gordon, R. 1991. High technology innovation and the global milieu: small and medium-sized enterprises in Silicon Valley. In: Perrin, J. and Maillat, D. (eds). *Milieux d'Innovateurs et Processus d'Innovation dans les Entreprises*. Paris: ERESA-Economica. (forthcoming)

Gorz, A. 1982. *Farewell to the Working Class*. London: Pluto Press.

Gough, J. 1986. Industrial policy and socialist strategy: restructuring and the unity of the working class. *Capital and Class*. 29: 58-82.

Gould, S. 1981. *The Mismeasure of Man*. New York: Norton.

Granovetter, M. 1985. Economic action and social structure: the problem of embeddedness. *American Journal of Sociology*. 91(3): 481-510.

Greater London Council. 1985. *The London Industrial Strategy*. London: GLC.

Greenfield, H. 1966. *Manpower and the Growth of Producer Services*. New York: Columbia University Press.

Habermas, J. 1971. *Knowledge and Human Interests*. Boston: Beacon Press.

Hadjimichalis, C. and Vaiou, D. 1990. Flexible labor markets and regional development in northern Greece. *International Journal of Urban and Regional Research*. 14(1): 1-23.

Håkånsson, H. (ed.) 1982. *International Marketing and Purchasing of Industrial Goods: an Interaction Approach*. Cirenchester: John Wiley.

Håkånsson, H. 1989. *Corporate Technological Behavior: Cooperation and Networks*. London: Routledge.

Håkånsson, H. and Johanson, J. 1988. Formal and informal cooperation strategies in international industrial networks. In: Contractor, F. and Lorange, F. (eds). *Cooperative Strategies in International Business*. Lexington, Mass.: Heath.

Håkånsson, H. and Östberg, C. 1974. Industrial marketing – an organizational problem? *Industrial Marketing Management*. 4: 113-23.

Hall, P. 1984. *The World Cities*. Third edition. London: Weidenfeld and Nicholson.

Hall, R. 1982. The Toyota Kanban system. In: Lee, S. and Schwardiman, G. (eds). *Management by Japanese Systems*. New York: Praeger. 141-51.

Hamelink, C. 1983. *Finance and Information*. Norwood, NJ: Ablex.

Hannah, L. 1980. Visible and invisible hands in Great Britain. In: Chandler, A. and Daems, H. (eds). *Managerial Hierarchies*. Cambridge, Mass.: Harvard University Press. 41-76.

Harcourt, G. 1972. *Some Cambridge Controversies in the Theory of Capital*. Cambridge: Cambridge University Press.

Hareven, T. (ed.) 1982. *Family and Kin in Urban Communities*. New York: New Viewpoints.

Harrigan, K. 1985. *Strategies for Joint Ventures*. Lexington: Lexington Books.

Harrison, B. 1989. The big firms are coming out of the corner. Paper presented to Conference on Industrial Transformation and Regional Development in an Age of Global Interdependence. Nagoya, Japan. September 18-21.

Harrison, B. and Bluestone, B. 1988. *The Great U-Turn: Corporate Restructuring, Laissez Faire and the Rise of Inequality in America*. New York: Basic Books.

Harrison, B. and Kelley, M. 1991. Outsourcing and the search for flexibility: the morphology of production subcontracting in US manufacturing. In: Storper, M. and Scott, A. (eds). *Pathways to Flexibility*. (forthcoming)

Hart, N. 1989. Gender and the rise and fall of class politics. *New Left Review*. 175: 19-47.

Hartmann, H. 1979. Capitalism, patriarchy and job segregation by sex. In: Eisenstein, Z. (ed.) *Capitalist Patriarchy and the Case for Socialist Feminism*. New York: Monthly Review Press. 206-47.

Hartmann, H., Kraut, R. and Tilly, L. 1986. *Computer Chips and Paper Clips: Technology and Women's Employment*. Volume I. Washington, D.C.: National Academy Press.

Hartsock, N. 1987. *Money, Sex and Power: Toward a Feminist Historical Materialism*. Boston: Northeastern University Press.

Haruo, S. 1981. Japanese automotive capital and international competition. Part 2. *Ampo: Japan-Asia Quarterly Review*. 13: 60-67.

Harvey, D. 1982. *The Limits to Capital*. Oxford: Basil Blackwell.

Harvey, D. 1985a. *Consciousness and the Urban Experience*. Baltimore: Johns Hopkins University Press.

Harvey, D. 1985b. *The Urbanization of Capital*. Baltimore: Johns Hopkins University Press.

Harvey, D. 1989. *The Condition of Postmodernity*. Oxford and Cambridge, Mass.: Basil Blackwell.

Hawrylyshyn, O. 1976. The value of household services: a survey of empirical estimates. *Review of Income and Wealth*. 22(2): 101-32.

Hay, E. 1988. *The Just-in-Time Breakthrough*. New York: Wiley.

Hayek, F. 1949. *Individualism and Economic Order*. London: Routledge.

Hayes, R. and Jaikumar, R. 1988. Manufacturing's crisis: new technologies, obsolete organizations. *Harvard Business Review*. September-October. 77-85.

Hays, S. 1959. *Conservation and the Gospel of Efficiency*. Cambridge, Mass.: Harvard University Press.

Hebdige, D. 1979. *Subculture: The Meaning of Style*. London: Methuen.

Heiman, M. 1988. *The Quiet Evolution: Power, Planning and Profits in New York State*. New York: Praeger.

Hepworth, M. 1990. *Geography in the Information Economy*. London: The Guilford Press.

Hergert, M. and Morris, D. 1988. Trends in international collaborative agreements. In: Contractor, F. and Lorange, P. (eds). *Cooperative Strategies in International Business*. Lexington, Mass.: Heath. 99-109.

Herman, E. 1981. *Corporate Control, Corporate Power*. New York: Cambridge University Press.

Hill, T. 1977. On goods and services. *Review of Income and Wealth*. 10(4): 315-38.

Himmelstein, D. and Woolhandler, S. 1990. The corporate compromise: a Marxist view of health policy. *Monthly Review*. 42(1): 14-29.

Hirschman, A. 1970. *Exit, Voice and Loyalty*. Cambridge, Mass.: Harvard University Press.

Hirschon, R. (ed.) 1984. *Women and Property, Women as Property*. London: Croom Helm.

Hirst, P. and Zeitlin, J. (eds). 1989. *Reversing Industrial Decline?* Oxford: Berg.

Hobsbawm, E. 1981. The forward march of labour halted? In: Jacques, M. and Mulhern, F. (eds). *The Forward March of Labour Halted?* London: New Left Books. 1-19.

Hobsbawm, E. 1990. *Nations and Nationalism Since 1780*. Cambridge: Cambridge University Press.

Hochschild, A. 1989. *The Second Shift: Inside the Two-Job Marriage*. New York: Viking.

Hodgson, G. 1974. *The Democratic Economy*. Harmondsworth: Penguin.

Hodgson, G. 1988. *Economics and Institutions*. Philadelphia: University of Pennsylvania Press.

Holmes, J. 1986. The organization and locational structure of production subcontracting. In: Scott, A. and Storper, M. (eds). *Production, Work, Territory*. London: Allen and Unwin. 80-106.

Holmes, J. 1987. Technical change and the restructuring of the North American automobile industry. In: Chapman, K. and Humphrys, G. (eds). *Technical Change and Industrial Policy*. Oxford and Cambridge, Mass.: Basil Blackwell. 121-56.

Holton, R. and Turner, B. 1989. *Max Weber on Economy and Society*. London: Routledge.

Horvat, B. 1982. *The Political Economy of Socialism*. Armonk: M.E. Sharpe.

Hounshell, D. 1984. *From the American System to Mass Production, 1800-1932*. Baltimore: Johns Hopkins University Press.

Hounshell, D. and Smith, J. 1988. *Science and Corporate Strategy: Dupont R&D, 1902-1980*. New York: Cambridge University Press.

Howells, J. 1988. *Economic, Technical and Locational Trends in European Services*. Avebury: Aldershot Press.

Howells, K. 1986. Mixing it: the environment and the unions. In: Weston, J. (ed.) *Red and Green: The New Politics of the Environment*. London: Pluto Press. 140-51.

Humphrey, J. 1987. *Gender and Work in the Third World: Sexual Divisions in Brazilian Industry*. London: Tavistock.

Hunt, E. 1979. The categories of productive and unproductive labor in Marxist theory. *Science and Society*. 43(3): 303-25.

Hymer, S. 1972. The multinational corporation and the law of uneven development. In: Bhagwati, J. (ed.) *Economics and World Order*. New York: The Free Press. 113-40.

Ichiyo, M. 1984. Class struggle on the shopfloor – the Japanese case, 1945-84. *Ampo: Japan-Asia Quarterly Review*. 16(3): 38-49.

Ikeda, M. 1979. The subcontracting system in the Japanese electronic industry. *Engineering Industries of Japan*. 19: 43-71.

Ikeda, M. 1987. Evolution of the Japanese subcontracting system. *Tradescope*. July. 2-6.

Illeris, S. 1989. *Services and Regions in Europe*. Aldershot: Avebury/Gower.

Illich, I. 1976. *Limits to Medicine*. London: Boyers.

Imai, K. 1987-88. The corporate network in Japan. *Japanese Economic Studies*. 16(2): 3-37.

Imai, K. and Itami, H. 1984. Interpenetration of organization and market: Japan's firm and market in comparison with the US. *International Journal of Industrial Organization*. 2: 285-310.

Ingham, G. 1984. *Capitalism Divided? The City and Industry in British Social Development*. London: Macmillan.

Irigaray, L. 1985. *This Sex Which is Not One*. English edition. Ithaca: Cornell University Press.

Itoh, M. 1987. Skilled labour in value theory. *Capital and Class*. 31: 39-58.

Jaikumar, R. 1986. Postindustrial manufacturing. *Harvard Business Review*. 64(6): 69-76.

Jameson, F. 1983. Postmodernism, or the cultural logic of late capitalism. *New Left Review*. 146: 53-93.

Jameson, F. 1989. Marxism and postmodernism. *New Left Review*. 176: 31-45.

Jessop, B. 1986. *The Capitalist State*. Oxford: Martin Robertson.

Jessop, B. 1990. Regulation theories in retrospect and prospect. *Economy and Society*. 19: 153-216.

Johnson, C. 1982. *MITI and the Japanese Miracle*. Stanford: Stanford University Press.

Jones, S. 1982. The organization of work: a historical dimension. *Journal of Economic Behavior and Organization*. 3(2-3): 117-38.

Jonscher, C. 1983. Information resources and information productivity. *Information Economics and Policy*. 1: 13-35.

Kamata, S. 1982. *Japan in the Passing Lane*. New York: Pantheon.

Kaplinsky, R. 1984. *Automation*. London: Longmans.

Katzenstein, P. (ed.) 1989. *Industry and Politics in West Germany*. Ithaca: Cornell University Press.

Katznelson, I. 1981. *City Trenches: Urban Politics and the Patterning of Class in the U.S.* New York: Pantheon.

Kelly, J. 1982. *Scientific Management, Job Redesign and Work Performance*. London: Academic Press.

Kelly, J. 1985. Management's redesign of work: labor process, labor markets, and product markets. In: Knights, D., Wilmott, H. and Colinson, D. (eds). *Job Redesign*. Aldershot: Gower. 30-51.

Kenney, M. 1986. *Biotechnology: The University-Industrial Complex*. New Haven: Yale University Press.

Kenney, M. and Florida, R. 1988. Beyond mass production: production and the labor process in Japan. *Politics and Society*. 16(1): 121-58.

Kerr, C. 1983. *The Future of Industrial Societies*. Cambridge, Mass.: Harvard University Press.

Kessler-Harris, A. 1982. *Out to Work: A History of Wage-Earning Women in the United States*. New York: Oxford University Press.

Kindleberger, C. 1984. *Multinational Excursions*. Cambridge, Mass.: MIT Press.

Kitching, G. 1983. *Rethinking Socialism*. London: Methuen.

Kocka, J. 1980a. *White Collar Workers in America, 1890-1940*. Beverly Hills: Sage.

Kocka, J. 1980b. The rise of the modern industrial enterprise in Germany. In:

Chandler, A. and Daems, H. (eds). *Managerial Hierarchies*. Cambridge, Mass.: Harvard University Press. 77-116.

Kolko, G. 1963. *The Triumph of Conservatism*. New York: Free Press.

Kornai, J. 1986. *Contradictions and Dilemmas: Studies in the Socialist Economy and Society*. Cambridge, Mass.: MIT Press.

Kotz, D. 1978. *Bank Control of Large Corporations in the United States*. Berkeley: University of California Press.

Kriedte, P. 1983. *Peasants, Landlords and Merchant Capitalists: Europe and the World Economy, 1500-1800*. Leamington Spa: Berg Publication.

Kuhn, A. and Wolpe, A. (eds). 1978. *Feminism and Materialism*. London: Routledge and Kegan Paul.

Kumar, K. 1978. *Prophecy and Progress: The Sociology of Industrial and Post-Industrial Society*. Harmondsworth: Penguin.

Kumazawa, M. and Yamada, J. 1989. Jobs and skills under the lifelong nenko employment practice. In: Wood, S. (ed.) *The Transformation of Work?* London: Unwin Hyman. 102-26.

Kuznets, S. 1966. *Modern Economic Growth: Rate, Structure, and Spread*. New Haven: Yale University Press.

Laclau, E. and Mouffe, C. 1985. *Hegemony and Socialist Strategy: Towards a Radical Democratic Politics*. London: Verso.

Lakoff, G. 1987. *Women, Fire and Dangerous Things*. Chicago: University of Chicago Press.

Lambooy, J. 1988. Intermediare dienstverlening en economisch complexiteit. *Economisch en Sociaal Tijdschrift*. 42(5): 617-29.

Larson, M. 1977. *The Rise of Professionalism*. Berkeley: University of California Press.

Lash, S. and Urry, J. 1987. *The End of Organized Capitalism*. Madison: University of Wisconsin Press.

Lazerson, M. 1988. Organizational growth of small firms: an outcome of markets and hierarchies. *American Sociological Review*. 53: 330-42.

Leaman, J. 1988. *The Political Economy of West Germany, 1945-85*. London: Macmillan.

Leante, J. 1989. Perestroika at a snail's pace. *Telos*. 80: 79-92.

Lebergott, S. 1966. Labor force and employment, 1800-1960. In: National Bureau of Economic Research (ed.) *Output, Employment and Productivity after 1800*. New York: Columbia University Press. 117-210.

Leijonhufvud, A. 1986. Capitalism and the factory system. In: Langlois, R. (ed.) *Economics as a Process*. Cambridge: Cambridge University Press. 203-21.

Lerner, G. 1986. *The Creation of Patriarchy*. New York: Oxford University Press.

Levi-Strauss, C. 1969. *The Elementary Structures of Kinship*. Boston: Beacon Press.

Levine, D. 1980. Aspects of the classical theory of markets. *Australian Economic Papers*. 19:1-15.

Levine, L. 1988. *Highbrow/Lowbrow: The Emergence of Cultural Hierarchy in America*. Cambridge, Mass.: Harvard University Press.

Levitt, T. 1976. The industrialization of service. *Harvard Business Review*. 76 (September-October): 63-74.

Lindblom, C. 1977. *Politics and Markets*. New York: Basic Books.

Lipietz, A. 1987. *Mirages and Miracles: The Global Crisis of Fordism.* London: Verso.

Littler, C. 1982. *The Development of the Labour Process in Capitalist Societies.* London: Heinemann Educational.

Littler, C. 1985. Taylorism, Fordism and job design. In: Knights, D., Willmott, H. and Collinson, D. (eds). *Job Redesign.* Aldershot: Gower. 1-9.

Littler, C. and Salaman, G. 1984. *Class at Work.* London: Batsford Academic.

Locksley, G. 1986. Information technology and capitalist development. *Capital and Class.* 27: 81-106.

Lockwood, J. 1958. *The Blackcoated Worker.* London: Allen and Unwin.

Logan, J. and Molotch, H. 1986. *Urban Fortunes: The Political Economy of Place.* Berkeley and Los Angeles: University of California Press.

Lukacs, G. 1922. *History and Class Consciousness.* English edition 1971. Cambridge, Mass.: MIT Press.

Lundvall, B. 1988. Innovation as an interactive process. In: Dosi, G., *et al.* (eds). *Technical Change and Economic Theory.* London: Frances Pinter. 349-69.

Lustig, J. 1982. *Corporate Liberalism: The Origins of Modern American Political Thought, 1890-1920.* Berkeley: University of California Press.

Luxenberg, S. 1985. *Roadside Empires: How the Chains Franchised America.* New York: Viking Penguin.

Lyon, D. 1988. *The Information Society: Issues and Illusions.* Cambridge: Polity Press.

Machlup, F. 1962. *The Production and Distribution of Knowledge in the United States.* Princeton: Princeton University Press.

MacKenzie, G. 1982. Class boundaries and the labor process. In: MacKenzie, G. and Giddens, A. (eds). *Social Class and the Division of Labor.* Cambridge: Cambridge University Press, 63-87.

MacNeil, I. 1978. Contracts: adjustments of long-term economic relations under classical, neoclassical and relational contract law. *Northwestern University Law Review.* 72: 854-906.

MacPherson, C. 1962. *The Political Theory of Possessive Individualism.* New York: Oxford University Press.

MacPherson, C. 1978. *Property: Mainstream and Critical Positions.* Oxford: Blackwell.

Mair, A., Florida, R. and Kenney, M. 1988. The new geography of automobile production: Japanese transplants in North America. *Economic Geography.* 64(4): 352-73.

Mäki, V. 1989. How to combine rhetoric and realism in the methodology of economics. *Economics and Philosophy.* 4: 89-109.

Mandle, J. 1978. *The Roots of Black Poverty: The Southern Plantation Economy of the Civil War.* Durham: Duke University Press.

Manwaring, T. and Wood, S. 1984. The ghost in the machine: tacit skills in the labor process. *Socialist Review.* 74: 57-86.

Marglin, S. 1974. What do bosses do? *Review of Radical Political Economy.* 6(2): 60-92.

Mark, J. 1982. Measuring productivity in service industries. *Monthly Labor Review.* June: 3-8.

Markusen, A., Hall, P., Deitrich, S. and Campbell, S. 1991. *The Rise of the Gun Belt.* New York: Oxford University Press.

Marris, R. 1979. *The Theory and Future of the Corporate Economy and Society.* Amsterdam: North-Holland.

Marris, R. and Mueller, D. 1980. The corporation, competition and the invisible hand. *Journal of Economic Literature.* 18: 32-63.

Marshak, P., Guy, N. and McMullen, J. (eds). 1987. *Uncommon Property: The Fishing and Fish-Processing Industries in British Columbia.* Toronto: Methuen.

Marshall, A. 1890. *Principles of Economics.* London: Macmillan.

Marshall, A. 1921. *Industry and Trade.* London: Macmillan.

Marshall, G., Rose, D., Newby, H. and Vogler, C. 1988. *Social Class in Modern Britain.* London: Unwin Hyman.

Martinelli, F. 1989. Une approche théoretique de la demande de services aux producteurs. In: Moulaert, F. (ed.) *La Production des Services et Sa Géographie.* Lille: Université de Lille 1. 45-66.

Martinez, J. and Jarillo, C. 1989. The evolution of research on coordination mechanisms in multinational corporations. *Journal of International Business Studies.* Fall: 489-514.

Marx, K. 1863. *Capital: Volume I.* English edition 1967. New York: International Publishers.

Marx, K. 1893. *Capital: Volume II.* English edition 1967. New York: International Publishers.

Marx, K. 1973. *Grundrisse.* Harmondsworth: Penguin and New Left Books.

Marx, K. and Engels, F. 1848. *Manifesto of the Communist Party.* English edition 1952. Moscow: Progress Publishers.

Massey, D. 1984. *Spatial Divisions of Labor: Social Structures and the Geography of Production.* London: Macmillan.

Maynard, A. and Williams, G. 1984. Privatisation in the National Health Service. In: Le Grand, J. and Robinson, R. (eds). *Privatisation and the Welfare State.* London: George Allen and Unwin. 95-110.

McCormick, K. 1985. The flexible firm and employment adjustment: fact and fable in the Japanese case. Mimeo. Unit for Comparative Research on Industrial Relations, University of Sussex.

McIntosh, M. 1978. The state and the oppression of women. In: Kuhn, A. and Wolpe, A. *Feminism and Materialism.* London: Routledge and Kegan Paul. 254-89.

McMillan, C. 1984. *The Japanese Industrial System.* London: Walter de Gruyter.

Meegan, R. 1988. A crisis of mass production? In: Allen, J. and Massey, D. (eds). *The Economy in Question.* London: Sage. 136-83.

Meillassoux, C. 1981. *Maidens, Meals and Money.* New York: Cambridge University Press.

Melman, S. 1951. The rise of administrative overheads in the manufacturing industries of the United States, 1899-1947. *Oxford Economic Papers.* 3: 62–113.

Merrington, J. 1975. Town and country in the transition to capitalism. *New Left Review.* 93: 71-92.

Meszaros, I. 1986. Marx's 'social revolution' and the divisions of labour. *Radical Philosophy.* 44: 14-23.

Meszaros, I. 1987. The division of labor and the post-capitalist state. *Monthly Review.* 39(3): 80-108.

Middleton, C. 1983. Patriarchal exploitation and the rise of English capitalism. In:

Gamarnikow, E., Morgan, D., Purvis, J. and Taylorson, D. (eds). *Gender, Class and Work*. London: Heinemann. 11-27.

Miles, I. and Gershuny, J. 1986. The social economics of information technology. In: Ferguson, M. (ed.) *New Communications Technologies and the Public Interest*. Beverly Hills: Sage. 19-36.

Miliband, R. 1969. *The State in Capitalist Society*. London: Weidenfeld and Nicholson.

Milkman, R. 1987. *Gender At Work: The Dynamics of Job Segregation by Sex During World War II*. Urbana: University of Illinois Press.

Miller, D. 1987. *Material Culture and Mass Consumption*. Oxford and Cambridge, Mass.: Blackwell.

Miller, D.L. 1989. *Market, State and Community*. Oxford: Clarendon Press.

Miller, J. and Vollman, T. 1985. The hidden factory. *Harvard Business Review*. 63(5): 142-50.

Miller, R.P. 1983. The Hoover in the garden: middle class women and suburbanization, 1870-1920. *Society and Space*. 1: 73-87.

Miller, R. 1984. A comment on productive and unproductive labor. *Studies in Political Economy*. 14: 141-53.

Millett, K. 1977. *Sexual Politics*. Garden City, NY: Doubleday.

Mills, C. 1951. *White Collar*. New York: Oxford.

Mintz, B. and Schwartz, M. 1985. *The Power Structure of American Business*. Chicago: University of Chicago Press.

Mitchell, J. 1975. *Psychoanalysis and Feminism*. Harmondsworth: Penguin.

Mitsui, I. 1987. The Japanese subcontracting system. Paper presented to the Workshop on Unemployment and Labour, University of Cambridge. 3 March.

Molyneux, M. 1979. Beyond the domestic labour debate. *New Left Review*. 116: 3-27.

Molyneux, M. 1990. The woman question in the age of perestroika. *New Left Review*. 183: 23-49.

Monden, Y. 1981. What makes the Toyota production system really tick? *Industrial Engineering*. Jan.: 37-44.

Monden, Y. 1983. *Toyota Production System*. Atlanta: Industrial Engineering and Management Press.

Montgomery, D. 1967. *Beyond Equality*. New York: Knopf.

Moore, B. 1966. *The Social Origins of Dictatorship and Democracy*. Boston: Beacon Press.

Moore, H. 1988. *Feminism and Anthropology*. Cambridge: Polity Press.

Moore, S. 1980. *Marx on the Choice Between Socialism and Communism*. Cambridge, Mass.: Harvard University Press.

Morgan, D. 1980. *Merchants of Grain*. Harmondsworth: Penguin.

Morgan, K. and Sayer, A. 1985. A "modern" industry in a "mature" region: the remaking of labour-management relations. *International Journal of Urban and Regional Research*. 9: 383-404.

Morgan, K. and Sayer, A. 1988. *Microcircuits of Capital*. Cambridge: Polity Press.

Morishima, M. 1982. *Why Has Japan Succeeded?* Cambridge: Cambridge University Press.

Morris, J. 1989. New technologies, flexible work practices and regional sociospatial differentiation: some observations from the United Kingdom. *Society and Space*. 6(3): 301-20.

Moulaert, F. (ed.) 1989. *La Production des Services et Sa Géographie*. Lille: Université de Lille 1.

Moulaert, F., Martinelli, F. and Djellal, F. 1989. The functional and spatial division of labor of information technology consultancy firms in Europe. Paper presented at the International Symposium on Regulation, Innovation and Spatial Development, University of Wales, Cardiff, 13-15 September.

Moulaert, F. and Swyngedouw, E. 1989. A regulation approach to the geography of flexible production systems. *Environment and Planning D: Society and Space*. 7(3): 327-46.

Mowery, D. (ed.) 1988. *International Collaborative Ventures in Manufacturing*. Cambridge: Ballinger.

Mueller, D. 1986. *Profits in the Long Run*. Cambridge: Cambridge University Press.

Murakami, Y. 1987. The Japanese model of political economy. In: Yamamura, K. and Yasuba, Y. (eds). *The Political Economy of Japan. Volume 1: The Domestic Transformation*. Stanford: Stanford University Press. 33-91.

Murphy, R. 1988. *Social Closure*. New York: Oxford.

Murray, F. 1983. The decentralization of production: the decline of the mass collective worker. *Capital and Class*. 19: 74-99.

Murray, R. 1972. Underdevelopment, the international firm and the international division of labor. In: *Towards a New World Economy: Papers and Proceedings of the Society for International Development*. Rotterdam: Rotterdam University Press. 159-248.

Murray, R. 1986. Public sector possibilities. *Marxism Today*. 29(7): 28-32.

Murray, R. 1987. Ownership, control and the market. *New Left Review*. 164: 87-112.

Murray, R. 1989. Life after Henry (Ford). *Marxism Today*. 32(10): 8-13.

Myrdal, G. 1954. *The Political Element in the Development of Economic Theory*. Cambridge, Mass.: Harvard University Press.

Nash, J. and Fernandez-Kelly, P. 1983. *Women, Men, and the International Division of Labor*. Albany: SUNY Press.

Navarro, V. 1976. *Medicine Under Capitalism*. New York: PRODIST.

Nelson, D. 1975. *Managers and Workers: Origins of the New Factory System in the United States*. Madison: University of Wisconsin Press.

Nelson, K. 1984. *Back Offices and Female Labor Markets: Office Suburbanization in the San Francisco Bay Area, 1965-1980*. Unpublished doctoral dissertation, Department of Geography, University of California, Berkeley.

Nelson, R. and Winter, S. 1982. *An Evolutionary Theory of Economic Change*. Cambridge, Mass.: Harvard University Press.

Nishiguchi, T. 1987. Competing systems of automotive components supply: an examination of the Japanese clustered control model and the Alps structure. Briefing paper for First Policy Forum, International Motor Vehicle Program, MIT. Niagara-on-the-Lake, Canada.

Noble, D. 1977. *America by Design*. New York: Knopf.

Noble, D. 1986. *Forces of Production: A Social History of Industrial Automation*. New York: Oxford University Press.

Nora, S. and Minc, A. 1980. *The Computerization of Society*. English edition. Cambridge, Mass.: MIT Press.

Nove, A. 1983. *The Economics of Feasible Socialism*. London: Allen and Unwin.

Nove, A. 1984. Some observations of intersystem comparisons. In: Zimbalist, A. (ed.) *Comparative Economic Systems: An Assessment of Knowledge, Theory and Method*. Amsterdam: Kluwer-Nijhtoff. 47-66.

Nove, A. 1989a. Socialism, capitalism and the Soviet experience. *Social Philosophy and Policy*. 6(2): 235-51.

Nove, A. 1989b. Central planning under capitalism and market socialism. In: Elster, J. and Moene, K. (eds). *Alternatives to Capitalism*. Cambridge: Cambridge University Press. 98-111.

Nusbaumer, J. 1987. *The Service Economy: Lever to Growth*. Boston: Kluwer.

Oakley, A. 1974. *The Sociology of Housework*. Oxford: Martin Robertson.

Ochel, W. and Wegner, M. 1987. *Service Economies in Europe*. London: Frances Pinter.

O'Connor, J. 1973. *The Fiscal Crisis of the State*. New York: St. Martin's Press.

Ohlin, B. 1939. *Interregional and International Trade*. Cambridge, Mass.: Harvard University Press.

Ohmae, K. 1989. The global logic of strategic alliances. *Harvard Business Review*. 69: 143-54.

Ohno, T. 1982. How the Toyota production system was created. *Japan Economic Studies*. 10(4): 83-103.

Okimoto, D. 1989. *Between MITI and the Market*. Stanford: Stanford University Press.

Okimoto, D., Sugano, T. and Franklin, B. 1984. *Competitive Edge: The Semiconductor Industry in the U. S. and Japan*. Stanford: Stanford University Press.

Ollman, B. 1971. *Alienation*. New York: Cambridge University Press.

O'Neill, J. 1989. Markets, socialism and information: a reformulation of a Marxian objection to the market. *Social Philosophy and Policy*. 6(2): 200-34.

Ong, A. 1987. *Spirits of Resistance and Capitalist Discipline: Factory Women in Malaysia*. Albany: SUNY Press.

Organization for Economic Cooperation and Development. 1980. *Women and Employment: Policies for Equal Opportunities*. Paris: OECD.

Organization for Economic Cooperation and Development. 1986. *Flexibility in the Labour Market: The Current Debate*. Paris: OECD.

Ouchi, W. 1980. Markets, bureaucracies and clans. *Administrative Science Quarterly*. 25: 120-42.

Page, B. 1992. The Restructuring of Regional Pork Production. Unpublished doctoral dissertation, Department of Geography. University of California, Berkeley.

Page, B. and Walker, R. 1991. From settlement to Fordism: the agroindustrialization of the Midwest. *Economic Geography*. 67 (forthcoming)

Pahl, R. 1984. *Divisions of Labour*. Oxford and Cambridge, Mass.: Basil Blackwell.

Palloix, C. 1976. The labour process from Fordism to neo-Fordism. In: *The Labour Process and Class Struggle*. London. 46-65.

Panzar, J. and Willig, R. 1981. Economies of scope. *American Economic Review*. 71(2): 268-72.

Parker, N. 1982. What's so Right about Adam Smith? *Radical Philosophy*. 30: 24-32.

Parker, R. 1969. *Management Accounting: An Historical Perspective*. London: Macmillan.

Parkin, F. 1979. *Marxism and Class Theory: A Bourgeois Critique*. London: Tavistock.

Parsons, T. and Bales, R. 1956. *Family, Socialization and Interaction Process*. Glencoe, Ill.: Free Press.

Patrick, H. and Rohlen T. 1988. Small scale family enterprises. In: Yamamura, K. and Yasuba, Y. (eds). *The Political Economy of Japan. Vol 1: The Domestic Transformation*. Stanford: Stanford University Press. 331-83.

Patterson, O. 1982. *Slavery and Social Death*. Cambridge, Mass.: Harvard University Press.

Perez, C. 1985. Microelectronics, long waves and world structural change. *World Development*. 13: 441-63.

Perrow, C. 1981. Markets, hierarchies and hegemony: a critique of Chandler and Williamson. In: Van de Ven, A. and Joyce, W. (eds). *Perspectives on Organization Design and Behavior*. New York: Wiley. 371-86.

Perrow, C. 1986. *Complex Organizations: A Critical Essay*. Third Edition. New York: Random House.

Peters, T. and Waterman, R. 1985. *In Search of Excellence: Lessons from America's Best-Run Companies*. New York: Harper and Row.

Petit, P. 1986. *Slow Growth and the Service Economy*. London: Frances Pinter.

Phillips, A. 1984. *Hidden Hands*. London: Pluto Press.

Phillips, A. and Taylor, B. 1980. Sex and skill: notes towards a feminist economics. *Feminist Review*. 6: 79-88.

Piaget, J. 1970. *Structuralism*. New York: Basic Books.

Piore, M. and Sabel, C. 1984. *The Second Industrial Divide*. New York: Basic Books.

Polanyi, K. 1944. *The Great Transformation*. New York: Rinehart.

Pollard, S. 1968. *The Genesis of Modern Management*. Harmondsworth: Penguin.

Pollard, S. 1981. *Peaceful Conquest: The Industrialization of Europe, 1760-1970*. New York: Oxford University Press.

Pollert, A. 1988. Dismantling flexibility. *Capital and Class*. 34: 42-75.

Porat, M. 1977. *The Information Economy*. Washington, D.C.: United States Department of Commerce.

Porter, G. and Livesay, H. 1971. *Merchants and Manufacturers*. Baltimore: Johns Hopkins University Press.

Porter, M. (ed.) 1986. *Competition in Global Industries*. Boston, Mass.: Harvard Business School Press.

Poulantzas, N. 1975. *Classes in Contemporary Capitalism*. London: New Left Books.

Powell, W. 1987. Hybrid organizational arrangements: new form or transitional development? *California Management Review*. 30(1): 67-87.

Powell, W. 1990. Neither market nor hierarchy: network forms of organization. In: Staw, B. and Cummings, L. (eds). *Research in Organizational Behavior*. 12: 295-336.

Pred, A. 1966. *The Spatial Dynamics of Urban Growth in the United States, 1800-1914*. Cambridge, Mass.: MIT Press.

Pred, A. 1973. *Urban Growth and the Circulation of Information, 1790-1840*. Cambridge, Mass.: Harvard University Press.

Pred, A. 1977. *City-Systems in Advanced Economies*. London: Hutchinson.

Pred, A. 1980. *Urban Growth and City Systems in the United States, 1840-60*.

Cambridge, Mass.: Harvard University Press.

Pred, A. 1981. Production, family and free-time projects: a time-geographic perspective on the industrial and societal changes in 19th century US cities. *Journal of Historical Geography*. 7: 3-36.

Pred, A. 1984. Structuration, biography formation and knowledge: observations on port growth during the late mercantile period. *Society and Space*. 2: 251-75.

Przeworski, A. 1985. *Capitalism and Social Democracy*. New York: Cambridge University Press.

Rainnie, A. 1984. Combined and uneven development in the clothing industry. *Capital and Class*. 22: 141-56.

Ramazanoglu, C. 1989. *Feminism and the Contradictions of Oppression*. London: Routledge.

Rattansi, A. 1982. *Marx and the Division of Labor*. London: Macmillan.

Remick, H. (ed.) 1984. *Comparable Worth and Wage Discrimination*. Philadelphia: Temple University Press.

Richardson, G. 1972. The organization of industry. *The Economic Journal*. 84: 883-96.

Riddle, D. 1986. *Service-Led Growth: The Role of the Service Sector in World Development*. New York: Praeger.

Riley, J. 1985. Factories automate before their time. *Computer Weekly*. May 23.

Robinson, J. 1932. *The Economics of Imperfect Competition*. New York: St Martin's Press.

Robinson, J. 1956. *The Accumulation of Capital*. London: Macmillan.

Roemer, J. 1982. *A General Theory of Exploitation and Class*. Cambridge: Cambridge University Press.

Roemer, J. (ed.) 1986. *Analytical Marxism*. New York: Cambridge University Press.

Rogers, E. and Larsen, J. 1984. *Silicon Valley Fever*. New York: Basic Books.

Rose, J. 1986. *Sexuality in the Field of Vision*. London: Verso.

Rose, S., Kamin, L. and Lewontin, R. 1984. *Not in Our Genes: Biology, Ideology and Human Nature*. Harmondsworth: Penguin.

Rosenberg, N. 1976. *Perspectives on Technology*. Cambridge: Cambridge University Press.

Rosenberg, N. 1982. *Inside the Black Box: Technology and Economics*. Cambridge: Cambridge University Press.

Rosenbloom, R. and Abernathy, W. 1982. The climate for innovation in industry: the role of management attitudes and practices in consumer electronics. *Research Policy*. 11(4): 209-25.

Ross, G. 1974. The second coming of Daniel Bell. In: Miliband, R. and Savile, J. (eds). *The Socialist Register*. London: Merlin Press. 331-48.

Rowbowtham, S. 1973. *Women's Consciousness, Men's World*. Harmondsworth: Penguin.

Rubery, J. 1980. Structured labor markets, worker organization, and low pay. In: Amsden, A. (ed.) *The Economics of Women and Work*. New York: St Martin's Press. 242-70.

Rubin, M. and Huber, M. 1986. *The Knowledge Industry in the United States, 1960-80*. Princeton: Princeton University Press.

Rueschmeyer, D. 1986. *Power and the Division of Labor*. Stanford: Stanford University Press.

Rumberger, R. 1981. The changing skill requirements of jobs in the US economy. *Industrial and Labor Relations Review.* 34(4): 578-90.

Russo, M. 1986. Technical change and the industrial district: the role of interfirm relations in the growth and transformation of ceramic tile production in Italy. *Research Policy.* 15: 329-43.

Rustin, M. 1986. Lessons of the London industrial strategy. *New Left Review.* 155: 75-84.

Ryan, M. 1981. *The Cradle of the Middle Class.* New York: Cambridge University Press.

Sabel, C. 1982. *Work and Politics.* New York: Cambridge University Press.

Sabel, C. 1989. Flexible specialization and the reemergence of regional economies. In: Hirst, P. and Zeitlin, J. (eds). *Reversing Industrial Decline?* New York: St Martin's Press. 17-70.

Sabel, C. 1991. Studied trust: building new forms of cooperation in a volatile economy. Unpublished paper. MIT.

Sabel, C. and Zeitlin, J. 1985. Historical alternatives to mass production: politics, markets and technology in nineteenth century industrialization. *Past and Present.* 108: 133-76.

Sako, M. 1989. Competitive cooperation: how the Japanese manage inter-firm relations. Mimeo. Industrial Relations Department, London School of Economics.

Samuel, R. 1977. Workshop of the world: steam power and hand technology in mid-Victorian Britain. *History Workshop Journal.* 3: 6-73.

Sassen, S. 1988. *The Mobility of Labor and Capital.* New York: Cambridge University Press.

Saxenian, A. 1988. *The Political Economy of Industrial Adaptation in Silicon Valley.* Unpublished doctoral dissertation, Department of Political Science, MIT.

Saxenian, A. 1989. In search of power: the organization of business interests in Silicon Valley and Route 128. *Economy and Society.* 18(1): 25-70.

Saxenian, A. 1991. The origins and dynamics of production networks in Silicon Valley. *Research Policy.* 20 (forthcoming)

Sayer, A. 1984. *Method in Social Science: A Realist Approach.* London: Hutchinson.

Sayer, A. 1985. Industry and space: a sympathetic critique of radical research. *Environment and Planning D: Society and Space.* 3: 3-29.

Sayer, A. 1989. Dualistic thinking and rhetoric in geography. *Area.* 21: 301-05.

Sayer, A. and Morgan, K. 1985. High technology industry and the international division of labour: the case of electronics. In: Breheny, M. and McQuaid, R. (eds). *The Development of High Technology Industries: An International Survey.* London: Croom Helm. 10-36.

Sayer, D. 1987. *The Violence of Abstraction: The Analytic Foundations of Historical Materialism.* Oxford: Basil Blackwell.

Schoenberger, E. 1988. From Fordism to flexible accumulation: technology, competitive strategies and international location. *Society and Space.* 6(3): 245-62.

Schoenberger, E. 1990. Some dilemmas of automation. *Economic Geography.* 66: 232-247.

Schonberger, R. 1982. *Japanese Manufacturing Techniques.* New York: The Free Press.

Schonberger, R. 1987. Frugal manufacturing. *Harvard Business Review.* 87(5): 95-100.

Schumpeter, J. 1939. *Business Cycles.* New York: McGraw-Hill.

Schumpeter, J. 1942. *Capitalism, Socialism, and Democracy.* New York: Harper and Row.

Schweikart, D. 1987. Market socialist capitalist roaders: a comment on Arnold. *Economics and Philosophy.* 3: 308-19.

Scott, A. 1986. Industrial organization and location: division of labor, the firm and spatial process. *Economic Geography.* 62(3): 215-31.

Scott, A. 1988a. *Metropolis: From the Division of Labor to Urban Form.* Berkeley and Los Angeles: University of California Press.

Scott, A. 1988b. *New Industrial Spaces.* London: Pion.

Scott, J. 1986. *Capitalist Property and Financial Power: A Comparative Study of Britain, the United States and Japan.* Brighton: Harvester/Wheatsheaf.

Scranton, P. 1983. *Proprietary Capitalism.* New York: Cambridge University Press.

Seccombe, W. 1974. The housewife and her labour under capitalism. *New Left Review.* 83: 3-24.

Seidman, H. 1980. *Politics, Position and Power.* New York: Oxford University Press.

Selucky, R. 1973. Marxism and self-management. In: Vanek, J. (ed.) *Self-Management.* Harmondsworth: Penguin. 47-61.

Selucky, R. 1979. *Marxism, Socialism, Freedom.* London: Macmillan.

Shaiken, H. 1984. *Work Transformed: Automation and Labor in the Computer Age.* New York: Holt, Reinhart and Winston.

Shapira, P. 1989. Steeltown to spaceworld? Industrial and regional restructuring strategies in Japanese heavy industry: Nippon Steel's Yawata Works and Kitakyushu city. Research Paper 8911. Regional Research Institute, West Virginia University.

Shapiro, D. 1989. Reviving the socialist calculation debate: a defence of Hayek against Lange. *Social Philosophy and Policy.* 6(2):139-59.

Sheard, P. 1983. Auto-production systems in Japan: organisational and locational features. *Australian Geographical Studies.* 21: 49-68.

Shelp, R., Stephenson, J., Truitt, N. and Wasow, B. 1984. *Service Industries and Economic Development.* New York: Praeger.

Shieh, G. 1990. *Manufacturing Bosses: Subcontracting Networks under Dependent Capitalism in Taiwan.* Unpublished doctoral dissertation, Department of Sociology, University of California, Berkeley.

Sidel, R. 1986. *Women and Children Last: The Plight of Poor Women in Affluent America.* New York: Viking.

Siltanen, J. and Stanworth, M. (eds). 1984. *Women and the Public Sphere.* London: Tavistock.

Silver, H. 1987. Only so many hours in a day: time constraints, labour pools and demand for consumer services. *The Service Industries Journal.* 7(4): 26-45.

Silverblatt, I. 1987. *Moon, Sun and Witches: Gender Ideologies and Class in Inca and Colonial Peru.* Princeton: Princeton University Press.

Simon, H. 1947. *Administrative Behavior.* New York: Macmillan.

Simon, H. 1957. *Models of Man.* New York: Wiley.

Singelmann, J. 1978. *From Agriculture to Services.* Beverly Hills: Sage.

Smith, A. 1776. *The Wealth of Nations.* 1937 edition. New York: Modern Library.

Soja, E. 1989. *Post-Modern Geographies.* London: Verso.

Sraffa, P. 1960. *Production of Commodities by Means of Commodities.* Cambridge: Cambridge University Press.

Stack, C. 1974. *All Our Kin: Strategies for Survival in a Black Community.* New York: Harper and Row.

Stanback, T. 1979. *Understanding the Service Economy.* Baltimore: Johns Hopkins University Press.

Stanback, T., Bearse, P., Noyelle, T. and Kanasek, R. 1983. *Services – The New Economy.* Totowa, NJ: Rowman and Allanheld.

Standing, G. 1991. Alternative routes to labour flexibility. In: Storper, M. and Scott, A. (eds). *Pathways to Industrial and Regional Development.* (forthcoming)

Stanworth, M. 1984. Women and class analysis: a reply to Goldthorpe. *Sociology.* 18: 159-70.

Starr, P. 1982. *The Social Transformation of American Medicine.* New York: Basic Books.

Ste. Croix, G. 1984. Class in Marx's conception of history, ancient and modern. *New Left Review.* 146: 94-111.

Steedman, I. 1977. *Marx After Sraffa.* London: New Left Books.

Steindl, J. 1952. *Maturity and Stagnation in American Capitalism.* Oxford: Basil Blackwell.

Steinmuller, E. 1988. International joint ventures in the integrated circuit industry. In: Mowery, D. *International Collaborative Ventures in US Manufacturing.* Cambridge: Ballinger. 111-46.

Stigler, G. 1951. The division of labor is limited by the extent of the market. *Journal of Political Economy.* 69: 213-25.

Stone, L. 1977. *The Family, Sex and Marriage in England, 1500-1800.* London: Weidenfield and Nicholson.

Stopford, J. and Wells, L. 1972. *Managing the Multinational Corporation.* New York: Basic Books.

Storper, M. 1988. The transition to flexible specialization in the US film industry: external economies, the division of labour, and the crossing of industrial divides. *Cambridge Journal of Economics.* 13: 273-305.

Storper, M. and Christopherson, S. 1988. Flexible specialization and regional industrial agglomerations: the case of the US motion picture industry. *Annals of the Association of American Geographers.* 77(1): 104-17.

Storper, M. and Harrison, B. 1991. Flexibility, hierarchy and regional development: the changing structure of industrial production systems and their forms of governance in the 1990s. *Research Policy.* 20 (forthcoming)

Storper, M. and Scott, A. 1988. The geographical foundations and social regulation of flexible production complexes. In: Wolch, J. and Dear, M. (eds). *The Power of Geography.* Boston: Allen and Unwin. 21-40.

Storper, M. and Walker, R. 1989. *The Capitalist Imperative: Territory, Technology and Industrial Growth.* Oxford and Cambridge, Mass.: Basil Blackwell.

Stowsky, J. 1987. The weakest link: Semiconductor production equipment, linkages, and the limits to international trade. Working Paper 27. Berkeley Roundtable on the International Economy, University of California, Berkeley.

Strange, S. 1986. *Casino Capitalism.* Oxford and Cambridge, Mass.: Basil Blackwell.

Studenski, P. and Krooss, H. 1952. *A Financial History of the United States.* New York: McGraw-Hill.

Sugimori, Y., Kusunoki, K., Cho, F. and Uchikawa, S. 1977. Toyota production system and Kanban system: materialization of just-in-time and respect-for-human system. *International Journal of Production Research*. 15(6): 553-64.

Suzaki, K. 1985. The application of Japanese competitive methods to US manufacturing. *Journal of Business Strategy*. 5(1): 10-19.

Sweezy, P. and Magdoff, H. 1987. *Stagnation and the Financial Explosion*. New York: Monthly Review Press.

Sweezy, P. and Magdoff, H. 1989. *The Irreversible Crisis*. New York: Monthly Review Press.

Szelenyi, I. 1986. The intelligentsia in the class structure of state-socialist societies. In: Burawoy, M. and Skocpol, T. (eds). *Marxist Inquiries: Studies of Labor, Class and State*. Chicago: University of Chicago Press. 287-327.

Tasker, P. 1987. *Inside Japan: Wealth, Work and Power in the New Japanese Empire*. London: Sidgwick and Jackson.

Taylor, M. and Thrift, N. (eds). 1982. *The Geography of Multinationals*. New York: St. Martin's Press.

Teaford, J. 1984. *The Unheralded Triumph: City Government in America, 1870–1900*. Baltimore: Johns Hopkins University Press.

Tedlow, R. 1990. *New and Improved: The Story of Mass Marketing in America*. New York: Basic Books.

Teece, D. 1980. Economies of scope and the scope of enterprise. *Journal of Economic Behavior and Organization*. 1: 223-47.

Teece, D. 1985. Multinational enterprise, internal governance and industrial organization. *American Economic Review*. 75: 233-38.

Teece, D. 1986. Profiting from technological innovation: implications for integration, collaboration, licensing and public policy. *Research Policy*. 15: 285-305.

Teece, D. 1988. Technological change and the nature of the firm. In: Dosi, G., Freeman, C., Nelson, R., Silverberg, G. and Soete, L. (eds). *Technical Change and Economic Theory*. London: Frances Pinter. 256-81.

Tentler, L. 1979. *Wage-Earning Women: Industrial Work and Family Life in the United States, 1900-1930*. New York: Oxford University Press.

Teubal, M., Yinnon, T. and Zuscovitch, E. 1991. Networks and market creation. *Research Policy*. 20 (forthcoming)

*Theory and Society*. 1987. Special issue. A capitalist road to communism. 15.

Thomas, H. and Logan, C. 1982. *Mondragon*. London: Allen and Unwin.

Thompson, E. 1964. *The Making of the English Working Class*. London: Penguin.

Thorelli, H. 1986. Networks: between markets and hierarchies. *Strategic Management Journal*. 7: 37-51.

Thrift, N. 1987. The fixers: the urban geography of international commercial capital. In: Henderson, J. and Castells, M. (eds). *Global Restructuring and Technical Development*. Beverly Hills: Sage. 203-33.

Thrift, N. and Leyshon, A. 1988. The gambling propensity: banks, developing country debt exposures, and the new international financial system. *Geoforum*. 19: 55-69.

Tidd, J. 1990. Technological Trajectories and Emerging Production Paradigms: Robotic Assembly as an Example of Flexible Manufacturing. D.Phil. thesis, University of Sussex, UK.

Tigar, M. and Levy, M. 1977. *Law and the Rise of Capitalism*. New York: Monthly Review Press.

Tomlinson, J. 1982. *The Unequal Struggle: British Socialism and the Capitalist Enterprise*. London: Methuen.

Tomlinson, J. 1986. *Monetarism: Is There an Alternative?* Oxford: Basil Blackwell.

Tomlinson, J. 1990. Market socialism. In: Hindess, B. (ed.) *Reactions to the Right*. London: Routledge. 32-49.

Touraine, A. 1971. *The Post-Industrial Society*. New York: Random House.

Tryon, R. 1917. *Household Manufactures in the United States, 1640-1860*. Chicago: University of Chicago Press.

Turnbull, P. 1986. The limits of Japanisation: just-in-time, labour relations, and the UK automotive industry. *Industrial Relations Journal*. 17: 193-206.

Tyebjee, T. 1988. A typology of joint ventures: Japanese strategies in the United States. *California Management Review*. 31(1): 75-86.

Urry, J. 1986. Capitalist production, scientific management and the service class. In: Scott, A. and Storper, M. (eds). *Production, Work, Territory*. Boston: Allen and Unwin. 41-66.

Urry, J. 1990. *The Tourist Gaze*. London: Sage.

Useem, M. 1983. *The Inner Circle: Large Corporations and Business Politics in the US and UK*. New York: Oxford University Press.

Vance, J. 1977. *This Scene of Man*. New York: Harper and Row.

Vance, J. 1986. *Capturing the Horizon: The Historical Geography of Transportation Since the Transportation Revolution of the Sixteenth Century*. New York: Harper and Row.

Vanek, J. 1970. *The General Theory of Labor-Managed Market Economies*. Ithaca: Cornell University Press.

Vanek, J. (ed.) 1973. *Self-Management*. Harmondsworth: Penguin.

Veblen, T. 1898. Why is economics not an evolutionary science? *Quarterly Journal of Economics*. 12: 373-97.

Veblen, T. 1899. *The Theory of the Leisure Class*. New York: Macmillan.

Veblen, T. 1908. On the nature of capital. *Quarterly Journal of Economics*. 22: 517-42.

Veblen, T. 1909. The limitations of marginal utility. *Journal of Political Economy*. 17(9): 620-36.

Veblen, T. 1921. *The Engineers and the Price System*. New York: Harcourt Brace and World.

Von Hippel, E. 1988. *The Sources of Innovation*. New York: Oxford University Press.

Von Tunzelman, G. 1978. *Steam Power and British Industrialization to 1860*. Oxford: Clarendon Press.

Walby, S. 1986. *Patriarchy at Work*. Oxford: Polity Press.

Walker, P. (ed.) 1979. *Between Capital and Labor*. Boston: South End Press.

Walker, R. 1981. A theory of suburbanization: capitalism and the construction of urban space in the United States. In: Dear, M. and Scott, A. (eds). *Urbanization and Urban Planning in Capitalist Societies*. New York: Methuen. 383-430.

Walker, R. 1985. Technological determination and determinism: industrial growth and location. In: Castells, M. (ed.) *High Technology, Space and Society*. Beverly Hills: Sage. 226-64.

Walker, R. 1988. The dynamics of value, price and profit. *Capital and Class*. 35: 147-81.

Walker, R. 1989a. In defense of realism and dialectical materialism: a friendly critique of Wright and Burawoy's philosophical Marxism. *Berkeley Journal of Sociology.* 34: 111-35.

Walker, R. 1989b. Machinery, labour and location. In: Wood, S. (ed.) *The Transformation of Work?* London: Unwin Hyman. 59-90.

Walker, R. 1991. Regulation and flexible specialization as theories of capitalist development: challengers to Marx and Schumpeter? *Regional Studies.* 25 (forthcoming)

Walker, R. and the Bay Area Study Group. 1990. The playground of US capitalism? The political economy of the San Francisco Bay Area in the 1980s. In: Davis, M., Hiatt, S., Kennedy, M., Ruddick, S. and Sprinker, M. (eds). *Fire in the Hearth: The Radical Politics of Place in America.* (*The Year Left* 4). 3-82.

Walker, R. and Greenberg, D. 1982. Post-industrialism and political reform in the city: a critique. *Antipode.* 14(1): 17-32.

Walsh, A. 1978. *The Public's Business: The Policy and Practice of Government Corporations.* Cambridge, Mass.: MIT Press.

Warf, B. 1989. Telecommunications and the globalization of financial services. *Professional Geographer.* 41(3): 257-71.

Weber, M. 1978. *Economy and Society.* Roth, G. and Wittich, C. (eds). Berkeley: University of California Press.

Weeks, J. 1981. *Capital and Exploitation.* Princeton: Princeton University Press.

Weiss, J. and Delbecq, A. 1988. Regional cultures and high-tecnology management: route 128 and Silicon Valley. In: Weiss, J. (ed.). *Regional Cultures, Managerial Behavior and Entrepreneurship.* Westport: Quorum/Greenwood. 9-22.

Whetten, J. 1981. Interorganizational relations: a review of the field. *Journal of Higher Education.* 52(1): 1-28.

Wilkinson, F. (ed.) 1981. *The Dynamics of Labor Market Segmentation.* London: Academic Press.

Williams, K., Cutler, T., Williams, J. and Haslam, C. 1987. The end of mass production? *Economy and Society.* 16(3): 405-39.

Williams, K., Williams, J. and Haslam, C. 1989. Why take the stocks out? Stock reduction in theory and in practice in Britain and Japan. Mimeo. Department of Economics, University of Wales, Aberystwyth.

Williams, R. 1982. *Dream Worlds: Mass Consumption in Late 19th Century France.* Berkeley: University of California Press.

Williamson, H. (ed.) 1951. *The Growth of the American Economy.* Second edition. New York: Prentice-Hall.

Williamson, O. 1975. *Markets and Hierarchies.* New York: The Free Press.

Williamson, O. 1980. The organization of work: a comparative institutional assessment. *Journal of Economic Behavior and Organization.* 1(1): 5-38.

Williamson, O. 1981. The modern corporation: origins, evolution, attributes. *Journal of Economic Literature.* 19: 1537-68.

Williamson, O. 1985. *The Economic Institutions of Capitalism: Firms, Markets, Relational Contracting.* London: Macmillan.

Williamson, O. 1986. *Economic Organization: Firms, Markets and Policy Control.* Brighton: Wheatsheaf.

Willis, P. 1977. *Learning To Labour.* Farnborough: Saxon House.

Willis, P. 1990. *Common Culture.* Milton Keynes: Open University Press.

Wilson, E. 1977. *Women and the Welfare State.* London: Tavistock.

Wolpe, A. 1988. *Within School Walls.* London: Routledge.

Wood, E. 1988. *Peasant-Citizen and Slave: The Foundations of Athenian Democracy*. London: Verso.

Wood, E. 1989. Rational choice Marxism: is the game worth the candle? *New Left Review*. 177: 41-88.

Woodward, J. 1965. *Industrial Organization: Theory and Practice*. London: Oxford University Press.

Wright, A. 1980. *Socialisms: Theories and Practices*. Oxford: Oxford University Press.

Wright, E. 1976. Class boundaries in advanced capitalist societies. *New Left Review*. 98: 3-41.

Wright, E. 1980. Varieties of Marxist conceptions of class structure. *Politics and Society*. 9(3): 299-322.

Wright, E. 1982. The status of the political in the concept of class structure. *Politics and Society*. 11(3): 321-41.

Wright, E. 1985. *Classes*. London: Verso.

Wright, E. 1989a. Rethinking, once again, the concept of class structure. In: Wright, E. (ed.) *The Debate on Classes*. London: Verso. 269-348.

Wright, E. 1989b. Women in the class structure. *Politics and Society*. 17(1). 35-66.

Yamamura, K. and Yasuba, Y. (eds) 1987. *The Political Economy of Japan*. Stanford: Stanford University Press.

Young, A. 1929. Increasing returns and economic progress. *Economic Journal*. 38: 527-42.

Young, I. 1981. Beyond the unhappy marriage: a critique of dual systems theory. In: Sargent, L. (ed.). *Women and Revolution*. Boston: South End Press. 43-69.

Zaretsky, E. 1976. *Capitalism, the Family and Personal Life*. London: Pluto Press.

Zysman, J. 1983. *Markets and Growth: Financial Systems and the Politics of Industrial Change*. Ithaca: Cornell University Press.

# Index